Melissa Marley

92-08

Yonni

Jen. D. Kid

Melissa Marley
was
here

Jen Loves David

Science

IDEAS AND APPLICATIONS

9

The Wiley Intermediate Science Program

Science Ideas and Applications 9

Science Ideas and Applications 9 Teacher's Guide

Science Ideas and Applications 10

Science Ideas and Applications 10 Teacher's Guide

Science Explorations 9

Science Explorations 9 Teacher's Guide

Science Explorations 10

Science Explorations 10 Teacher's Guide

Science
IDEAS AND APPLICATIONS

SERIES EDITORS
H. Murray Lang
E. Usha R. Finucane

AUTHOR TEAM

Carol A. Caulderwood
St. Ignatius of Loyola High School

Thomas R. Donovan
Pope John Paul II Secondary School

E. Usha R. Finucane
Bloor Collegiate Institute

Jane L. Forbes
Acton High School

H. Murray Lang
Faculty of Education, University of Toronto

Ron Raymer
North Toronto Collegiate Institute

Ian R. Skinner
Harwood Secondary School

Jean Bullard
Qualicum Secondary School

John Wiley & Sons
Toronto   New York   Chichester   Brisbane   Singapore

Canadian Cataloguing in Publications Data
Main entry under title:

Science ideas & applications 9

For use in secondary schools.
Includes index.
ISBN 0-471-79779-0

1. Science - Juvenile literature. I. Lang, H.
Murray (Harold Murray), 1922-

Q161.2.S25 1988   500   C87-095094-0

Designer: Julian Cleva
Typesetter: CompuScreen Typesetting Ltd.
Printer: The Bryant Press Limited

Printed and bound in Canada.

10 9 8 7 6 5 4 3

Contents

Acknowledgements

This textbook has been reviewed by science teachers at several stages in the developmental process. In addition, each unit in the textbook was class tested in typical science classrooms. Our thanks to Mary Kay Winter for organizing the class testing.

We would like to express our appreciation to the following reviewers and class testers who, as science teachers and educators, have provided us with many useful suggestions and comments on the manuscript at various stages of its development. Many of their recommendations have been incorporated in the textbook, but for the final content, we alone are responsible.

Daniele DiFelice, Bloor Collegiate Institute
David Freeman, Albert Campbell Collegiate

Tamas Holcz, Gloucester High School
Richard Krol, Newtonbrook Secondary School
Kathy Nealon, Francis Libermann High School
Henry Pasma, Cawthra Park Secondary School
Earl Shieff, Bendale Secondary School
Peter Williams, Ward 1 Core Team Consultant, Toronto Board of Education
Nora Wood, Bloor Collegiate Institute
Tom Wright, Acton High School
Douglas Yack, Michael Power/St. Joseph High School

We would also like to thank our developmental editors, Jonathan Bocknek and Mary Kay Winter, who provided valuable support and suggestions during the writing of this book.

Science Ideas and Applications 9 and its companion textbook, **Science Ideas and Applications 10**, have been written with a specific goal in mind—to provide learning resources that are completely appropriate for students who are most likely to begin their working careers with the completion of grade 12, or who will receive post-secondary technician/technology training.

Our textbooks are designed to provide a general introduction to science that will help students prepare for their role as citizens in an increasingly technological society. We have strived to prepare materials that will be relevant to the daily lives of your students. Our very title reflects this goal of relating science ideas to practical applications. Please note below the features of the textbook that we think you will find most useful in implementing your course.

FEATURES OF THE TEXT

- **Science Ideas and Applications 9** employs a mosaic approach to the study of science, an approach that emphasizes the breadth of science and ensures that students are exposed to a variety of science disciplines. The textbook contains seven units, each representing a separate topic of scientific enquiry.
- Each chapter opens with a motivational introduction, followed by a brief summary of the *Key Ideas* of the chapter.
- Reading level has been carefully controlled so that material is appropriate for the grade 9 students for whom this textbook has been developed.

- Activities have been included in the textbook. Distinctive logos identify each activity as either a *Laboratory Experiment*, a *Thought Experiment*, or a *Teacher Demonstration*.
- Instructions for the safe use of equipment and materials are emphasized throughout the textbook. The *Safety Rules* section should be read carefully at the onset of the course and referred to as needed. In each activity, notes of *Caution* occur whenever there are procedures or techniques that require special care or attention.
- *Self-check* questions, occurring at the end of most numbered sections in the textbook, are provided to allow the student or the class to review short sections of content and to consolidate ideas and applications covered.
- *Challenges*, interspersed throughout the textbook, stimulate students to think about ideas and/or applications presented and to extend their knowledge beyond the limits of the required course material.
- *Ideas and Applications*, likewise found throughout the textbook, provide examples, in addition to those included in narrative passages, of the relevance to life and society of a science idea.
- Each unit contains a feature called *Science and Technology in Society*. As the title suggests, this feature focusses on the science behind, and the societal implications of, a particular technological invention or event.
- Each unit contains a *Science at Work* feature that spotlights the work of a Canadian scientist or group of scientists, and the relation of that work to a key concept or concepts in the unit.

- Newly defined terms, in **boldface** type, are also found in the *Words to Know* section at the end of each chapter. These terms are defined in the *Glossary* at the end of the textbook.
- The end of each chapter contains the following:
  * *Chapter Objectives* to provide students with a checklist of the ideas and applications they should know by the time they have completed their study of the chapter (these are keyed to appropriate sections if the students require review)
  * *Words to Know*, a listing of words which the student should now be able to define and/or use in a sentence
  * *Tying It Together*, a series of chapter review questions
  * *Applying Your Knowledge*, a selection of questions that involve applications of the ideas covered
  * *Projects for Investigation*, questions which involve library research

- Each unit concludes with a two-page self-review and/or class review called *How Well Have You Understood. . . .* This feature contains a matching exercise, 10 multiple choice questions, 10 true/false questions, and a case study.
- Appendices on *The Metric System* and *Graphing* may be found at the end of the textbook.
- The accompanying *Teacher's Guide* contains many helpful teaching suggestions, including a rationale for teaching each unit, ways to help stimulate interest, advance preparation and planning, answers to all questions, additional activities and questions, and lists of resource materials.

We hope you and your students have a successful year investigating science ideas and applications.

The Authors

The title of our book, **Science Ideas and Applications**, shows that the ideas of science are often put to use (applied) in our daily lives. As you go through this course, you will look at the materials of your daily life in terms of their properties and changes. These ideas may help you use with understanding the changes involved in heating, cooling, mixing, dissolving, cooking, and burning fuels. You will learn the skills of handling such materials and changes safely. Such an understanding may save your life.

You will also find out about living things. These include the cells you are made of, the plants that make your life possible and pleasant, and the food that keeps you alive. The final unit of this book shows how some of the ideas of science are applied to sports and leisure—activities that go with you beyond school for your whole lifetime.

Many of the applications of science ideas suggest ways of making a living. If you enjoy science and the applications of science to the machines and instruments of modern life, you may also enjoy an occupation that puts science to work. The pictures here show several examples of people whose jobs use (directly or indirectly) skills and knowledge you will gain in this course. You will also find references to other science-related jobs throughout the book.

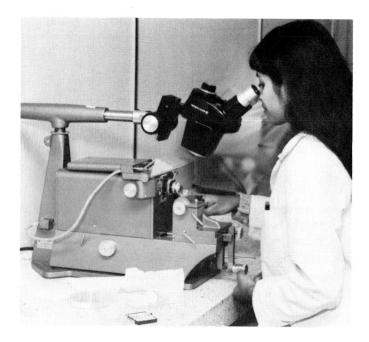

But what if you're not interested in a science-related job. Does science have anything else to offer you? To answer this question, think about the headlines shown below.

RESEARCHERS SEARCH FOR CANCER CURE

Scientists Announce New Pest-Resistant Grain

Investigators Probe Cause of Chemical Explosion

Why is a cure for cancer needed? How can food shortages be avoided? What causes dangerous explosions and how can they be prevented? These are the kinds of questions that a knowledge of science will help you understand. We hope that, by the time you finish your studies, you will feel you have a better understanding of the world and events that affect you every day.

The Authors

Your school laboratory, like your kitchen, need not be dangerous. In both places, understanding how to use materials and equipment and following proper procedures will help you maintain an accident-free environment.

Care has been taken to ensure a safe environment for all activities in this textbook. Take special note of the word CAUTION associated with certain activities. This alerts you to the fact that you will be working with potentially dangerous equipment, such as Bunsen burners and scalpels, or using chemicals that require special care.

Follow the guidelines and general safety rules listed below. Your teacher will give you specific information on additional safety procedures recommended in your province and on routines that apply to your school. You will also be informed about the location and proper use of all safety equipment.

1. Read through each activity before you begin.
2. Clear the laboratory bench of all materials except those you are using in the activity.
3. Learn the location and use of the safety equipment available to you, such as safety goggles, protective aprons, fire extinguishers, fire blankets, eyewash fountains, and showers. Find out the location of the nearest fire alarm.
4. Do not begin an activity until you are instructed to do so.
5. Do not taste any material unless you are asked to do so by your teacher.

6. When you are instructed to smell a chemical in the laboratory, follow the procedure shown here. Only this technique should be used to smell chemicals in the laboratory. Never sniff a chemical by placing it close to your nose.

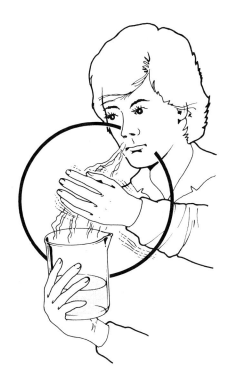

7. Use flames only when instructed to do so. Read the special Bunsen burner safety rules.
8. When using dangerous chemicals, wear safety goggles.

9. When heating materials, wear safety goggles. Make sure the test tubes you use are Pyrex and are clean and not cracked. Always keep the open end of the test tube pointed away from other people and yourself. Move the test tube through the flame so heat is distributed evenly.
10. Handle hot objects carefully. If you suffer a burn, immediately apply cold water or ice.
11. If any part of your body comes in contact with a harmful chemical, wash the area immediately and thoroughly with water. If your eyes are affected, do not touch them, but wash them immediately and continuously for at least ten minutes.
12. Wash your hands after you handle chemicals, biological specimens, and micro-organisms that your teacher has instructed you to use.
13. Clean up any spilled chemicals immediately, following instructions given by your teacher.
14. Never pour harmful substances into the sink. Dispose of them as instructed by your teacher.
15. Clean all apparatus before putting it away.
16. Always unplug electric cords by pulling on the plug, not the cord.
17. Watch for sharp or jagged edges on all apparatus. Do not use broken or cracked glassware. Place broken glass only in special marked containers.
18. Report to your teacher all accidents (no matter how minor), broken equipment, damaged or defective facilities, and suspicious looking chemicals.

BUNSEN BURNER SAFETY RULES

If a Bunsen burner is used in your science classroom, make sure you follow the procedures listed below. (Note: Hot plates should be used in preference to Bunsen burners whenever possible.)

1. Do not wear scarves or ties, long necklaces, or earphones suspended around your neck. Tie back loose hair, and roll back and secure loose sleeves before you light a Bunsen burner.
2. Obtain instructions from your teacher on the proper method of lighting and using a Bunsen burner.
3. Never heat a flammable material (for example, alcohol) over a Bunsen burner.
4. Be sure there are no flammable materials nearby before you light a Bunsen burner.
5. Never leave a lighted Bunsen burner unattended.
6. Always turn off the gas at the valve, not at the base of the Bunsen burner.

WHAT TO DO IF A FIRE OCCURS

1. Shut off all gas supplies at the desk valves.
2. Notify your teacher immediately. Since every second is vital, move quickly to provide help in an emergency.
3. Pull the fire alarm.
4. Here are some other recommendations for action.
   (a) If clothing is on fire, roll on the floor to smother the flames. Use a fire blanket to smother the flames if a fellow student's clothing has caught fire.
   (b) A small fire can be smothered by using sand or a small container such as an inverted large beaker or can.
   (c) Make sure you know how to operate the fire extinguishers in order to assist your teacher.
   (d) If the fire is not quickly and easily put out, leave the building in a calm manner.

Science is a special way of asking questions. The answers to these questions help to explain things and solve problems. Explaining and problem solving give us knowledge about the world around us.

In earlier science courses, you were probably introduced to the *scientific method* of solving problems. Table 1 shows the steps of the scientific method.

Put yourself in the place of a young child who does not yet know through experience that sandwiches exposed to air will dry out. First read the step of the scientific method in column two. Then read the way the child might use the step to solve this simple problem. (Note: You might not always follow all these steps in this order. Also, sometimes the results of an experiment might be different from what you expected. Then you would have to go back a step and suggest a new hypothesis or idea for why you got the results you did.)

Perhaps the most important step in the scientific method is the last one, communicating. New knowledge should always be shared with other people. It is important for you to be able to communicate your knowledge simply and effectively. In this way, you benefit from other people's knowledge, and they benefit from yours.

Think of a problem from your own experience that could best be solved using the scientific method.

Table 1 *Using the Scientific Method to Solve a Problem*

STEP NUMBER	THE SCIENTIFIC METHOD	USING THE SCIENTIFIC METHOD
1	*Observe* an unexplained object or event.	My sandwiches at lunch are dried out instead of fresh.
2	Identify a *problem*.	I don't like dry sandwiches. I want to prevent this from happening.
3	Pose a *question*.	What causes my sandwiches to dry out?
4	Collect *data* (information).	I compare my lunch with my friends' lunches. Are their sandwiches fresh? Is my bread packed differently?
5	Suggest a *hypothesis*.	My sandwiches dried out because they were exposed to the air. My friends have the same bread and it stays fresh. But their sandwiches are tightly sealed in plastic wrap. Mine have fallen out of their paper bag and are loose in my school bag.
6	Conduct *experiments* to test this hypothesis.	I cut my sandwich in half, and wrap one half in plastic. I leave the other half unwrapped. At lunch, I look at, feel, and taste both sandwiches to find out which one is fresher.
7	Draw some *conclusions* from the results of the experiments.	Only the wrapped half remains fresh. I conclude that my lunch should be carefully wrapped in plastic so that the bread isn't exposed to air. In future, I'll wrap my sandwiches securely.
8	*Communicate* the conclusions to others.	I'll share what I've discovered with my friends.

Physical Properties and Physical Change

The parts of an automobile include many different substances. Some parts are metals. Other parts are glass. And still other parts are plastics. What other substances are used to make automobiles? Why do you think these substances are used? What other substances could be used to make cars? What advantages and/or disadvantages would these substances have?

Each substance has its own set of characteristics or properties which cause it to behave in specific ways. In this unit, you will compare properties of different substances. You will also find out how certain properties of substances are useful. And finally, you will develop a model to explain why substances have these properties.

Properties of Matter

Key Ideas

- Matter occupies space and has mass.
- Matter can be described by its properties.
- The various properties of substances allow us to identify and compare them.

These students are trying to identify substances by examining their characteristics or **properties**. The student on the right is able to describe properties that he can see. What properties must the blindfolded student depend on?

How can you describe the various properties of different substances? How can you identify and compare different substances? In this chapter, you will learn several methods to help you answer these questions.

Matter, Volume, and Mass

Each substance in Figure 1.1, and in the world around you, is made of matter. **Matter** is the name given to the "stuff" that everything is made of. Each type of matter differs from the others, yet all have two things in common. Matter occupies space and has mass.

Matter Occupies Space

All matter occupies space. Together, the length, width, and height of an object determine the space it occupies. The space that matter occupies is called its **volume**. For example, the length, width, and height of this book determine its volume.

Volumes of liquids are usually measured in litres (L) or millilitres (mL). See Figure 1.2. Solids are more likely to be measured in cubic metres (m³) or cubic centimetres (cm³). For a review of the units used to measure volume, along with their prefixes and meanings, turn to Appendix A.

Matter Has Mass

How do you measure how much of you there is? The amount of matter in an object is its **mass**. You are more likely to measure your mass in kilograms (kg). Many of the substances you will work with this year will be measured in grams (g). Some common objects sold by mass are shown in Figure 1.3. For a more complete description of the units used to measure mass, turn to Appendix A.

Challenge

Describe one job for which you would need the ability to measure mass. In what type of work would you need to make measurements of volume?

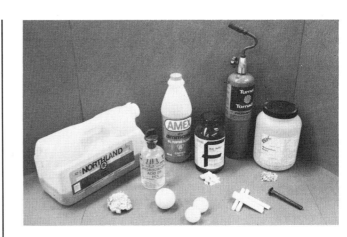

Figure 1.1 *Each of these substances is made of matter.*

Figure 1.2 *These products are packaged using the volume units, millilitres (mL) and litres (L).*

Figure 1.3 *These products are packaged using the mass units, milligrams (mg), grams (g), and kilograms (kg).*

In the four parts of this activity, you will measure the volume and mass of different solids and liquids.

PART 1

Problem

How can you measure the volume of a regular solid?

Materials

ruler
regular objects

Procedure

1. In your notebook, make a table similar to Table 1.1.
2. Use your ruler to measure the length, width, and height of a regular object (Figure 1.4). Record your values, along with the units, in your table.
3. Repeat step 2 for other regular objects provided by your teacher.

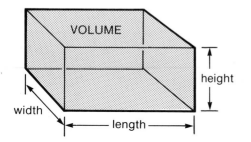

Figure 1.4 *How to calculate the volume of a regular object*

Observations

Table 1.1

OBJECT	LENGTH	WIDTH	HEIGHT	VOLUME
Rectangular block				
Etc.				

(Note: To calculate the volume of each regular solid, multiply length X width X height. Be sure to include the proper units when you record the volume in your table.)

Questions

1. If the length, width, and height of an object are measured in cm, what will be the units of its volume?
2. If the length, width, and height of an object are measured in m, what will be the units of its volume?

PART 2

Problem

How can you measure the volume of an irregular solid?

Materials

irregular objects   graduated cylinder
overflow container   water

Procedure

1. Fill the overflow can with as much water as possible.
2. Hold the graduated cylinder as shown in Figure 1.5, and gently lower an object into the overflow can. Collect the overflowing water in the graduated cylinder.

Figure 1.5 *How to measure the volume of an irregular object*

3. Measure the volume of water in the cylinder (Figure 1.6).
4. Repeat steps 1, 2, and 3 for two other irregular objects.

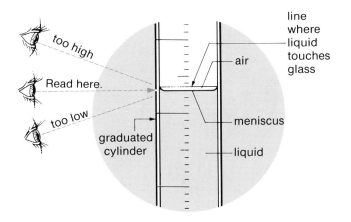

too high

Read here.

too low

graduated cylinder

air

line where liquid touches glass

meniscus

liquid

Figure 1.6 *The scale on a graduated cylinder allows you to read the volume of a liquid directly. Be sure to read from the bottom of the curved surface (the meniscus).*

Observations

What is the volume of each of the irregular objects?

Questions

1. How would you measure the volume of a sample of apple juice?
2. How would you measure the volume of a lump of clay?

PART 3

Problem

How can you measure the mass of a solid?

Materials

several solid objects
balance
set of masses

Procedure

1. Make a table similar to Table 1.2 in your notebook. (See page 6.)
2. List the objects in the first column. Pick up each object and estimate its mass. Write your estimate in the second column. Be sure to include the units.
3. Two common types of balances are shown in Figure 1.7. A set of masses is needed for the balance in Figure 1.7 (b). Your teacher will give you instructions for your type of balance. Carefully measure and record the mass of each of the solid objects.

Figure 1.7 *Two common balances used in the science classroom*

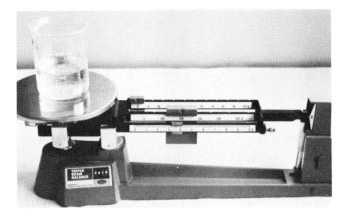

(a) *triple-beam balance*

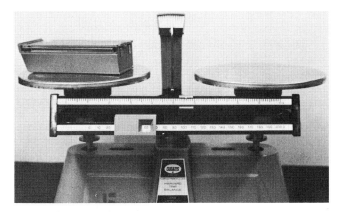

(b) *equal-arm balance*

Observations

Table 1.2

OBJECT	ESTIMATED MASS	MEASURED MASS

Questions

1. Which object has the most mass?
2. Which object has the least mass?
3. What unit did you use for most of your mass measurements?
4. Were your estimates generally higher or lower than the measured mass?

PART 4

Problem

How can you measure the mass of a liquid?

Materials

balance
set of masses
150 mL beaker
graduated cylinder
water

Procedure

1. Make a table similar to Table 1.3 in your notebook.
2. Measure the mass of the beaker. Record the value on line B of your table.
3. Pour exactly 100 mL of water into the graduated cylinder. Carefully pour all the water into the beaker.
4. Measure the mass of the beaker and water. Record this value on Line A of your table.
5. Calculate the mass of the water by subtracting line B from line A. Record this value in your table at line C.

Observations

Table 1.3

A. Mass of beaker and water	g
B. Mass of dry beaker	g
C. Mass of water (A – B = C)	g

Questions

1. Explain in your own words the three steps you must follow to measure the mass of a liquid.
2. From your own result, calculate the mass of 1 mL of water.

Challenge

Obtain an object with a regular shape. Measure it and calculate its volume. Ask a friend to find its volume by displacement of water. Compare your results. (Note that 1 mL = 1 cm^3.)

The Three States of Matter

Matter exists in three states: solid, liquid, and gas (Figure 1.8). Have you ever stopped to think about how you identify these states? We can identify the **states of matter** by examining the mass, volume, and shape when a substance is placed in a different container. Solids and liquids always have the same volume no matter what container they are placed in. A solid usually keeps its own shape, unless it consists of small particles like sand. A liquid always takes the shape of its container. A gas always takes the shape of its container and always fills its container. These properties of the three states of matter are summarized in Figure 1.9.

Figure 1.8 *Matter exists in three states: solid, liquid, and gas.*

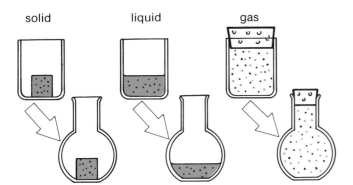

	SOLID	LIQUID	GAS
Mass	Stays the same	Stays the same	Stays the same
Volume	Stays the same	Stays the same	Changes to fill its container
Shape	Stays the same	Changes to fit its container	Changes to fit its container

Figure 1.9 *Each state of matter behaves in a different way when placed in a new container.*

Self-check

1. Name two properties of all matter.
2. How would you calculate the volume of a regular object?
3. Tell if the following statements about mass are true or false. In your notebook, correct any false statements.
   (a) Mass is measured using a graduated cylinder.
   (b) The gram, milligram, and kilogram are mass units.
   (c) Of the units listed in (b) above, the milligram is the largest quantity.
   (d) Chocolate bars are measured and sold in grams.
   (e) Flour is measured and sold in milligrams.
4. Describe the method you would use to measure the mass of the following items:
   (a) a gold ring
   (b) cereal in a box
5. Describe the method you would use to measure the volume of the following items:
   (a) a box of tissues
   (b) orange juice in a can
   (c) a lump of gold
6. Decide whether each of the following is a solid, a liquid, or a gas.
   (a) This substance keeps the same mass, volume, and shape when placed in a new container.
   (b) This substance keeps the same mass but changes its volume and shape to suit its container.
   (c) This substance keeps the same mass and volume but changes its shape to suit its container.

 **Activity 1B**

Describing The Properties of Matter

Observation is important in science and in daily life. You observe things in order to describe them or identify them. Observations may be made using all five senses: sight, touch, hearing, taste, and smell (Figure 1.10).

When preparing a meal, cooks observe the smell and appearance of the food, as well as its taste. How the food feels in the mouth may also be important.

Automobile mechanics use sight to inspect items such as fan belts and fluid levels in a car. They use hearing to listen for unusual sounds made by an engine. The sense of smell may be used to check for burning oil or leaking engine coolant.

Figure 1.10 *The chef in the kitchen and the mechanic in the shop must make accurate observations using their senses.*

A reasoned conclusion based on observations is called an **inference**. After making observations, a cook may *infer* that the soup needs more basil or that it is time to stop cooking the rice. A mechanic may infer that the strange sounds made by a car are due to a worn wheel bearing or a problem in the braking system. In this activity you will be making your own inferences based on observations.

Problem

How can you identify something you can't see?

Materials

substances in plastic containers labelled A, B, C, etc. (for example, sand, fur, detergent, flour, candle wax) blindfold

Procedure

1. Make a table similar to Table 1.4 in your notebook.
2. Choose three plastic containers. Record the letters in your table.
3. With your eyes blindfolded, shake a container and listen to any sounds made by its contents.
4. While blindfolded, open the container, and smell and feel the substance inside it.
5. After closing the container, uncover your eyes and record all observations in your table.
6. Repeat steps 3, 4, and 5 for the other two containers.
7. Try to identify each substance based on your observations.

Observations

Table 1.4

SUBSTANCE	OBSERVATIONS		
	Sound	Feel	Smell
A.			
B.			
C.			

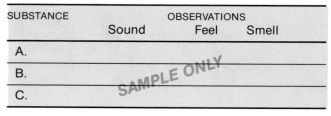

1. Which senses did you find most useful for identifying substances?
2. Explain the difference between observing and seeing.
3. How did you use inferences in this activity?

Challenge

How could you describe the substance from which this sculpture is made to someone who has not seen it? Try writing a brief description.

Describing Matter

You can describe a substance in two ways. Those properties which may be observed directly, such as colour, state, hardness, texture, and smell, are called **qualitative properties**. For example, water is a clear, colourless liquid. Copper wire is a shiny, red-brown, smooth solid. These descriptions of water and copper wire are based on their qualitative properties.

Those properties which may be measured or quantified in some way are called **quantitative properties**. You already know two quantitative properties: volume and mass. In the next section, you will find out about several other important quantitative properties that can help you describe matter more accurately.

Challenge

Name four jobs not mentioned in this chapter that require accurate observations. Explain how these observations are important in each job.

Self-check

1. Describe the difference between observations and inferences.
2. In your notebook, identify the following examples as observations or inferences.
   (a) The sky appears blue in colour.
   (b) A few clouds appear in the sky.
   (c) There appear to be more clouds in the sky.
   (d) It will probably rain.
3. The following descriptions are based on qualitative properties of substances. Try to identify the substance described in each case.
   (a) clear, colourless, odourless, tasteless liquid
   (b) grey, shiny, smooth solid
   (c) colourless, odourless, tasteless gas
   (d) white, shiny, odourless, salty-tasting solid
   (e) reddish-brown, shiny, smooth solid
4. (a) If you had difficulty identifying any of the substances in question 3, here's a hint. The five substances (not in the correct order) are air, water, copper, steel, and salt.
   (b) Agree or disagree with the following statement, and give reasons for your answer: "Knowing some of the quantitative properties of the substances in question 3 would have helped me identify them more easily and with more certainty."

Describing Matter Quantitatively

Some properties of matter may be accurately measured, rather than just observed. These measured properties of a substance are called their quantitative properties because they involve amounts or quantities. Quantitative properties can also be used to compare and identify substances. In the rest of this section, you will find out about several different quantitative properties.

Melting Point and Boiling Point

Melting is the change from the solid state to the liquid state (Figure 1.11). Ice melts if left at room temperature (20°C). Lard must be heated above room temperature for it to melt. Very high temperatures, such as those reached in a blast furnace, are needed to melt most metals. The temperature at which a substance changes from a solid to a liquid is the **melting point** of the substance.

When it is cooled, a liquid substance solidifies or freezes. In other words, it changes from the liquid state to the solid state. The temperature at which a liquid freezes is called its **freezing point**. The freezing point and the melting point of a substance are the same (Table 1.5).

Boiling refers to the rapid change of a substance from a liquid state to the gas state. You can tell when a liquid is boiling because of the rapid production of bubbles of gas. Substances boil at a definite temperature known as **boiling point**. Each substance has a specific value for its boiling point. Boiling point, together with melting point and freezing point, are therefore quantitative properties of substances (Table 1.5).

Figure 1.11 *Solid substances such as (a) ice and (b) lard melt at different temperatures.*

Activity 1C

PART 1

Problem

How can you use a thermometer to measure temperature?

Materials

thermometer	hot water
2 beakers (150 mL)	cold water

Procedure

1. Examine your thermometer and note the level of liquid in the tube. This level indicates the temperature in your classroom.
2. Place the bulb of the thermometer in the cold water. Watch the level of the liquid in the thermometer. When it has finished changing, it indicates the temperature of the cold water.
3. Repeat step 2 using hot water.

Observations

1. Record your temperature readings as
   (a) room temperature
   (b) temperature of the cold water
   (c) temperature of the hot water.

Questions

1. What would happen to the temperature of the water in the two beakers if they were left in the classroom
   (a) for 5 min?
   (b) for 30 min?

PART 2

Problem

At what temperature does ice melt, and at what temperature does water boil?

Table 1.5 *Melting, freezing, and boiling points of some common substances. Although the temperature of the melting point of a substance is the same temperature as its freezing point, we commonly speak of "melting points of solids" and "freezing points of liquids." Which of these substances are liquids at room temperature?*

SUBSTANCE	MELTING POINT/ FREEZING POINT (°C)	BOILING POINT (°C)
Ammonia	-77.7	-33.4
Chlorine	-103	-34.6
Copper	1083	2300
Ethanol	-114.4	78.3
Glycerine	18	290
Tin	232	2260

Materials

crushed ice   stirring rod
beaker   clamp
retort stand   hot plate
ring clamp   watch
thermometer

Procedure

1. Make a table similar to Table 1.6 in your notebook. You will need at least 25 lines in the table.
2. Assemble the apparatus shown in Figure 1.12. Draw a labelled diagram of this apparatus in your notebook.
3. Stir the crushed ice using the stirring rod. Use the thermometer to measure the temperature of the ice. Record this temperature in your table at 0 min.

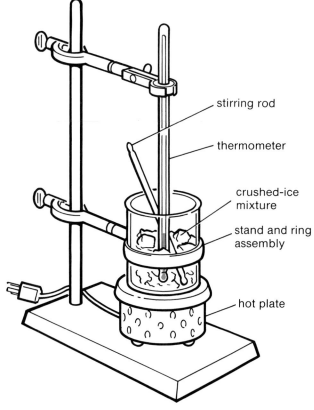

Figure 1.12 *Apparatus for Activity 1C, Part 2*

> **CAUTION** Do not use your thermometer to stir your crushed ice mixture. Use the stirring rod only. Wear safety goggles.

4. Heat the ice in the beaker using the hot plate. Stir the ice mixture occasionally with the stirring rod. Measure and record the temperature of the ice at intervals of 1 min.
5. Between temperature readings, observe any changes that occur in the beaker. Record these on your table.
6. Repeat steps 4 and 5 until the melted ice has boiled for at least 5 min. Be sure to note the temperature at which (a) the ice melts and (b) the water boils.
7. Repeat steps 4 and 5 until the water has boiled for 5 min.
8. Keep your data table in a safe place. You will need it again for Activity 2C in Chapter 2.

Observations

Table 1.6 *Data for Activity 1C, Part 2*

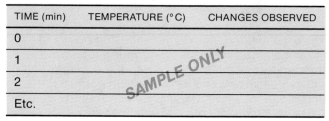

TIME (min)	TEMPERATURE (°C)	CHANGES OBSERVED
0		
1		
2		
Etc.		

Questions

1. What is the melting point of your sample of ice?
2. What changes indicate that a liquid is boiling?
3. Describe what happened to the temperature as the water boiled.
4. What did you find to be the boiling point of water?
5. Compare your values for melting point and boiling point with those of your classmates. Are the values exactly the same? Explain why there might be slight differences.

Density

Look at Figure 1.13. Which is heavier—a kilogram of feathers or a kilogram of lead? Once you think about it, the answer is obvious. But this old riddle continues to fool people of all ages.

Figure 1.13

The riddle points out an important difference between feathers and lead. Equal volumes of these two substances have very different masses. A volume of feathers large enough to stuff a sleeping bag might have a mass of only 1 kg. The same volume of lead would have a mass of 100 kg or more. This difference between feathers and lead results from a property of matter that depends on the relationship between mass and volume. This property of matter is called density. Density is the amount of mass a substance has in a certain volume (Figure 1.14).

Density is the mass (in kilograms) of one cubic metre of a substance (Figure 1.15). Thus, the SI unit for density is the kilogram per cubic metre (kg/m^3). Because a cubic metre is a large unit of volume, smaller units are often used in the laboratory to measure density. The density of solids is usually given in g/cm^3. The density of liquids may be given in g/mL (1 cm^3 = 1 mL). Thus, using water as an example, we could express its density as 1000 kg/m^3, 1 g/cm^3, or 1 g/mL.

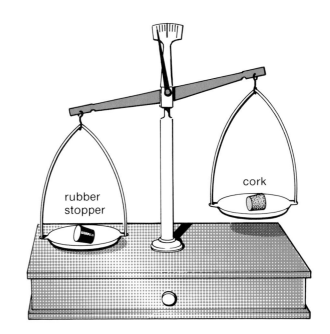

Figure 1.14 *Although both have the same volume, the rubber stopper has more mass than the cork. This shows that the density of rubber is greater than the density of cork.*

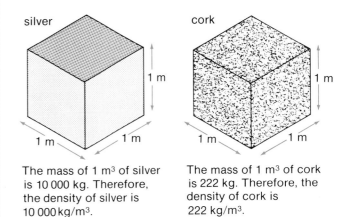

The mass of 1 m^3 of silver is 10 000 kg. Therefore, the density of silver is 10 000 kg/m^3.

The mass of 1 m^3 of cork is 222 kg. Therefore, the density of cork is 222 kg/m^3.

Figure 1.15 *Density is the mass of a cubic metre of a substance.*

Conductivity

Conductivity is a measure of how well a substance allows electricity to pass through it. Substances such as copper that can allow electricity to pass through them very well are called electrical **conductors**. Because copper is such a good conductor, it is used in electrical wiring. Other metals also conduct electricity, but not as well as copper does. Substances such as plastic that do not conduct electricity are called electrical **insulators**. Plastic is often used for the coating around copper wire. As an insulator, the plastic blocks the flow of electricity from the wire.

For an accurate measurement of how much electric current is flowing, you can use a **galvanometer** (Figure 1.16). An electric current causes the galvanometer needle to move. Good conductors allow an electric current to flow. They give a high reading on the scale. Poor conductors do not allow much electricity to flow. Thus, the galvanometer shows a lower reading for poor conductors. Insulators, which allow no current to pass, show a reading of zero on the scale. Is the substance tested in Figure 1.16 a good conductor, poor conductor, or insulator?

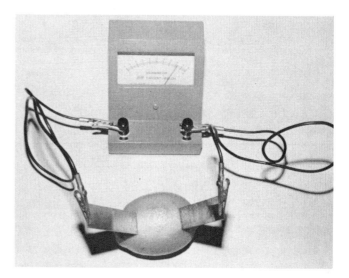

Figure 1.16 *A galvanometer*

14

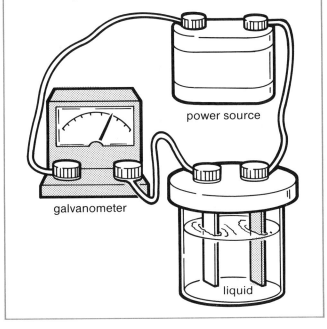
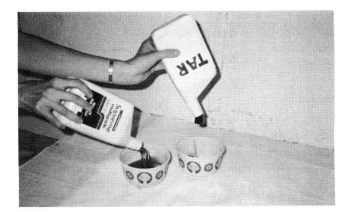

Figure 1.17 *Oil flows more readily than tar.*

Ideas and Applications

The oil in an automobile must have a low viscosity so that it can be pumped through the engine. But it must be viscous enough to lubricate the motor. The correct grade of engine oil must be chosen.

As the engine runs, it gets hotter. The oil also gets hotter, so its viscosity decreases. This could cause damage to the engine. Without proper lubrication, metal parts would wear against one another, and in time they would no longer fit. To prevent this, multigrade oils are used. These oils maintain a safe viscosity over the normal operating temperature of the engine.

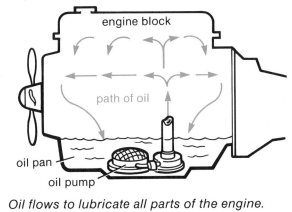

engine block

path of oil

oil pan

oil pump

Oil flows to lubricate all parts of the engine.

Viscosity

The samples of oil and tar in Figure 1.17 were poured at the same time, in the same way, and at the same temperature. You can see that the oil flows much more readily than the tar. The two substances differ in the quantitative property, viscosity. **Viscosity** is the measure of how much a liquid resists flow.

You can compare the viscosity of liquids by placing different liquids in graduated cylinders and timing how long it takes a small ball to fall through a column of each liquid. Predict which liquid will have the highest and lowest viscosities. Then see if your predictions were correct.

Problem

How can you compare the viscosity of liquids?

Materials

liquids with different viscosities
funnel
small ball or ball bearing
100 mL graduated cylinder
1 L beaker

Procedure

1. Make a table similar to Table 1.7 in your notebook.
2. Using a funnel, pour 100 mL of one of the liquids into a graduated cylinder.
3. Hold a small ball at the mouth of the cylinder. Then let the ball drop through the liquid (Figure 1.18). Record the time it takes for the ball to reach the bottom of the cylinder.
4. Repeat this procedure for the other liquids.

Observations

Table 1.7 Comparing Viscosity

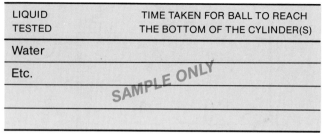

LIQUID TESTED	TIME TAKEN FOR BALL TO REACH THE BOTTOM OF THE CYLINDER(S)
Water	
Etc.	

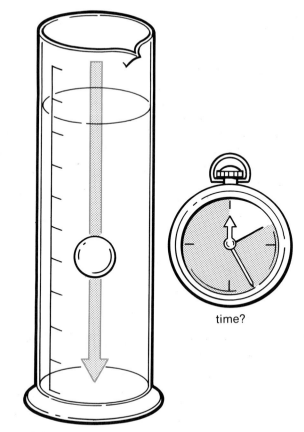

Figure 1.18 How long will it take the ball to reach the bottom of the cylinder?

Questions

1. (a) In which liquid did the ball take the longest time to reach the bottom of the cylinder?
   (b) Is this liquid, therefore, the most viscous or the least viscous? Explain your answer.
2. (a) How do your predictions compare with your results?
   (b) If your predictions were correct, explain how you were able to make the correct conclusion. If your predictions were incorrect, explain what you didn't take into account.
3. People who apply roofing tar usually heat the tar before applying it. Why do you think this is done?

Safety engineers at Transport Canada perform well-designed experiments to test safety designs and the properties of the materials used to make cars and trucks. This helps Transport Canada ensure that vehicles are as safe as possible for their occupants. The dummy shown here will be placed in the impact sled which is being prepared for a test crash. The sled will slam into a solid barrier at a speed of 48 km/h. This test situation is similar to an actual car hitting a parked car at 96 km/h. The engineers use modern instruments and high-speed photography to analyze the test results. Once the dummy has been put through the test crash, the effects on its head and chest are measured to see if a person in a similar crash situation would have suffered any injuries.

Self-check

1. In your notebook, write the following statements and state whether each is true or false. Correct any false statements.
   (a) Properties are characteristics of a substance.
   (b) Qualitative properties involve measured amounts.
   (c) Colour is a quantitative property of a substance.
   (d) Melting point is a quantitative property.
   (e) All types of matter have the same melting point.

2. Define melting point, freezing point, and boiling point.
3. (a) What is density?
   (b) Which would require a larger package, 50 g of chocolate or 50 g of popcorn? Explain.
4. (a) Define conductivity.
   (b) Name two substances that are conductors.
   (c) Name two substances that are insulators.
5. (a) Define viscosity.
   (b) Name two liquids that have high viscosity.
   (c) Name two liquids that have low viscosity.

How Are Electric Guitars Made?

The next time you listen to your favourite electric guitar music, you might think about what you learned in science class. The sound of an electric guitar depends on the physical properties of the substances used to make the guitar. Each substance has certain properties. These properties contribute to the types of sounds the guitar can make.

The final product is unique. It produces a sound that results from the physical properties of all its parts: wood, metal, acrylic—and, of course, the talent and skill of the player!

rut

pickups

bridge

electronics

The strings contain the metals chromium, nickel, and steel. Steel is an important part of the strings. Steel has two properties that are needed. It is magnetic, so it can affect the pickups. Also, it will not rust.

The pickups on the guitar detect the vibrations of the strings. The pickup closest to the neck detects sound of low pitch (bass response). The pickup near the bridge detects higher-pitched sound (treble response). Controls allow the player to modify the sound.

The last finish sprayed on the guitar must be long lasting and attractive. The best choice is acrylic lacquer.

tuner

neck

fingerboard

strings

body

Start Here

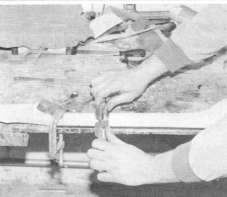

Maple is often used for the neck because it is a strong wood. The neck must be strong enough to withstand the pull of the strings without bending. The best necks use one piece of wood that runs the length of the guitar. A metal rod inside the neck helps to keep it from bending.

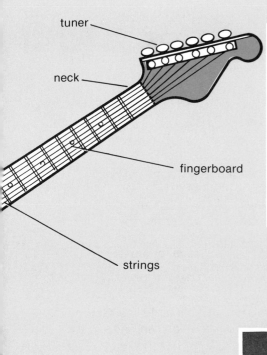

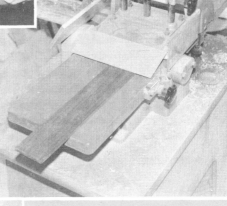

Ebony, a very hard wood, is a good choice for the fingerboard of guitars.

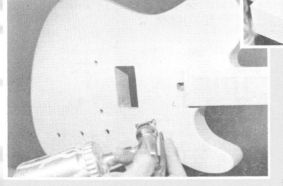

The body is made of wood. The wood must be of good quality and prepared properly. The moisture content of the air (humidity) must be controlled while the guitar is being made. The type of wood affects the quality of sound. A guitar made of dense maple will have a harsh, metallic sound. Red alder and butternut, which are less dense woods, produce a more mellow sound.

Even the finish applied to the guitar affects the sound. Some finishes deaden the sound. Other finishes change colour or crack with time.

Comparing Properties of Substances

A mechanic performs a series of tests to examine an automobile. See Figure 1.19 (a). Lab technicians perform many different tests when they examine samples sent to the laboratory. See Figure 1.19 (b). Both must keep accurate records of the tests performed and of their results. In this activity, you will compare several properties of liquids. Be sure to keep accurate records of the results.

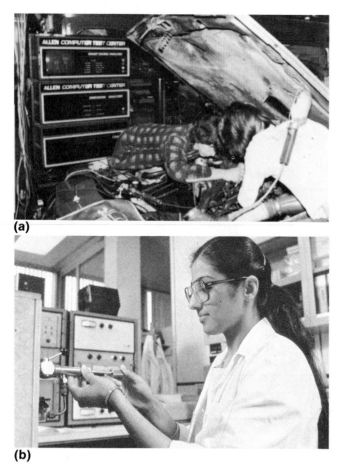

(a)

(b)

Figure 1.19. *Auto mechanics and lab technicians must perform many tests and keep accurate records.*

Activity 1E

Problem

How can you compare many properties of substances?

Materials

brine	gauze
ethanol	retort stand
water	ring clamps
ethylene glycol (50%)	test-tube clamps
iodine	hot plate
graduated cylinder	4 watch glasses
4 beakers (150 mL)	tweezers
balance	marking pen
thermometer	tape for labels

Procedure

1. In your notebook, make a table like Table 1.8. Record all your observations and measurements in this table.
2. Label four clean, empty beakers "brine," "ethanol," "water," and "ethylene glycol."
3. Put the beaker labelled "brine" on a balance. Determine the mass of the beaker and record it in your table.
4. Measure 100 mL of brine in the graduated cylinder and carefully pour it into the beaker. Record the mass of the beaker and liquid. Subtract the mass of the empty beaker, and record the mass of 100 mL of liquid.
5. Describe the qualitative properties of brine and record them in the table.
6. Repeat steps 4 and 5 for ethanol, water, and ethylene glycol.
7. Note any differences in viscosity and record these in the table.
8. Add an ice cube to each beaker. Observe and record in your table the position of the ice cube in each liquid. Did it float or sink? Remove the ice cube.
9. Wrap the bulb of a thermometer with gauze (Figure 1.20). Dip the bulb end of the thermometer in the brine. Remove it from the liquid while observing what happens to the

Table 1.8 *Observations for Activity 1E*

OBSERVATIONS	BRINE	ETHANOL	WATER	ETHYLENE GLYCOL
(a) Colour				
(b) Is the liquid clear?				
(c) Odour				
(d) Comparison of viscosity				
(e) Mass of beaker and liquid				
(f) Mass of empty beaker				
(g) Mass of 100 mL of liquid				
(h) Ice cube position in liquid				
(i) Boiling Point (°C)				
(j) Did the liquid freeze?				
(k) Evaporation time				
(l) What was left on the watch glass?				
(m) Does iodine dissolve?				

temperature. Record in your table the lowest temperature reading. Then repeat this procedure for the other three liquids.

10. Attempt to freeze each liquid by measuring equal volumes into plastic or paper cups and placing the cups overnight in the freezer compartment of a refrigerator. Observe what happens and record this in your table.

Figure 1.20 *Thermometer bulb wrapped with gauze*

11. Set out four watch glasses. Place 1 mL of one of the liquids in each watch glass (Figure 1.21). Time how long it takes for each liquid to completely evaporate. Examine the watch glass to see if anything remains after the liquid has evaporated. Record your observations.

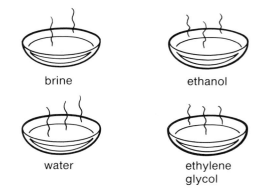

brine   ethanol

water   ethylene glycol

Figure 1.21 *Which liquid do you think will evaporate first?*

12. Using tweezers, add a crystal of iodine to each liquid. Observe and record what happens.

13. Your teacher may demonstrate how to measure the boiling point of each liquid (Figure 1.22). Record the boiling point for each.
14. Your teacher may demonstrate to you the electrical conductivity of these liquids. Record the conductivity in your table.

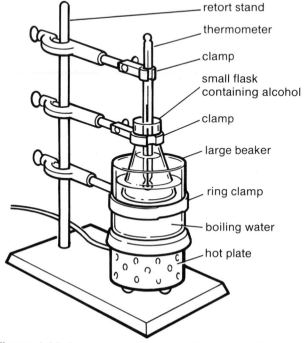

retort stand
thermometer
clamp
small flask containing alcohol
clamp
large beaker
ring clamp
boiling water
hot plate

Figure 1.22 *Apparatus for measuring the boiling point of alcohol*

Questions

1. Did any of the four liquids appear different from the others? Describe any differences.
2. Which liquid was different from the others in viscosity?
3. List the four liquids in order of density, from least dense to most dense.
4. (a) Which liquids were less dense than ice? Which liquids were more dense?
   (b) Explain how you know this.
5. (a) What happened to the temperature when the liquid evaporated from the gauze around the thermometer?
   (b) Which liquid reached the lowest temperature?
   (c) Which of the liquids evaporates the fastest?
   (d) Try to explain why your answers to (b) and (c) are the same.
6. (a) Which liquid had the highest boiling point?
   (b) Which liquid had the lowest boiling point?
   (c) Why must special care be taken around boiling alcohol?
7. (a) Which liquids did not freeze when placed in a freezer?
   (b) Explain this using your knowledge of freezing points.
8. Which liquid would be best for use as automobile antifreeze? Give two reasons why you chose this liquid over the others.
9. Which of the liquids tested would you use to make a solution of iodine crystals? Explain.

Chapter Objectives

NOW THAT YOU HAVE COMPLETED THIS CHAPTER, CAN YOU DO THE FOLLOWING?	FOR REVIEW TURN TO SECTION
1. Define matter.	1.1
2. Accurately measure volume, mass, and temperature.	1.1
3. Identify the three states of matter.	1.1
4. Recognize the difference between an observation and an inference.	1.2
5. Distinguish between qualitative properties and quantitative properties.	1.2
6. Use your senses to observe and describe qualitative properties of a substance.	1.2
7. Describe several quantitative properties of substances.	1.3
8. Explain how quantitative properties may be used to distinguish one substance from another.	1.4

Words to Know

property
matter
volume
mass
state of matter
inference
qualitative property
quantitative property

melting point
freezing point
boiling point
density
conductivity
conductor
insulator
viscosity

Tying It Together

1. Use the following words in sentences that help to illustrate their meaning.
   (a) matter
   (b) mass
   (c) volume
   (d) observation
   (e) inference
   (f) property
2. Name the equipment and units used to measure the following:
   (a) the mass of a chocolate bar
   (b) the volume of a piece of modelling clay
   (c) the volume of water
   (d) the temperature at which water boils

3. (a) List the five senses used for observation.
   (b) Explain why one of these senses must not be used in the science laboratory.
4. In your notebook, identify the following as observations or inferences.
   (a) The young boy ran out of the store.
   (b) The boy was carrying a bag.
   (c) The boy was chased by a man.
   (d) The boy must have stolen the contents of the bag.
   (e) The boy is a thief.
5. Explain the differences between qualitative and quantitative properties of a substance.
6. Identify the following as either qualitative or quantitative properties of substances:
   (a) colour
   (b) density
   (c) state
   (d) melting point
   (e) viscosity
   (f) mass
7. (a) Explain why electrical wiring is usually coated with plastic or rubber.
   (b) Why is copper used to make power lines?
   (c) If electrical equipment falls in water, it can cause a severe shock to a person who is some distance away. What property of water makes this possible?

8. Identify the property that is described in each of the following:
   (a) how much a liquid resists the flow of a marble through it
   (b) solid, liquid, or gas
   (c) the temperature at which a substance melts
   (d) how well a substance allows electricity to pass through it
   (e) the mass in a certain volume of substance
   (f) the smoothness of a substance's surface
9. Explain why the following people perform many tests, rather than just one when doing the following activities:
   (a) a mechanic identifying a problem with a car engine
   (b) a doctor examining a patient
   (c) a scientist identifying an unknown substance
10. Explain why accurate records are important in each of the cases described in question 9.

Applying Your Knowledge

1. Describe how the senses are used to make observations in each of the following situations:
   (a) a detective at the scene of a crime
   (b) a person hiking through a forest
2. Describe how temperature, mass, and volume measurements are important in the following:
   (a) cooking a meal
   (b) running a supermarket
   (c) driving and maintaining a truck
   (d) a situation of your choice
3. A piece of hardwood floats lower in the water than a piece of softwood of the same size and shape.
   (a) Use your knowledge of density to explain why this occurs.
   (b) Which would have a greater volume, 1 kg of softwood or 1 kg of hardwood? Explain.
4. Give an example of a substance which suits each of the following descriptions:
   (a) a liquid of high viscosity
   (b) a liquid of low viscosity
   (c) an electrical insulator
   (d) an electrical conductor

5. Describe one use for each substance named in question 4.
6. (a) Ethylene glycol is used as antifreeze in car engines. What properties of ethylene glycol make it useful as automobile antifreeze?
   (b) Ethylene glycol in the cooling system of your car also helps prevent the coolant from boiling in summer. What property of ethylene glycol makes this possible?
7. Ice and glass may appear similar. What properties could you use to distinguish ice from glass?

Projects for Investigation

1. Compare the viscosities of several grades of motor oil (Figure 1.23) by using the procedures outlined in Activity 1D. Research the uses of each grade and explain each in terms of its viscosity.

Figure 1.23 *Each grade of oil has specific uses based on its viscosity.*

Physical Change

Courtesy: Canada Blower and Pump Ltd.

Key Ideas

- There are many kinds of physical changes.
- Matter exists in three states: solid, liquid, and gas.
- Changes in state occur when the temperature is changed.
- An understanding of freezing and boiling points is useful in everyday life.

Steel is a strong, hard, solid substance that can be used to make cooking utensils, bridges, and other kinds of support structures. The picture here shows sheets of steel being cut by a laser. Cutting steel in this way is possible because of the extreme heat produced by the laser. The laser beam is focussed into a narrow beam that melts the steel in order to cut it. The beam is so fine that it leaves the steel looking as if it had been cut with a saw.

Melting is an example of a physical change. **Physical change** involves changes in properties that do not change the type of substance. For example, when steel melts, the melted (liquid) steel is still the same substance as solid steel. What other physical changes are there? What causes these changes? In this chapter, you will investigate physical changes and find out how they affect our lives.

Temperature and Changes in Volume

The ping-pong ball shown in Figure 2.1 was dented, but not punctured. It regained its original shape when placed in boiling water for a short time. Heating softened the shell of the ball and caused the air inside to expand, increasing its volume. This forced the surface of the ball back to its original shape. This change in the ping-pong ball is a **physical change**. Heating changes the volume of air in the ball without changing the type of substance.

Figure 2.1 *How did heating fix this ping-pong ball?*

Activity 2A

Does heating affect the volume of matter: What effect does cooling have on the volume of matter? Predict the answers to these questions. Then test your predictions by performing this activity.

PART 1

Problem

How does heat affect the volume of a solid?

> **CAUTION** Wear safety goggles when using a Bunsen burner. Do not touch the hot metal.

Materials

cold water
ball and ring apparatus
Bunsen burner

Procedure

1. Slide the ball through the ring. Then slide it back (Figure 2.2).
2. Heat the ball for 2 min in the flame of the Bunsen burner. Attempt to slide the hot ball through the ring.
3. Cool the ball by putting it in the cold water. Attempt to slide the ball through the ring again.

Figure 2.2 *Apparatus for Activity 2A, Part 1*

Observations

1. How easily did the ball fit through the ring before it was heated?
2. What happened when you tried to fit the ball through the ring after it was heated?
3. After it was cooled, how easily did the ball fit through the ring?

Questions

1. What happened to the ball when it was heated? How do you know?
2. Was this change permanent? Explain.
3. Do you think the ball would go through the ring if both were heated? Explain.
4. (a) Do you think the ball would go through the ring if the ball alone was cooled in a freezer? Explain.
   (b) Do you think the ball would go through the ring if the ring alone was cooled in a freezer? Explain.
5. What can you infer about the effect of heat on the volume of solids?

PART 2

Problem

How does heat affect the volume of liquids?

Materials

coloured water
flask
stopper and tube assembly
ring clamp
retort stand

clamp
marker pen
hot plate

Procedure

1. Assemble the apparatus as shown in Figure 2.3. Draw a neat, labelled diagram of the apparatus in your notebook.
2. Use a marker pen to mark the level of water in the tube.
3. Heat the water in the flask. Observe what happens to the level of water in the tube.
4. Allow the flask to cool and observe any change in the level of water in the tube.

Observations

1. What happened to the level of water in the tube when the flask was heated? Mark on your diagram the new level of water in the tube.
2. What happened to the water level as the flask cooled?

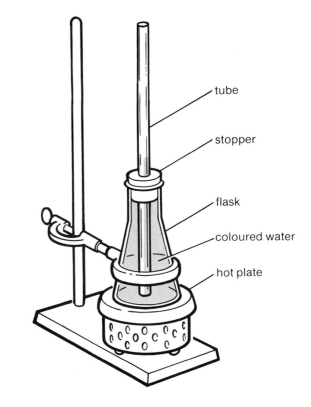

tube

stopper

flask

coloured water

hot plate

Figure 2.3 *Apparatus for Activity 2A, Part 2*

Questions

1. What caused the water level to change?
2. Was this change permanent? Explain.
3. What would happen to the water level in the tube if the flask were placed in the refrigerator? Explain.
4. What can you infer about the effect of heat on the volume of liquids?

PART 3

Problem

How does heat affect the volume of a gas?

CAUTION Due to the possibility of implosion, safety goggles should be worn by everyone as the activity is being demonstrated by your teacher. (Implosion is a bursting inwards, with a loud clap.)

Materials

coloured water
Florence flask
stopper and tube assembly
ring clamp
beaker
retort stand
Bunsen burner

Procedure

1. Your teacher will assemble the apparatus as shown in Figure 2.4. Draw a neat, labelled diagram of the apparatus in your notebook.
2. Your teacher will then carefully heat the air in the flask for 2 min. Observe what happens.
3. The air in the flask will be allowed to cool. Observe what happens.

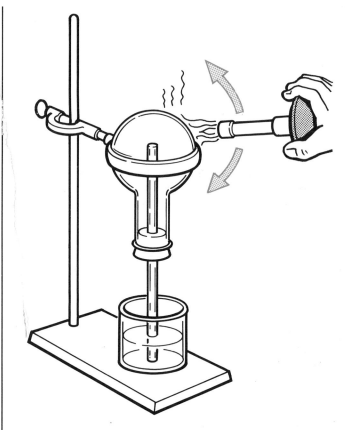

Figure 2.4 *Apparatus for Activity 2A, Part 3*

Observations

1. What happened when the air was heated?
2. What happened when the air in the flask was allowed to cool?
3. Draw the apparatus as it appeared after the air had cooled.

Questions

1. What happened to the volume of the air in the flask when it was heated?
2. Why did the water move into the flask?
3. What happens to the volume of air when it is heated?
4. What can you conclude about the effect of heat on the volume of gases?

Ideas and Applications

Different materials expand and contract by different amounts when the temperature changes. The bimetallic strip shown here is made of layers of two metals. When heated, both metals expand. But since one expands more than the other, the bar bends. The metal which expands the most forms the outside of the curve.

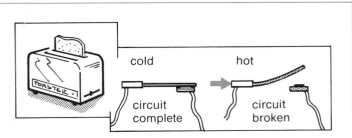

(a) *A bimetallic strip thermostat in a toaster*

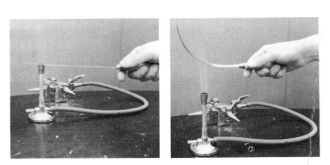

A bimetallic strip

Temperature sensing devices, called **thermostats**, use bimetallic strips or coils in their operation. Thermostats may be found in such devices as toasters, home heating systems, and indicator lights on automobiles. These use the curving of bimetallic strips to turn electric switches on or off. The cooling system of automobiles uses a bimetallic-coil thermostat. The coil opens and closes a valve, controlling the flow of coolant between the radiator and engine.

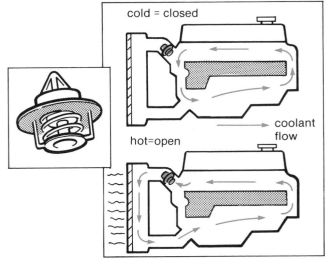
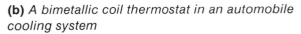

(b) *A bimetallic coil thermostat in an automobile cooling system*

What other devices can you think of that use thermostats in their operation?

The Effect of Heat on Solids

When solids are heated, their volume becomes larger. When they are cooled, their volume becomes smaller. The increase in volume is called **expansion**. The decrease in volume is called **contraction**. In all kinds of construction, the expansion and contraction of substances is very important.

Building materials expand in summer because of the heat. They contract in the cold of winter. Each of the photographs in Figure 2.5 on page 30 shows expansion and contraction of solid substances. In Figure 2.5 (a), spaces have been left between pieces of rail so that there is room for expansion to take place. Expansion joints are also built into bridges during their construction. See Figure 2.5 (b).

(a)

(b)

Figure 2.5 *How is expansion and contraction involved in these cases?*

overflow in hot conditions. What problems could occur if the overflowing coolant were allowed to escape? What problems could occur if there were no space for the expansion of the coolant as it is heated?

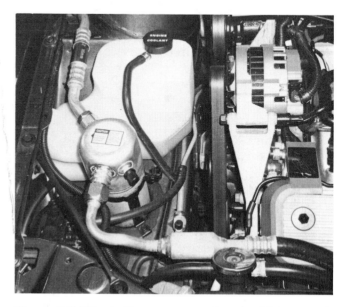

Figure 2.6 *Why is there more coolant in this reservoir when the engine is hot?*

The Effect of Heat on Liquids

Thermometers use the expansion and contraction of liquids to measure changes in temperature. As the liquid in a thermometer is heated, it expands. Liquid is pushed up the tube, indicating a higher temperature. At colder temperatures, contraction of the liquid occurs. The liquid moves down the tube, indicating a lower temperature.

Most new cars have a reservoir for collecting radiator coolant that overflows as the engine heats up (Figure 2.6). This overflow occurs because the coolant expands when heated. As the engine cools, the coolant contracts and flows back into the radiator. No coolant is lost due to

The Effect of Heat on Gases

Automobile tires contain air, which is a mixture of gases. High speed travel causes the air in the tires to become heated. The air would expand, but there is no place for it to go, so the air pressure in the tire increases. In extreme cases this may result in a blowout (Figure 2.7).

When the air in a hot-air balloon is heated, it expands (Figure 2.8). This increase in volume makes it less dense than the air around the balloon. Since it is less dense, it floats and rises into the sky. Can you think of two methods the pilot could use to make the balloon come down again? Explain how these methods involve changes in density.

Figure 2.7 *A tire blow-out*

Figure 2.8 *Expanding hot air makes this balloon rise.*

Self-check

1. (a) Define physical change.
   (b) List two examples of physical changes.
2. Describe the effect that an increase in temperature has on (a) volume and (b) mass.
3. Explain the reason for each of the following:
   (a) Spaces are left between sections of railway lines.
   (b) The liquid in a thermometer rises up the tube when its bulb is placed in a hot substance.
   (c) A hot-air balloon rises.
4. (a) Describe how a thermostat works.
   (b) Describe three common uses of thermostats.

Changes of State

A change in temperature can cause matter to change from one state to another. A **change of state** involves a change of a solid, liquid, or gas to a different state of matter. For example, water can change from solid ice to liquid water to steam, which is a gas. Ice, water, and steam have different properties. Yet they are the same type of matter. Changes of state are physical changes.

Identifying the Changes of State

Heat must be added to a substance to change it from a solid to a liquid or to a gas. As a gas changes to a liquid or a solid, heat is given off from it. In Figure 2.9 all the changes of state are shown.

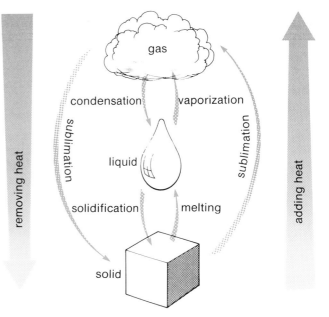

Figure 2.9 *Changes of state of matter*

- **Melting** is the change from solid to liquid. Heat must be added to a substance to make it melt. Even metals melt at high temperatures. In industry, melting may be used as a means of cutting steel.

- **Solidification** is the change from a liquid to a solid. When water solidifies, we call the process "freezing." We prefer to say that substances such as wax or molten metals "solidify."
- **Vaporization** is the change from liquid to gas. When vaporization occurs rapidly at a high temperature, we call it "boiling." Slow vaporization is known as "evaporation." Heat is required to change a substance from a liquid to a gas. When you perspire on a hot day, the evaporation of the sweat takes heat away from your body. You can speed up the evaporation and increase the cooling effect by fanning yourself.
- **Condensation** is the change from a gas to a liquid. When a gas is cooled, it may condense. Condensation causes the dew which forms on cool summer evenings. It also causes the mist which forms on the bathroom mirror while you are showering.
- **Sublimation** is the change of state from a solid directly to a gas or the change of state from a gas directly to a solid. For example, moth crystals sublime, or change directly from a solid to a gas, without becoming liquid. You can observe this sublimation because the smell of the moth crystals is noticeable. Water vapour (a gas) can sublime directly to form solid crystals of ice known as "frost."

Challenge

Put an ice cube in a beaker of corn oil and observe what happens as it melts. Try to explain what you observe using your knowledge of density.

Ice floats in liquid water (Figure 2.10). This happens because ice is less dense than water. No matter is added or removed during freezing, so the mass remains the same. Thus, any change in density must be due to a change in volume. Predict what happens to the volume of ice when it melts. Then perform the activity to test your prediction.

Figure 2.10 *Ice floats in water.*

Problem

How does melting affect the volume of water?

Materials

graduated cylinder
ice
water

Procedure

1. Make a table similar to Table 2.1 in your notebook.
2. Pour 50 mL of water into a graduated cylinder. Add ice to bring the volume up to 100 mL.
3. Allow the ice in the cylinder to melt completely. Once the ice has melted, read the volume of water in the graduated cylinder. Record this volume in your table.

4. Share results with other members of your class. Record their data in your table.
5. Calculate the average volume for the water after melting. Record the average in your table.

Observations

Table 2.1 *Data for Activity 2B*

GROUP	VOLUME AFTER MELTING (mL)
1	
2	
Etc.	

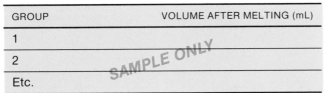

Questions

1. Was the volume greater before or after melting? Explain.
2. What happens to the volume of ice when it melts?
3. What happens to the volume of water when it freezes?
4. Explain why ice is less dense than water.

Ideas and Applications

Have you ever seen a special effects machine that can produce "smoke" at concerts or plays? The machine uses dry ice, a fan, and two changes of state to produce this effect. Dry ice is carbon dioxide in the solid state. Above −79° C, dry ice sublimes, forming carbon dioxide gas, which is still very cold. When ordinary air comes in contact with this cold gas, the water vapour in the air condenses. This forms a mist made up of very small droplets of liquid water.

A "smoke machine" blows air over blocks of dry ice. The air is cooled, and water vapour condenses to form vast streams of small water droplets. The droplets remain in the air and look like smoke.

Water in the Liquid and Solid State

As water freezes, it expands. If water in a pipe freezes, the expansion can cause the water pipe to burst. For this reason, pipes carrying water to outside taps must be drained before winter.

The cells of your body contain water. Exposure to the cold can cause this water to freeze. As the water freezes, it expands. It breaks open cells and causes a condition known as "frostbite." What precautions can people take to avoid frostbite in winter?

Since ice is less dense than water, it remains on the surface of lakes in winter. The layer of ice acts as an insulating barrier. It allows fish and other organisms to survive and protects them during the harsh winter months (Figure 2.11). If thick enough, the ice also allows us to travel over a frozen lake surface on cross-country skis, on skates, or by snowmobile.

Figure 2.11 *Ice fishing on a frozen lake surface*

Challenge

What effect does freezing have on the mass of water? Design an experiment to test your prediction.

Problem

How does heating affect water?

Materials

completed data table from Activity 1C, Part 2

Procedure

1. Following the instructions given by your teacher, draw a graph of temperature vs. time using the data from Activity 1C, Part 2.

Observations

1. What was happening to the temperature of the water in those sections of the graph which are sloping?
2. What was happening to the temperature of the water in those sections of the graph which are level?
3. What changes occurred to the water in those sections of the graph where the temperature remained constant?

Questions

1. How can you identify the melting point and boiling point of water from your graph?
2. How does heating affect water at 0° C?
3. How does heating affect cold water?
4. How does heating affect water at 100° C

Challenge

Does the temperature of a substance remain constant as it solidifies? Design an experiment to test this. Get your teacher's permission before testing your experiment.

How Does Temperature Affect Changes of State?

If a substance is heated, the heat may cause the temperature to rise. But if the substance is at its freezing point or boiling point, the heat may cause it to change state. Figure 2.12 on page 35 shows how temperature rises as a substance is heated over a period of time.

Challenge

Obtain three beakers filled with crushed ice. Add no salt to one beaker, two scoops of salt to another, and four scoops of salt to the third. Stir the ice and salt mixtures. As the ice melts, measure the temperature in each beaker. In which one is the melting point the lowest? Repeat this experiment, mixing a different substance with the crushed ice. You could try antifreeze, methyl hydrate, sugar, baking soda, or any substance suggested by your teacher.

Applications of Changes of State

Many substances have specific melting and boiling points. Pure water, for example, freezes at 0°C and boils at 100°C. These specific temperatures may be changed by dissolving other substances in the water. Dissolving means mixing substances together so thoroughly that the parts cannot be told apart. Sugar and water can be mixed together in this way. (You will learn more about this in Unit Two.) Water containing dissolved substances remains liquid even below 0°C and only boils when it is heated above 100°C.

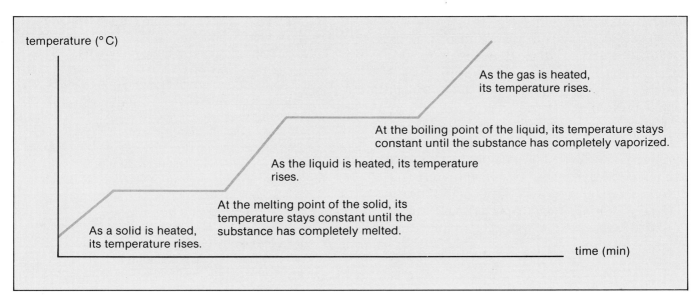

Figure 2.12 *Temperature versus time graph (heat added)*

Text inside Figure 2.12:

temperature (°C)

As the gas is heated, its temperature rises.

At the boiling point of the liquid, its temperature stays constant until the substance has completely vaporized.

As the liquid is heated, its temperature rises.

At the melting point of the solid, its temperature stays constant until the substance has completely melted.

As a solid is heated, its temperature rises.

time (min)

You have probably seen salt spread on snowy roads in winter (Figure 2.13). The mixture of snow and salt has a melting point lower than 0°C. This causes the mixture to be a liquid even at temperatures below 0°C and makes driving safer. However, if the temperature dips below -15°C, the mixture itself will freeze, creating extremely dangerous driving conditions. In very cold parts of the country, sand is used instead of salt. Although sand does not melt the snow, it makes the snow less slippery.

Dissolved substances are also used in nature to lower melting points. Sugar in the sap of trees prevents freezing of this liquid in winter. Should freezing occur in the buds of these trees, expanding ice could damage the cells.

Ethylene glycol (antifreeze) is added to the cooling system of an automobile. It lowers the freezing point of the liquid, thus preventing damage due to freezing in winter. Different mixtures of antifreeze and water are used, depending on the low temperatures expected.

Figure 2.13 *Why is salt spread on icy road surfaces in winter?*

A mixture of 50 parts antifreeze to 50 parts of water allows a car to run at temperatures as low as -35°C. If there are 70 parts antifreeze to 30 parts of water it can run at temperatures as low as -64°C. What other liquids used in automobiles must be prevented from freezing?

The addition of antifreeze also helps to prevent the cooling system from boiling in the summer. The mixture has a higher boiling point than pure water. For this reason, antifreeze is added to the cooling system throughout the year. What do you think would happen if pure water were used in the cooling system of an automobile?

Self-check

1. Identify the following changes of state:
   (a) change from solid to gas
   (b) change from solid to liquid
   (c) change from gas to liquid
   (d) change from liquid to solid
   (e) change from liquid to gas
   (f) change from gas to solid
2. Give one example of each of the changes of state. Try to use an example not already given in this book.

3. (a) Which changes of state require heat to be added?
   (b) Which changes of state require heat to be removed?
4. Describe the effect of freezing on the following:
   (a) mass of water   (c) density of water
   (b) volume of water
5. (a) Why does ice float on water?
   (b) How does this property of ice affect people in Canada? Give three examples.
6. The graph of temperature vs. time for substance X is shown in Figure 2.14. Use this graph to answer the following questions:
   (a) What is the melting point of substance X?
   (b) What is the boiling point of substance X?
   (c) In what state is substance X at 10°C?
   (d) In what state is substance X at 100°C?
   (e) Is substance X being heated or cooled?
7. What effect do dissolved substances have on
   (a) the freezing point of a liquid?
   (b) the boiling point of a liquid?
8. Sap in the buds of plants contains water. Explain why this water does not freeze in winter.
9. Explain why salt is not used on roads in areas with temperatures below -15°C in winter.
10. Explain why antifreeze is left in the cooling system of a car in summer as well as in winter.

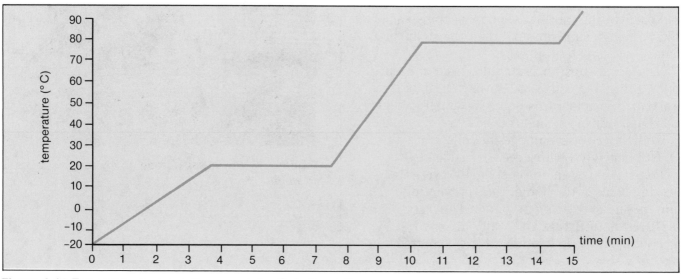

Figure 2.14 *Temperature versus time graph for Substance X*

Chapter Objectives

NOW THAT YOU HAVE COMPLETED THIS CHAPTER, CAN YOU DO THE FOLLOWING?	FOR REVIEW TURN TO SECTION
1. Define physical change and list several examples.	2.1
2. Describe how changes in temperature affect the volume and density of matter.	2.1
3. Describe several practical applications of the effect of temperature on the volume of matter.	2.1
4. Name and define the six changes of state.	2.2
5. Give an example of one application of each change of state.	2.2
6. Explain how changes in volume and density as water freezes affect our lives.	2.2
7. Demonstrate that changes of state involve the addition or removal of heat.	2.2
8. Identify the melting and boiling point of a substance from a graph of temperature vs. time.	2.2
9. Describe the effect that dissolved substances have on the boiling points and melting points of substances.	2.2
10. List several useful applications of changes in the freezing point and boiling point of water.	2.2

Words to Know

physical change
expansion
contraction
change of state
melting
solidification

vaporization
condensation
sublimation

Tying It Together

1. Define physical change and list three examples.
2. In your notebook identify each of the following statements as true or false. Correct any false statements.
   (a) Matter expands when heated.
   (b) All types of matter expand at the same rate when heated.
   (c) The volume of matter decreases when it is cooled.
   (d) The mass of matter increases when it is heated.
   (e) The density of matter increases when it is heated.
3. (a) Explain why hot air rises.
   (b) What happens to this air as it cools?
4. Explain why changes in volume and density are considered physical changes.
5. Explain why changes of state are considered physical changes.
6. Define the following changes of state. For each, tell whether heat must be added to or removed from the substance.
   (a) melting
   (b) condensation
   (c) sublimation
   (d) vaporization
   (e) solidification
7. Identify the changes of state involved in the following situations:
   (a) The refrigerator changes water into ice.
   (b) Water droplets form on the cold surface of a mirror.

(c) Ice cubes decrease in size if left for a long time in the freezer.

(d) Puddles of water disappear on a hot day in summer.

(e) A block of dry ice slowly disappears as it turns to carbon dioxide gas.

(f) A fog forms around cold carbon dioxide gas.

8. Complete the following statements as you write them in your notebook.

Word choices: more, volume, less, density, increases, decreases

(a) Water _____ in volume as it freezes to ice.

(b) The _____ of water decreases, when it freezes.

(c) Because ice is _____ dense than water, ice floats on liquid water.

9. Water was left in the pipes of a cottage over the winter and the pipes burst. Explain the cause of this problem.

10. A pure liquid has a freezing point of 20° C and a boiling point of 80° C. If another substance were dissolved in this liquid,

(a) what might the new freezing point be? Explain.

(b) would you expect the boiling point of the mixture to be 80° C? Explain.

Applying Your Knowledge

1. Explain the cause of the following events:

(a) Spaces are often left between girders in buildings.

(b) Heating a tight metal lid on a jar of jam allows it to be easily removed.

(c) Hydro wires hang lower in summer than in winter.

(d) Coolant in a hot car overflows into a container and then flows back into the radiator as the car cools.

(e) A bimetallic strip in a thermostat bends when heated.

(f) Aerosol cans explode when heated.

2. Substance X is heated from –10° C to 80° C. If the melting point of substance X is 10° C and its boiling point is 50° C, draw a graph of temperature vs. time for substance X.

3. Urea pellets, commonly used as fertilizer, can also be spread on walkways to melt ice. Explain why the ice melts.

4. The guide which accompanies most cars tells the owner to use the correct mixture of water and antifreeze.

(a) Explain why pure water should not be used.

(b) Should more or less antifreeze be added if you are travelling to colder climates? Explain.

(c) Salt would have a similar effect on the freezing point of water. Why is this dissolved substance not recommended in this case.?

5. A cook notices that vegetables cook faster if salt is added to the water. Explain a possible reason for this.

6. Ice cream melts at –5° C. You want to keep ice cream in a cooler, but the ice, at 0° C, is too warm. How could you lower the melting point of the ice below –5° C in order to keep the ice cream frozen?

Projects for Investigation

1. Heat pumps are machines that transfer heat from one place to another. Air conditioners and refrigerators are heat pumps. Find out how these and other heat pumps use changes of state to transfer heat from one place to another.

2. Research the method used by the pilot of a hot-air balloon to control the height to which the balloon rises. How does the pilot bring the balloon back to the ground? Explain how these methods involve changes in density.

3. The cooling system of an automobile is sealed with a cap to raise the pressure. Find out why the cooling system is kept under pressure. (Hint: The reason is connected with boiling point.)

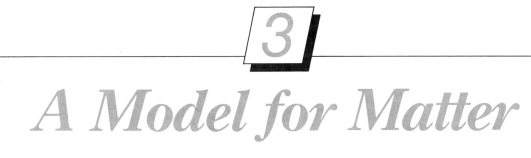

3

A Model for Matter

Key Ideas

- Models can help us explain the behaviour of matter.
- The model of matter is called the particle model.
- The particle model is useful for explaining physical properties of matter and changes in matter.

Would you like to be able to drive a car like the one shown here? That is what the auto makers would like to find out when they display a vehicle such as this. In fact, this car is not for sale. It is only a model. The model vehicle is used to explain the behaviour of new design features, such as gull-wing doors or an improved engine. Your reaction to the model helps the auto makers predict what automobile features the public will want in the future.

Models are often used in science to explain behaviour and to predict future behaviour. In this chapter, you will develop a model for matter. You will use this model to explain why matter behaves in certain ways. Then you will test this model to see if it can explain other ways in which matter behaves.

Designing Models

Why are some substances solid? Why are other substances liquids or gases? What causes substances to expand when they are heated and contract when they are cooled? What really happens when liquid water changes to ice and then changes back to the liquid state again?

Even in ancient times, people wondered about the answers to these questions. Then, as now, no one could see what matter was made of. Scientists tackled this problem by designing a model of matter. This model was not based on what matter looks like. Instead, the model was based on how matter behaves.

The first step in designing a model is to gather accurate observations. This is not always an easy task in science. Many things cannot be seen. For example, you cannot see the force that attracts or repels magnets. You cannot see the gases in the air we breathe. If you cannot see something, you must rely on its behaviour to help you design a model for it. In Activity 3A, you will have a chance to design models for several items based only on their behaviour.

Activity 3A

Problem

How can you design an accurate model based only on behaviour?

Materials

numbered boxes containing unknown items (bolts, marbles, etc.)
modelling clay

Procedure

1. Design a table to record all your observations.
2. Pick up a numbered box. Record the number in your table.
3. Without opening the box, observe as much about the object inside as possible. These questions will help you: Does the object roll or is it flat? Is the object heavy or light? How large is the object in relation to the box? Is there one object or many?
4. When you have made as many observations as you can, try to infer what you think the object(s) in the box could be.
5. Design and build a model of the object(s) using modelling clay. This model should behave in the same ways as you have observed the object in the box to behave. Draw a diagram of your completed model in your table.
6. Your teacher will have a list of the actual objects in each box. Record this in your table.

Observations

1. Make sure all your observations are recorded in your table.

Questions

1. Compare your model with the observations you made. In which ways does your model behave like the actual object?
2. In which ways is your model different from the actual object it represents?
3. What other observations could you have used to make your model more accurate?

When you open a bottle of perfume, the smell of the perfume spreads throughout the room. The perfume becomes mixed with the air. This gradual mixing together of substances is called **diffusion**. Diffusion occurs in both liquids and gases.

PART 1

Problem

How can you observe diffusion in a liquid?

Materials

potassium permanganate crystals	water
petri dish	tweezers

> **CAUTION** Potassium permanganate stains skin and clothing. Be careful when handling this substance.

Procedure

1. Fill a petri dish half-full with water. Allow it to sit on a flat surface for about 1 min to let the water become perfectly still. See Figure 3.1 (a).
2. Using the tweezers, carefully place one crystal of potassium permanganate in the centre of the dish. See Figure 3.1 (b).
3. Observe the crystal and the water in the plate for 5 min.

Observations

1. Record your observations in your notebook. Draw the petri dish as it appeared after 5 min.

Questions

1. Did the potassium permanganate move in any particular direction?
2. Draw a diagram showing what the petri dish would look like
   (a) after 10 min
   (b) if it were left undisturbed for 24 h.

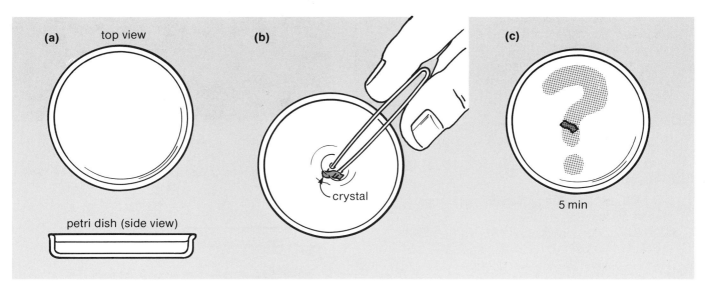

Figure 3.1 *What will happen to the potassium permanganate crystal?*

PART 2

Problem

Does temperature affect the rate of diffusion?

Materials

potassium permanganate crystals
hot and cold water
petri dishes
tweezers

> **CAUTION** Potassium permanganate stains skin and clothing. Be careful when handling this substance.

Procedure

1. Label one dish "hot" and the other "cold." Fill these dishes half-full with hot or cold water and allow them to sit for about 1 min until the water becomes perfectly still.
2. Use the tweezers to place one crystal of potassium permanganate in the centre of each dish at the same time.
3. Allow the potassium permanganate to diffuse for 5 min and observe the dishes at this time.

Observations

1. Record your observations in your notebook.

Questions

1. In which dish had the potassium permanganate spread more?
2. What effect does increasing temperature have on the rate of diffusion?
3. Try to explain why diffusion occurs.

Using a Model to Explain Diffusion

What causes diffusion? We can explain this behaviour by using a model. In the **particle model** of matter, all matter is assumed to be made of tiny particles. The particles of one substance are all the same. For example, all the particles of water are alike, and all the particles of perfume are alike. These particles are constantly moving about in many directions. This motion with no definite direction is called **random motion** (Figure 3.2).

Figure 3.2 *The dust particles can be seen moving in many directions. The particles in matter are smaller in size than dust particles, but they also display this random motion.*

Challenge

Do all substances diffuse at the same rate? Design an experiment to test this. Get your teacher's permission before performing your experiment. Then use the particle model to explain your observations.

Particles move about randomly. They collide with other particles and are bounced about in all directions. They slowly spread out and mix with any other particles that are around them. They diffuse, as shown in Figure 3.3.

Figure 3.3

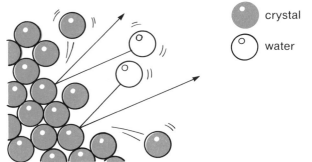

crystal

water

(a) *Particles of water collide with the crystal causing some crystal particles to break free.*

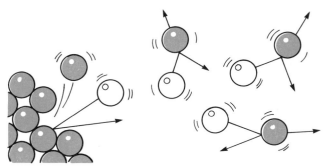

(b) *The particles in the water continue to be bounced about by water particles. This causes them to move out from the crystal.*

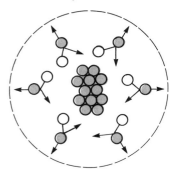

(c) *Since the motion occurs in all directions, the substance spreads out in a circular pattern.*

Diffusion occurs faster at higher temperatures. This can also be explained using the particle model (Figure 3.4). The particle model states that the particles of matter move faster at higher temperatures. Since the particles move faster, they spread out and mix faster.

In the next section, you will find out more about the particle model. You will also investigate more about the behaviour of matter.

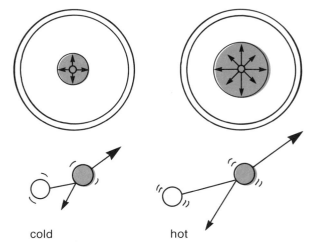

cold   hot

Figure 3.4 *Since particles move faster at higher temperature, they also spread faster by diffusion.*

Self-check

1. Describe two situations in which models are used to explain or predict behaviour.
2. Why are models used in science?
3. If you cannot see something, what observations can you use to help you make a model of it?
4. (a) Define diffusion.
   (b) Why do substances spread out by diffusion?
   (c) How is diffusion affected by temperature?
5. Draw a diagram to represent the motion of a particle of matter.

Using the Particle Model

The particle model has proven useful in explaining diffusion and the effect of temperature on diffusion. This model can also be used to explain other ways matter behaves. For this, we need to add two more parts to the model.

The particles of substances do not touch each other. They are separated by empty space. The particles also attract each other. The closer together they are, the more strongly attracted they are. All five points of the particle model are listed in Figure 3.5.

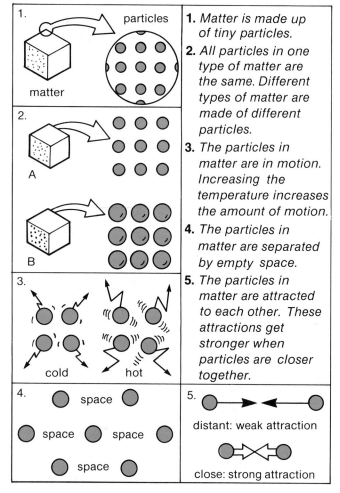

Figure 3.5 *The particle model for matter*

1. *Matter is made up of tiny particles.*
2. *All particles in one type of matter are the same. Different types of matter are made of different particles.*
3. *The particles in matter are in motion. Increasing the temperature increases the amount of motion.*
4. *The particles in matter are separated by empty space.*
5. *The particles in matter are attracted to each other. These attractions get stronger when particles are closer together.*

Expansion and Contraction of Matter

All matter—solid, liquid, and gas—expands when heated and contracts when cooled. Can we now use the particle model to explain the cause of this behaviour?

The particle model states that matter is made of particles in motion. This motion increases with an increase in temperature. As the particles are heated, their motion increases, and they collide more with their neighbours. These collisions force the particles to move farther apart. This results in an increase in volume (Figure 3.6).

When matter cools, the particles slow down, resulting in fewer collisions. This allows the particles to move closer together and causes contraction.

Figure 3.6

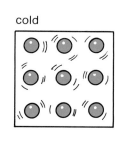
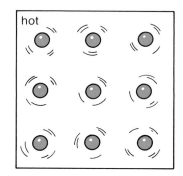

(a) *In a cold substance, the particles are close and moving slowly.*

(b) *In a hot substance, the faster moving particles move farther apart.*

The States of Matter

The solid state occurs at the lowest temperature. When heated, solid matter changes to a liquid and then to gas. The particle model states that heating causes the particles of matter to move faster. Thus, the particles of matter in a gas move faster than those in a liquid and much faster than those of a solid.

What makes a solid keep its rigid shape even when it is placed in a different container? The particles in a solid are arranged in a closely packed pattern. There are strong attractions between them (Figure 3.7). Although the particles vibrate, they cannot change places.

The attractions are weaker in a liquid. The particles of a liquid can move past each other so the liquid takes the shape of its container (Figure 3.7). The attractions are still strong enough to hold the liquid particles together.

Gas particles are so far apart and the attractions are so weak that they behave almost independently of each other. The particles of gas spread randomly, always taking the shape and volume of the container (Figure 3.7). This information is summarized in Table 3.1.

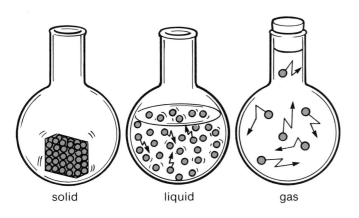

solid   liquid   gas

Figure 3.7 *Particles in the solid, liquid, and gas states*

Table 3.1 *The States of Matter and the Particle Model*

STATE OF MATTER	SOLID	LIQUID	GAS
Particle motion	Slowest	Faster	Fastest
Particle spacing	Close together	Farther apart	Farthest apart
Attraction of particles for each other	Strong	Weaker	Weakest

Ideas and Applications

When a substance is cooled, its particles are closer together. For most substances, this means that the solid state is denser than the liquid. If the solid and liquid are mixed together, the solid sinks below the liquid. But water has one very unusual property. The solid state (ice) is less dense than the liquid state. Thus, ice floats on water. This property occurs because of the way the water particles fit together in the rigid structure of ice. The spaces between the particles are smaller in liquid water. Since the particles fit closer together, the liquid is denser.

Life in Canada would be very different if water did not behave this way. If ice sank to the bottom of lakes, they would freeze solid in winter, and the heat of summer would only melt the surface. How might that affect the living things in the lakes? How would it affect our summer recreation? How do you think Canada's climate might be affected if lakes remained frozen all summer?

Changes of State

According to the particle model, heat causes the particles of a substance to move about more. Particles in a solid, which are held in a fixed position, vibrate faster when they are heated. When they vibrate hard enough, the attractions can no longer hold them in a rigid pattern. See Figure 3.8 on page 46. As the pattern collapses, the substance becomes a liquid, flowing to take the shape of its container. For water, the temperature at which this occurs is 0°C. This temperature is the melting point of water. Since particles of different substances have different attractions, their melting points are different.

Vaporization occurs when heat causes the particles to move even faster, thus weakening the attractions between them. Eventually some particles move fast enough to break free from their neighbours. These particles escape from the liquid and move independently (Figure 3.8). At the boiling point, many gas particles escape at one time in the form of large bubbles.

When heat is removed from a gas, the particles slow down and move closer together. There are greater attractions between particles, and the gas condenses into the liquid state. Continued cooling causes the liquid to solidify. This occurs as its particles form into rigid patterns because of the increasing strength of the attractions.

Figure 3.8 *The particle model and changes of state*

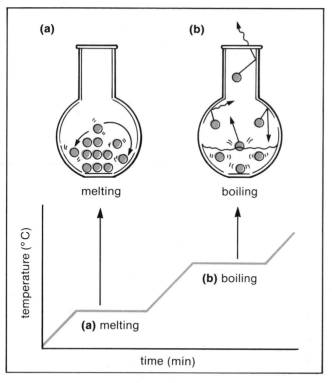

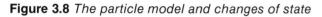

(a) *When the attractions holding solids together are overcome, the rigid pattern breaks down and the solid becomes a liquid.*

(b) *When the attractions holding liquid particles together are overcome, some particles break free and escape as a gas.*

You have used the particle model to explain the behaviour of matter. Now, try to use it to explain the results of this activity.

Problem

What causes the change in volume when water and alcohol are mixed?

Materials

water
ethyl alcohol
2 graduated cylinders

Procedure

1. Set up a table in your notebook similar to Table 3.2.
2. Pour exactly 50 mL of water into one graduated cylinder and exactly 50 mL of alcohol into another (Figure 3.9). Record these volumes in your table. What volume would you expect the mixture to have when the two liquids are combined?
3. Add the 50 mL of alcohol to the water and observe the final volume of the mixture. Record this in your table.

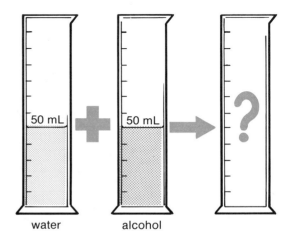

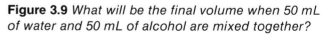

Figure 3.9 *What will be the final volume when 50 mL of water and 50 mL of alcohol are mixed together?*

Observations

Table 3.2 *Changes in Volume on Mixing*

CYLINDER CONTENTS	VOLUME (mL)
A Exact volume of water alone	
B Exact volume of alcohol alone	
C Total of the two above volumes (A + B = C)	
D Measured volume of the mixture	

SAMPLE ONLY

Questions

1. How did the actual volume of the mixture compare with the expected value?
2. Describe a possible reason for this change in volume. (Hint: Look at Figure 3.10.)

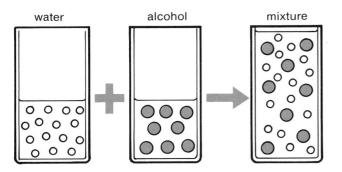

water   alcohol   mixture

Figure 3.10

Challenge

When water and alcohol are mixed, is there any change in the total mass? Design an experiment to find out. Using the particle theory, predict what the mass of the mixture will be. Then, with your teacher's permission, test your prediction.

Self-check

1. In your notebook state if each of the following statements is true or false. Correct any false statements.
   (a) Matter is made of particles.
   (b) All substances are made of the same kind of particles.
   (c) The particles of matter are in motion.
   (d) Heating slows down the motion of particles.
   (e) Particles all move in the same direction.
   (f) Particles of matter are separated by spaces.
   (g) The spaces between particles get smaller when matter is heated.
   (h) Attractions exist between the particles in matter.
   (i) Heating matter weakens the attractions between particles.
2. (a) Use the particle model to explain why matter expands when heated.
   (b) Use the particle model to explain why matter contracts when cooled.
3. (a) In which state is the temperature lowest?
   (b) In which state are the particles moving fastest?
   (c) In which state are the particles closest together?
   (d) In which state are the attractions weakest?
4. Use the particle model to explain the cause of the following changes of state:
   (a) Solids melt when heated.
   (b) Liquids vaporize when heated.
   (c) Gases condense when cooled.
   (d) Liquids freeze when cooled.
5. When 30 mL of water are mixed with 30 mL of alcohol, the total volume is found to be less than 60 mL. Explain this change in volume, using the particle model.

Challenge

The diesel engine compresses air to the point that it gets hot enough to ignite the fuel. Give a possible reason why compressing air releases heat. Explain your reason in terms of the particle theory.

47

Chapter Objectives

NOW THAT YOU HAVE COMPLETED THIS CHAPTER, CAN YOU DO THE FOLLOWING?	FOR REVIEW TURN TO SECTION
1. Describe how models are used to explain and predict behaviour.	3.1
2. Design a model based on observed behaviour.	3.1
3. Define and give examples of diffusion.	3.1
4. List the five points in the particle model of matter.	3.1
5. Use the particle model to explain diffusion and the effect of heat on diffusion.	3.1
6. Use the particle model to explain expansion and contraction of matter.	3.2
7. Compare the motion of the particles in solids, liquids, and gases.	3.2
8. Use the particle model to explain changes of state.	3.2

Words To Know

diffusion
particle model
random motion

Tying It Together

1. Complete the following statements as you write them in your notebook. Word choices: particle, predict, explain
   (a) Models are used to ▓▓▓▓ behaviour and to ▓▓▓▓ future behaviour.
   (b) In this chapter you used the ▓▓▓▓ model to ▓▓▓▓ the behaviour of matter known as diffusion.
   (c) You also used the model to ▓▓▓▓ the cause of the change in volume when water and alcohol were mixed.
2. A bottle of coloured gas is opened in a room with no air currents. Sketch the room as seen from above
   (a) 1 min after the bottle was opened
   (b) 1 h after the bottle was opened.
3. State the five points in the particle model of matter.
4. (a) What is diffusion?
   (b) Use the particle model to explain why substances spread by diffusion.
   (c) Diffusion occurs in all directions. Explain why, using the particle model.
   (d) Use the particle model to explain why diffusion occurs faster at higher temperatures.
5. In your notebook, state whether the following statements are true or false. Correct any false statements.
   (a) The particle model states that the different types of matter have different types of particles.
   (b) Matter expands when heated because the particles get larger in size.
   (c) The density of matter decreases when the matter is heated because the particles are spread further apart.
   (d) Matter contracts when it is cooled because the particles move closer together.
   (e) When matter is cooled, its density increases.

6. Name the state of matter which
   (a) has the highest temperature
   (b) has the least motion of its particles
   (c) has particles which are furthest apart
   (d) has the strongest attractions between its particles.
7. Use the particle model to explain why it easier to compress a gas than a liquid or a solid.
8. Explain the following, using the particle model:
   (a) why solids have a hard, rigid shape
   (b) why liquids take on the shape of their container but keep the same volume
   (c) why gases take on the shape and volume of their container
   (d) what happens when solids melt
   (e) how liquids vaporize
   (f) why gases condense when they are cooled
   (g) what happens when liquids freeze
9. Use the particle model to explain why the total volume decreases when alcohol and water are mixed together to form a solution.

Applying Your Knowledge

1. Name one group of people, other than scientists, who use models in their work. Describe the models used, and explain how they are used on the job.
2. A chemical leak occurs in a pipeline built through a pond of still water. Describe how you could use your knowledge of diffusion to locate the source of the leaking chemical.
3. Snowflakes form from water vapour in the air by sublimation. Use the particle theory to explain this change of state.
4. Hot-air balloons rise when the air they contain is heated. Explain in terms of the particle theory why hot air is less dense than normal air.
5. Someone suggests that the decrease in volume when alcohol and water are mixed is due to evaporation. How could you test this idea?

6. Every model has its limitations. To illustrate this,
   (a) identify one model used in daily life
   (b) describe ways in which the model is similar to the real item it represents
   (c) describe ways in which the model differs from the real item it represents.
7. Use the particle model to explain the following. Diagrams may be helpful.
   (a) Why can a gas be compressed into a smaller space?
   (b) Why do different substances have different densities?
   (c) Why does an elastic band get hot when it is stretched repeatedly?

Projects for Investigation

1. Items cook rapidly in a microwave oven. Microwaves put energy into the water particles inside the food. Research how microwaves work and use the particle model to explain why microwaves cook food so quickly.
2. Before the particle model was proposed, there were other models to explain the behaviour of matter. One model was proposed by the Greek philosopher Aristotle. Aristotle was not a scientist, but he was so well respected that his ideas about matter were not challenged for nearly 2000 years. Find out about Aristotle's model for matter and about the early scientists who were finally able to change people's thinking about matter. You might try looking in an encyclopedia under *alchemy*, *Robert Boyle*, *Joseph Priestley*, and *Antoine Lavoisier*.

Unit One: Physical Properties and Physical Changes

MATCH

In your notebook, write the letters (a) to (j). Beside each letter, write the number of the word in the right column that corresponds to each description in the left column.

(a) characteristic used to describe or identify a substance	1. inference
	2. balance
	3. physical property
(b) used to explain the behaviour of matter	4. solid
	5. physical changes
(c) keeps its shape and volume when placed in a new container	6. observation
	7. expansion
	8. particle model
(d) study using the five senses	9. liquid
	10. condensation
(e) changes in properties that do not change the type of substance	11. graduated cylinder
	12. gas
(f) increase in volume caused when matter is heated	
(g) used to measure mass	
(h) a reasoned conclusion based on observations	
(i) used to measure volume	
(j) has fast moving particles that are spaced far apart and attract each other weakly	

MULTIPLE CHOICE

In your notebook, write the numbers 1 to 10. Beside each number, write the letter of the best choice.

1. This equipment is used to measure volume.
   (a) graduated cylinder
   (b) overflow can
   (c) ruler
   (d) All of these can be used to measure volume.

2. Which product would be packaged in kilogram units?
   (a) vitamin pills
   (b) soft drink
   (c) flour
   (d) chocolate bar

3. A substance is observed to keep its same volume as it takes on the shape of a new container. In what state is this substance?
   (a) solid
   (b) liquid
   (c) gas
   (d) It's impossible to tell from this information.

4. Which of the following is an example of an inference?
   (a) My car has stopped at an intersection.
   (b) Another car is approaching the intersection from the left.
   (c) The other driver is signalling a left turn.
   (d) The other driver will make a left turn.

5. Which list of qualitative properties best describes iron?
   (a) brown, shiny, smooth solid
   (b) colourless, odourless liquid
   (c) brown, dull, powdery solid
   (d) grey, shiny, smooth solid

6. In which state are the particles of matter moving fastest, spaced farthest apart, and most weakly attracted to each other?
   (a) solid
   (b) liquid
   (c) gas
   (d) All states are exactly the same.

Questions 7 to 10 refer to the graph in Figure 2.14 in Chapter 2.

7. What is the melting point of this substance?
   (a) -10° C
   (b) 20° C
   (c) 50° C
   (d) 80° C

8. At 15° C, in which state is this substance?
   (a) solid
   (b) liquid
   (c) gas
   (d) It is a mixture of solid and liquid.

9. If 10 g of this substance was cooled from 90° C to 50° C, it would become
   (a) 9 g of liquid
   (b) 10 g of gas
   (c) 9 g of solid
   (d) 10 g of liquid
10. When heated from 15° C to 30° C, what will happen to the volume of this substance?
   (a) increase
   (b) decrease
   (c) remain the same
   (d) change to fit the volume of its container

TRUE/FALSE

Write the numbers 1 to 10 in your notebook. Beside each number, write T if the statement is true and F if the statement is false. For each false statement, rewrite the statement so it becomes true.

1. Matter is anything that has mass and volume.
2. Quantitative properties may be observed directly, but qualitative properties must be measured.
3. Density is a quantitative property of matter.
4. Ice floats on water because water is less dense than ice.
5. Viscosity is a measure of how well a substance allows electricity to pass through it.
6. Volume is measured with a balance.
7. Diffusion occurs because the particles of matter spread out in all directions.
8. Diffusion occurs more slowly at high temperatures because the particles of matter move more slowly when heated.
9. Changes of state involve breaking and forming attractions between particles.
10. Gases are easier to compress than liquids because the particles are closer together in a gas.

FOR DISCUSSION

Read the paragraphs and answer the questions that follow.

Paramedics are specialized ambulance attendants. They must be excellent observers. They must also be able to use measuring equipment quickly and accurately.

Upon arriving at the scene of an accident, paramedics first observe the scene. They locate victims and observe any wounds they may have suffered. A small flashlight is used to shine in the victim's eyes. Observing the reaction to the light helps the paramedics determine whether the victim has suffered a concussion or has gone into shock. They may also feel for areas of tenderness or for broken bones.

Always alert, paramedics must be aware of smells such as alcohol on the breath of a victim or gas fumes leaking from damaged vehicles. A stethoscope may be used to listen for fluids in the victim's lungs or abdomen. If fluid is detected, this can indicate that there are internal injuries.

Paramedics also make use of measuring skills. They may use a pressure cuff to determine blood pressure. A thermometer is used to measure the victim's temperature. A watch comes in handy for determining pulse rate and breathing rate.

All of the data collected by paramedics greatly assists doctors in treating the patient upon arrival at the hospital.

1. What two skills do paramedics need on the job?
2. Describe how paramedics use their senses to observe the scene of an accident.
3. List four measurements a paramedic might make at the scene of an accident.
4. Why is it important to make quick, accurate observations and measurements?
5. Observation and measurement skills are also important in the home.
   (a) List two ways in which you use each of your senses in the home.
   (b) List three measurements used regularly in the home and describe how each is used.

Mixtures

You may have heard the expression, "oil and water don't mix." Have you ever tried mixing them? Keeping the two mixed together can be more difficult than separating them.

For some power tools, motor oil is regularly mixed with gasoline. The two substances mix completely. If you needed to separate them again, you would need time, suitable equipment, and some specialized knowledge. You would also have to know the necessary safety precautions.

What other kinds of mixtures are there? How are they made and what are they used for? In this unit, you will be working with many different mixtures and discovering their properties. You will also learn about some methods you can use to separate the parts of a mixture.

4

Mechanical Mixtures and Solutions

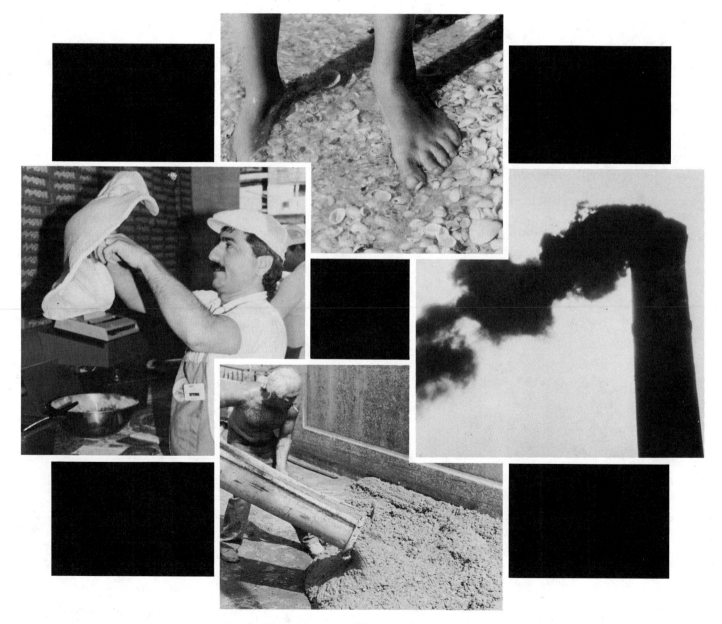

Key Ideas

- There are many useful mixtures of different substances.
- Some mechanical mixtures can be separated into parts by filtration.
- Solutions can be made with a variety of solutes and solvents.
- Solvents and bleaches remove stains from fabrics.

*T*hink of the dozens of different substances you come into contact with each day—the air you breathe, the water you drink, the cereal you had at breakfast (and the spoon you used to eat it with), the concrete sidewalk and earth you walked on. The list goes on and on. All of these are mixtures of one kind or another. This means simply that they contain two or more substances.

There seems to be an endless number of different mixtures. But mixtures can be grouped into only a few categories. In this chapter you will learn about two kinds of mixtures: those whose parts can be seen and those whose parts cannot be seen.

Mechanical Mixtures

Mixtures are made up of two or more substances. Each of the mixtures shown in Figure 4.1 is a special kind of mixture called a mechanical mixture. **Mechanical mixtures** are made up of easily identifiable parts. For example, when you add milk to corn flakes, you have made a mechanical mixture. Try to name the parts of the other mechanical mixtures shown in Figure 4.1.

Figure 4.1 *You can find mechanical mixtures almost anywhere.*

Activity 4A

Problem

What are some properties of mechanical mixtures?

Materials

sand (about 20 g)
magnifier
iron filings (about 10 g)
stirring rod
2 pieces of paper
water
250 mL beaker

Procedure

1. Make an observation table in your notebook like Table 4.1.
2. Place the iron filings on one piece of paper. Place half of the sand on the other piece of paper.
3. Examine each substance by looking at it through the magnifier. Write a description in your table.
4. Thoroughly mix the two substances together, using the stirring rod.
5. Examine the mixture using the magnifier. Record your observations in your table.
6. In your table, describe the appearance of the water.
7. Use the other half of the sand to mix with the water. Pour it into the beaker and mix with the stirring rod.
8. Hold the beaker at eye level and observe the mixture closely.

Observations

Table 4.1

SUBSTANCE	DESCRIPTION
Sand	
Iron filings	
Mixture of sand and iron	
Water	
Mixture of sand and water	

SAMPLE ONLY

Questions

1. (a) In what ways were these two mechanical mixtures alike?
   (b) In what ways were they different?
2. (a) How could you separate the parts of the sand and iron mixture?
   (b) What properties of sand and iron would make this separation possible?
3. (a) How do you think you could separate the mixture of sand and water?
   (b) What properties of sand and water would make this separation possible?

Properties of Mechanical Mixtures

You have seen that mechanical mixtures are made up of two or more easily identified parts and that each part retains its own properties. Another way of saying this is that mechanical mixtures are always **heterogeneous**. This, then, is one property of mechanical mixtures. They are heterogeneous.

Another property of mechanical mixtures is that they may be either opaque or translucent. For example, if you place a mixture of sand and iron in a test tube and hold it up to the light, you will find it quite *opaque*. That is, it does not allow any light to pass through it. On the other hand, many mechanical mixtures do allow some light to pass through. They are referred to as *translucent*. A mixture of sand and water is an example of a translucent mechanical mixture. Such mixtures are never completely clear. In other words, mechanical mixtures are never transparent.

A third property of mechanical mixtures is that you can usually separate the parts fairly easily. You will be discovering one way of doing so in section 4.2.

Problem

What other kinds of mixtures are there?

Materials

5 beakers (150 mL)
5 stirring rods
balance
masking tape
hand magnifier
5 pieces of paper (10 cm x 10 cm)
5 pieces of aluminum foil (10 cm x 10 cm)
10 g clay
10 g dextrose
10 g calcium hydroxide
10 g starch
15 white beans
water

Procedure

1. Make a table similar to Table 4.2 in your notebook. (See page 58.)
2. Label each of the five pieces of paper "starch," "dextrose," "clay," "calcium hydroxide," and "beans" (Figure 4.2).
3. Using five short pieces of masking tape, make labels identical to the labels you made in step 2. Stick one of these to each of the beakers.

Figure 4.2 *These five substances will be mixed with water.*

4. Place the paper labelled "starch" on the balance and measure 10 g of starch onto it.
5. Repeat step 4 for each of the other four substances. It is not necessary to measure the mass of the beans.
6. With the aid of a hand magnifier, observe closely the appearance of each of the substances. Record your observations in your table.
7. Place each of the samples into its own labelled beaker.
8. Pour 50 mL of water into each beaker. Stir for about 30 s, using a different stirring rod for each mixture.
9. Observe each of the mixtures and record your observations.
10. Cover each of the beakers with a piece of aluminum foil and store them as directed by your teacher. You will be using them again in Activity 4C.

Observations

Table 4.2 *Data for Activity 4B*

SUBSTANCE	APPEARANCE BEFORE MIXING	APPEARANCE AFTER MIXING AND STIRRING
Starch		
Dextrose		
Clay		
Calcium hydroxide		
Beans		

SAMPLE ONLY

Questions

1. Which mixture or mixtures were heterogeneous after you finished stirring?
2. Which mixture or mixtures were not heterogeneous after you finished stirring?
3. In which mixture or mixtures was there evidence that some (or all) of the solid mixed fully with the water?
4. Which of these mixtures is not a mechanical mixture? Explain your answer.

A Concrete Idea

WANTED: Multi-purpose building material with the following properties:

- available in almost any size, shape, texture, and colour
- water-resistant
- able to withstand attack by insects and rodents
- able to resist fire and heat
- both plentiful and inexpensive

This may seem like a lot to ask of a building material. But, fortunately, at least one material meets all the above requirements. The material is concrete.

Concrete is a mechanical mixture made up of three parts: paste, aggregate, and water. The paste is itself a mixture of water, air, and a special kind of cement called Portland cement. The aggregate, also a mixture, is made up of sand, gravel, and crushed stone. When the paste, aggregate, and water are mixed together, the paste binds the aggregate together. The aggregate serves as an inexpensive "filler," which gives the concrete its bulk. The concrete's strength depends on the amount of water added and how the concrete dries.

While it is wet, concrete must be mixed evenly for the best consistency. It needs to be workable, so it can be formed into the required shape. When it has hardened, it must be strong. Prepared properly, concrete will resist the effects of weather.

At one time, concrete had to be thick to be strong. Over the years, experiments and experience have resulted in different concrete mixtures. These can be combined with steel reinforcement and special techniques to provide maximum strength with a minimum of material.

What is the role of concrete in your life? Think about the school you go to or the building you live in. Chances are that concrete played a role at least in the early development of these structures. Think about the roads you walk or bicycle on. What about the floor under your feet or the walls you lean against? Think also about patio slabs, stairways, curbs, and bridges. All these—and more—involve the use of this multi-purpose building material.

Concrete is by far the most widely used of all building materials. As a result, there is usually a demand for engineers, technicians, tradespeople, and other workers associated with the concrete industry.

Self-check

1. What does the word "heterogeneous" mean?
2. Which of the following is heterogeneous? Give reasons for your answer.
   (a) vegetable soup mix
   (b) cooked vegetable soup
   (c) vanilla milkshake
   (d) water
   (e) chocolate chip muffin
   (f) granola
3. List three properties of mechanical mixtures, and give an example for each.
4. Is it possible to have a mixture that is not heterogeneous? Give an example.

Filtration

Filtration is a method that can be used to separate some mixtures. Filter paper is paper that has very small openings through it. If you pour a mixture through the paper, particles that are larger than the openings in the paper will not pass through. Particles smaller than the openings in the paper will pass through the paper and can be collected.

The part of the mixture that passes through the filter paper is called the **filtrate**. The part that remains behind in the filter paper is called the **residue**.

Folding a piece of filter paper twice prepares it for filtration (Figure 4.3). Forming it into a cone shape allows you to fit it into the funnel (Figure 4.4). Wetting the cone then helps it to stick to the funnel.

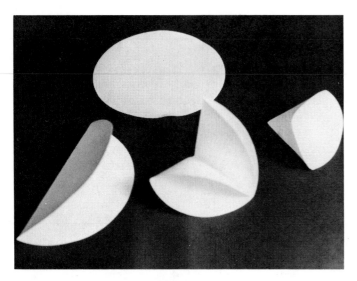

Figure 4.3 *Prepare the filter paper by folding it twice and then forming it into a funnel.*

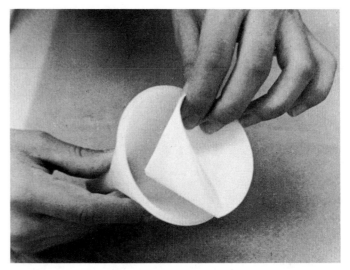

Figure 4.4 *Fit the cone of filter paper into the funnel, and wet it to make it stick.*

PART 1

Problem

How can filtration separate the parts of a mixture?

Materials

the mixtures, in beakers, from Activity 4B
funnel   5 pieces of filter paper
retort stand   5 beakers (150 mL)
ring clamp   masking tape
stand   aluminum foil

Procedure

1. Make a table similar to Table 4.3. Record your observations in this table.
2. Use masking tape to label each of the empty beakers with the names of the mixtures listed in Table 4.3. Label them in the same way you labelled the five beakers from Activity 4B.
3. Set up the apparatus as shown in Figure 4.5 with filter paper in the funnel. Place the empty beaker labelled "starch" under the funnel.

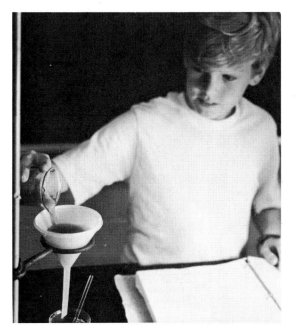

Figure 4.5 *Filtering a mixture*

4. Pour the starch mixture carefully into the funnel, making sure the level of the liquid in the funnel is below the top of the filter paper.
5. Continue to add the mixture slowly as the filtration proceeds.
6. When the filtration of the starch mixture is complete, remove the filter paper from the funnel, and set it aside for observation of the residue. Clean the funnel and prepare another piece of filter paper for filtration.
7. Repeat the filtration procedure for each of the other four mixtures. Clean the funnel each time, and save the paper containing the residue.
8. Examine each of the residues, and describe them in your observation table.
9. Hold each of the beakers containing the filtrates up to the light, and examine them closely. Describe the appearance of each in your observation table.
10. You will use the five filtrates in Part 2 of this activity. If they must be kept until another day, cover them with aluminum foil.

Observations

Table 4.3 *Data for Activity 4C, Part 1*

MIXTURE FILTERED	APPEARANCE OF FILTRATE	APPEARANCE OF RESIDUE
Starch		
Dextrose		
Clay		
Calcium hydroxide		
Beans		

Questions

1. What similarities are there in the appearance of the filtrates?
2. Which filtrate is less clear than the others?
3. How many different substances do you detect in each of the filtrates?
4. For which mixture or mixtures was there no residue?

PART 2

In this part of the activity, you will learn about an important approach to solving problems: *controlled experimentation*. To help you understand what a controlled experiment is, look at Figure 4.6 and read the following paragraph.

Suppose your school is building a new gym and must decide whether to install wood or rubber flooring. Since the basketball teams want the bounciest floor, your science class is given a sample of each kind of flooring to test. Some class members decide to test the flooring by dropping two balls and measuring how high they bounce. Figure 4.6 shows how four students conducted the test. Study each illustration carefully to decide whether the student in each conducted a fair or an unfair test.

There are many possible factors that could affect the results of any experiment. In the flooring test described above, these factors include the type of ball that was dropped and the height from which it was dropped. In a fair experiment, the only factor that should have changed was the type of flooring. All the other factors should have been kept the same. They are said to be *controlled*. Another name for a fair experiment is a *controlled experiment*.

Problem

How can a controlled experiment help us understand the nature of the filtrates?

Materials

5 filtrates, in beakers, from Part 1
distilled water, 50 mL in a 150 mL beaker
150 mL beaker
12 test tubes
dropping bottle of bromthymol blue indicator
dropping bottle of iodine solution
6 straws
masking tape for labels
marking pen

Procedure

1. Make a table in your notebook similar to Table 4.4.
2. Using each of the five filtrates in turn, pour about one-third of the filtrate into two test tubes each. Leave the remaining one-third in the beaker.
3. Pour about one-third of the distilled water into two test tubes, and leave the remaining one-third in the beaker. Make sure all your test tubes are labelled properly.

Figure 4.6 *An experiment to test flooring for a new gym*

Table 4.4 *Data for Activity 4C, Part 2*

SUBSTANCE TESTED	EFFECT OF BROMTHYMOL BLUE	EFFECT OF IODINE SOLUTION	EFFECT OF CARBON DIOXIDE
Starch			
Dextrose			
Clay			
Calcium hydroxide			
Beans			
Water			

SAMPLE ONLY

4. Do this step for each of the six substances being tested.
    (a) To the first test tube, add two drops of bromthymol blue indicator solution. In your table, record any colour change in the filtrate.

> **CAUTION** Iodine is a corrosive liquid. Be careful to keep from spilling it on your hands, clothing, books, or table. If any iodine is spilled, wipe it up quickly and rinse with plenty of water. Inform your teacher immediately.

    (b) To the second test tube, add two drops of iodine solution. Record any changes.
5. Using a straw, blow your breath gently through the remainder of the filtrate in each beaker. Record your observations.

Observations
(See Table 4.4.)

Questions

1. Bromthymol blue is an indicator that turns yellow in acids and blue in bases.
    (a) Which of the filtrates are acids?
    (b) Which are bases?
2. Iodine is an indicator that shows whether starch is present by turning blue-black or purple.
    (a) Was there any starch in any of the filtrates?
    (b) What does this tell you about starch?
3. Your breath contains carbon dioxide. Carbon dioxide forms a white substance when it meets a solution of calcium hydroxide. What does this show about the filtrates?
4. (a) A control is a test to which all the other tests in a controlled experiment can be compared. What was the control in this experiment?
    (b) Why is using a control helpful in an experiment such as this?

Self-check

1. How does a filter separate the two different substances of a mixture?
2. How does a tea bag serve as a filter? What materials does it separate? On what property does the separation depend?
3. Many coffee makers use a filter. What is the residue in this process? What is the filtrate?
4. What is a controlled experiment?
5. Why are controlled experiments useful?

Solutions

When sugar is mixed with water, the sugar **dissolves** in the water. This means that it mixes so thoroughly that the parts of the mixture can't be seen. The mixture is said to be **homogeneous** (Figure 4.7) and is called a **solution**. The material that dissolves is called the **solute**. The material into which the solute dissolves is called the **solvent**. In a solution of sugar and water, the solute is sugar and the solvent is water.

Reexamine Figure 3.1 from Chapter 3. In this solution, potassium permanganate is the solute and water is the solvent. Even if the water is not stirred, the solute particles can still spread throughout the solvent by diffusion. Think of some other examples of solutions, and name the solute and the solvent for each.

The solute and solvent affect the properties of a solution. For example, the boiling point of a solution of salt and water is different from the boiling point of pure water. Yet some of the properties of the starting materials are still present in a solution. You can still taste the salt in a salt solution. The particles of salt and the particles of water have not changed. They are simply mixed together, but the particles are so tiny that we cannot easily separate them.

Challenge

In Activity 4B, there was one solution. Which mixture was the solution? Name the solute and the solvent.

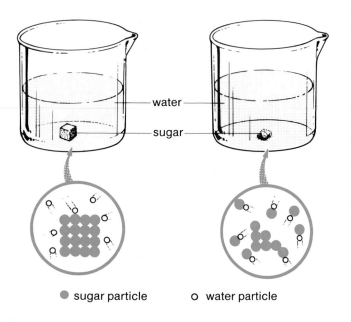

water

sugar

● sugar particle   ○ water particle

Figure 4.7 *Before dissolving, the sugar particles are close enough together so that the solid is visible. After dissolving, the individual sugar particles are so far apart from each other that they cannot be seen.*

Activity 4D

Suppose you continue to add solute to a solution. Is there a limit to how much solute will dissolve?

Problem

How much salt will dissolve in 50 mL of water?

Materials

graduated cylinder
beaker
water
salt (about 25 g on a small piece of paper)
stirring rod
scoopula or spoon
balance

Figure 4.8 *Is there a limit to the amount that can dissolve in a solvent?*

Procedure

1. Prepare a table in your notebook similar to Table 4.5.
2. Find the mass of the salt and paper, and record the mass in your table.
3. Measure 50 mL of water into the graduated cylinder and then pour it into the beaker.
4. Using the spoon or scoopula, place a small amount of salt into the water and stir to dissolve.
5. When the salt has completely dissolved, continue adding a small quantity of salt, stirring until dissolved. Look for traces of undissolved salt in the water (Figure 4.8). Eventually you will find that the salt in the bottom will not dissolve, no matter how much you continue to stir.
6. When no more salt will dissolve, measure the mass of the salt that remains on the paper. Record this amount.
7. Calculate the mass of the salt that dissolved.

Observations

Table 4.5 *Calculations for Activity 4D*

Mass of salt and paper before dissolving	_____ g
Mass of salt and paper when no more salt dissolves	_____ g
Mass of salt that dissolved in 50 mL of water	_____ g

Questions

1. Was there a limit to how much solute would dissolve?
2. How much salt dissolved in the 50 mL of water?
3. Try to explain why there might be a limit to how much solute will dissolve in a solvent. (Hint: Think about what you learned in Chapter 3.)

Saturated Solutions

In a mechanical mixture, there is no limit to the amounts of the various parts of the mixture. There is a limit, however, to the amount of solute which can dissolve in a solvent. When a solution contains the maximum amount of solute, the solution is said to be **saturated**.

The differences between solutions and mechanical mixtures are summarized in Table 4.6.

Table 4.6 *Comparison of Solutions and Mechanical Mixtures*

SOLUTIONS	MECHANICAL MIXTURES
Are homogeneous	Are heterogeneous
May be transparent	Are never transparent
Have parts which are not easily separated	Have parts that can be easily separated
Have a limit to the amount of solute that can dissolve	Can be mixed in any proportion

Ideas and Applications

The particle theory can be used to explain why some substances are insoluble. If the particles of a solute have no attraction for the solvent particles, they will not diffuse out into the solvent. For example, particles of cornstarch do not mix with the water because they have no attraction for water particles. Particles of sugar, however, are greatly attracted by water particles, so the solubility of sugar in water is high.

Solubility

Some substances do not dissolve in a solvent. In other words, they are **insoluble**. Cornstarch is an example of a substance that is insoluble in water (Table 4.7). Many substances do dissolve in water, but for all of them, there is a limit. The **solubility** of a solute is the amount that will dissolve in 1 L of the solvent at a stated temperature. The solubility of some common solutes is shown in Table 4.7. Notice that the temperature must be specified because solubility depends on temperature.

Table 4.7 *Solubility of Some Common Substances*

SOLUTE	MAXIMUM AMOUNT THAT CAN DISSOLVE IN 1 L OF WATER AT 20°C
Sugar	2000 g
Salt	360 g
Baking soda	70 g
Cornstarch	0 g

Challenge

Use the particle theory to explain why most solids dissolve faster at higher temperatures.

Challenge

Suppose you were given a saturated solution of sugar in water. What could you do to make it an unsaturated solution?

Some substances do not dissolve in water. Are there solvents that will dissolve these substances? In this activity, you will use both water and corn oil as solvents to find out.

Problem

How can solubility differ in different solvents?

Materials

4 beakers (100 mL)
water
table salt
corn oil
petroleum jelly
masking tape
graduated cylinder
4 stirring rods

Figure 4.9 *Even if a substance is insoluble in one solvent, it can still be soluble in another.*

Procedure

1. Prepare a table in your notebook, similar to Table 4.8.
2. Using masking tape, label the beakers A, B, C, D.
3. Place 25 mL of water into each of beakers A and B.
4. Place 25 mL of corn oil into beakers C and D.
5. Into beakers A and C place 1 mL of petroleum jelly.
6. Into beakers B and D place 1 mL of salt.
7. Stir the contents of each beaker (Figure 4.9).
8. Observe the contents of each of the four beakers. Look for evidence of dissolving.

Observations

Table 4.8

BEAKER	CONTENTS OF BEAKER	EVIDENCE OF DISSOLVING
A		
B		
C		
D		

Questions

1. In which liquid was salt soluble?
2. In which liquid was petroleum jelly soluble?
3. Using the particle theory, explain why the solute did or did not dissolve in each of the four beakers.

Self-check

1. Which of the following statements is/are true? For any which are not true, write the corrected sentence in your notebook.
   (a) A mechanical mixture can be formed by diffusion.
   (b) Solutions are homogeneous mixtures.
   (c) There is no limit to the amount of solute which can dissolve in a solvent.
2. For each of the following, write a single word that expresses the same idea:
   (a) gradual spreading out of particles
   (b) uniform distribution
   (c) having different parts which can be seen
3. Explain the meaning of "solubility."
4. What is meant by the statement, "It is a saturated solution"?

Ideas and Applications

You have been working with and reading about solutions in which a solid solute is mixed with a liquid solvent. These may be the most common kinds of solutions, but they are by no means the only ones. For example, the properties of solid iron can be improved by adding other substances to it. The most important of these additives is carbon. Adding carbon to melted iron forms a solution. When the solution cools, it becomes a solid metal solution called an alloy. The alloy formed by mixing carbon with iron is steel. Steel is stronger and less brittle than iron. Thus, steel may be used in constructing skyscrapers or bridges. Iron alone would be unsuitable for such objects.

Dissolving and Bleaching Stains

The problem of removing stains from clothing is an important one in the household. It is also the basis for a large cleaning industry. If you know what kind of stain is causing the problem, you will be able to use the correct solvent to remove it (Figure 4.10).

Oil and grease stains are the most common type of hard-to-remove stains. Other difficult stains are those made by gravy, mayonnaise, shoe polish, chocolate, and lipstick. For such stains, a spot remover such as *perchloroethylene* can be effective because it dissolves the grease. This chemical is used widely in the dry-cleaning industry.

Some stains are at least partly soluble in water. These include stains from coffee, tea, mustard, some cosmetics, felt pens, and grass. Rubbing liquid detergent into the stain can be effective in getting the stain to dissolve, and treatment with a grease-dissolving spot remover may also help.

Figure 4.10 *Supermarket shelves display a bewildering choice of cleaning products.*

The third type of stain has a protein base. Examples are cream, milk, eggs, gravy, and meat juice. Detergents may be able to remove these, but some of these stains are really not soluble. They may require the addition of an enzyme to break down the substances that make up the stain.

There are many factors that could determine success in removing a stain. In extreme cases the stain may be removed, but the colour of the fabric may also disappear. The fabric itself may also be damaged by the solvent.

Bleaches

When a stain cannot be removed by the use of solvents or detergent, a bleach can be used. Bleach removes the colour from the stain, making it invisible. There are two kinds of bleach. One, a chlorine-based bleach, is the stronger of the two. It is usually used only on white fabrics. For coloured fabrics, a milder, oxygen-based bleach is used.

Challenge

Choose some of the common stains mentioned in the text and try a variety of solvents with each one. In addition to regular cleaning solvents, hair spray or nail polish remover may be effective. A good sturdy fabric to use is white cotton. Other fabrics may be damaged more easily by some of the solvents.

CAUTION Do not breathe the vapour from these solvents. Use them in a well-ventilated location.

Activity 4F

PART 1

Problem

What is the effect of bleaches on coloured food?

Materials

2 evaporating dishes
5 mL chlorine-based liquid bleach
5 mL oxygen-based bleach (such as hydrogen peroxide)
4 mL coloured food (such as ketchup)
2 stirring rods

CAUTION Safety goggles must be worn during this investigation. Do not breathe the vapours. Place the dishes in a well-ventilated location.

Procedure

1. Place the chlorine-based bleach into one of the evaporating dishes. Place the oxygen-based bleach into the other.
2. Place half the coloured food into each dish. Use the stirring rods to mix the contents.
3. Observe any colour change in the two dishes.

Observations

1. In your notebook, describe what you observe in the two dishes.

Questions

1. What differences were there in the reactions of the two bleaches with the food?

PART 2

Problem

What is the effect of bleaches on fabrics?

Materials

2 samples of each of various coloured fabrics
(8 cm x 8 cm) (e.g., cotton, silk, denim,
polyester, rayon, etc.)
evaporating dishes (2 for each sample of fabric)
chlorine-based liquid bleach
oxygen-based liquid bleach
masking tape

Procedure

1. Make a table in your notebook, similar to Table 4.9.
2. With the masking tape, label half the dishes "chlorine base" and half "oxygen-base." Place 10 mL of the correct bleach into each dish.
3. Choose one type of fabric. Place a corner of one piece of the fabric into the chlorine-based bleach and a corner of the other piece into the oxygen-based bleach (Figure 4.11).
4. Repeat step 3 for all the other types of fabrics.
5. Allow the samples to soak for 15 min, and then observe any changes. If possible, let them soak overnight and make final observations.

Figure 4.11 *Testing the effect of bleach on coloured fabric*

Observations

Table 4.9

FABRIC TESTED	REACTION WITH CHLORINE-BASED BLEACH	REACTION WITH OXYGEN-BASED BLEACH
First fabric		
Second fabric		
Third fabric		
Etc.		

SAMPLE ONLY

Questions

1. Which bleach had the greater effect on the colours of the fabrics?
2. Which fabrics, if any, were damaged by the bleach?
3. Which fabrics, if any, were unaffected by either of the bleaches?

Challenge

Design an experiment that combines both parts of Activity 4F. In your experiment, make stains on various kinds of fabric, and try to find out if the stains are removed equally well from all the fabrics.

Chapter Objectives

NOW THAT YOU HAVE COMPLETED THIS CHAPTER, CAN YOU DO THE FOLLOWING?	FOR REVIEW, TURN TO SECTION
1. Explain the meaning of "mechanical mixture" and give several examples of mechanical mixtures.	4.1
2. Describe how you would separate the liquid and solid parts of a mixture.	4.2
3. Explain what is the residue and what is the filtrate when a mixture is filtered.	4.2
4. State the name for a homogeneous mixture.	4.3
5. List four properties of solutions that make them different from mechanical mixtures.	4.3
6. Explain dissolving in terms of the particle theory.	4.3
7. Name three common types of stains and some methods used to remove them.	4.4
8. Describe the effects of different bleaches on fabrics.	4.4

Words to Know

mixture
mechanical mixture
heterogeneous
filtration
filtrate
residue
dissolve
homogeneous
solution

solute
solvent
saturated
insoluble
solubility

Tying It Together

1. State whether each of the following statements is true or false. Correct all false statements, and write the correct sentence in your notebook.
   (a) All solutions are colourless.
   (b) Solutions appear homogeneous.
   (c) The residue always passes through the filter paper.
   (d) All mechanical mixtures can be separated by filtration.
   (e) Water can dissolve everything.

2. Give an example of a mechanical mixture which consists of
   (a) a solid and a liquid
   (b) a solid and a gas
   (c) two solids
   (d) a liquid and a gas.
3. Describe briefly how you would separate a mixture of
   (a) sand and iron filings
   (b) sand and water
   (c) sand, water, and wood chips
   (d) dust and air.
4. List four characteristics of mechanical mixtures.
5. State two ways in which a solution looks different from a mechanical mixture.
6. What is a saturated solution? Give an example.
7. Why is temperature specified when the solubility of a substance is given?
8. Use the particle model of matter to explain how salt dissolves in water.
9. Name one substance that does not dissolve in water, and name a solvent in which it does dissolve.
10. What is the meaning of "homogeneous"?
11. What is the main difference in the action of a solvent and the action of a bleach?

Applying Your Knowledge

1. Sugar will dissolve faster in tea if it is stirred. Explain this in terms of the particle model for matter.
2. A television commercial states that Brand A gasoline gives 1 km/L better performance than Brand B. Some students decide to test this claim.
   (a) What factors other than gasoline might affect the performance of the car?
   (b) How can they keep these other factors constant while testing the claim?
3. Imagine a mechanical mixture consisting of iron filings and gold dust. What method could be used to separate the two metals?

Projects for Investigation

1. Filters may be used in at least three places in an automobile engine. Find out what mixtures in the engine require filtering. For each of these, what is the filtrate and the residue?
2. Read the labels on the cleaning solutions that you have at home.
   (a) How many of them indicate what is in the product?
   (b) Is there one ingredient that seems to appear more frequently than others? Which is it?
   (c) How many of these labels include safety precautions?
3. Steel, like many other alloys, was discovered by accident. Research the discovery, history, and use of one of the following metal alloys: stainless steel, bronze, brass, Wood's metal, pewter, solder.

4. There is a limit to the amount of salt that can be dissolved in a given amount of water. Plan an experiment to find out if there is also a limit to the amount of propanol that can be dissolved in water. (Propanol is the main ingredient in rubbing alcohol.) Consider the following questions while planning your experiment.
   (a) Can 1 mL of propanol be dissolved in 10 mL of water? Which is the solute and which is the solvent?
   (b) Can 1 mL of water be dissolved in 10 mL of propanol? Which is solute and which is the solvent? With your teacher's permission, perform your experiment and report your findings.

Solutions and Crystals

Key Ideas

- Temperature affects the solubility of most substances.
- Supersaturated solutions contain more solute than would normally dissolve in the solvent at that temperature.
- Crystals form when solutions evaporate.
- Crystals form from supersaturated solutions when a seed crystal is added.

*T*he substances shown in both these photographs "grew" from solutions. The large, icicle-like formations were grown in a cave. The small crystals of bluestone were grown in a school classroom. Later in this chapter, you will have a chance to grow your own bluestone crystals.

How do crystals form? And what is the connection between crystal formation and solutions? In this chapter, you will continue to investigate solutions in order to answer these questions.

73

Solubility and Temperature

You already know that many substances are soluble in water. For example, the solute sugar is soluble in the solvent water. You also know that there is a limit to the amount of sugar that can dissolve in water. When this limit is reached, a saturated solution of sugar and water results.

Can the solubility of a solvent such as water be changed? In other words, is there a way to make a solvent hold more solute than you would expect? In Activity 5A, you will answer this question by finding out how temperature affects the amount of potassium nitrate that can dissolve in water.

Ideas and Applications

Potassium nitrate, the substance you will be testing in Activity 5A, is a yellowish-white solid that has a wide variety of uses. For example, potassium nitrate (also known as saltpetre) is the main component in gunpowder. Thus, it may not surprise you to learn that this substance is also used in making matches. You may also have heard of potassium nitrate being involved with the following: agricultural fertilizers, the preservation (pickling) of meats, the manufacturing of glass, the production of some medicines, and the manufacture of rocket fuels. (Potassium nitrate helps rocket fuels burn more efficiently.)

In this activity, you will first make a saturated solution. Then you will make it unsaturated. And, finally, you will make it saturated again.

Problem

How does temperature affect the solubility of a substance?

Materials

graduated cylinder
test tube
rubber stopper
thermometer
scoopula or 2 mL measuring spoon
potassium nitrate
test tube rack
250 mL beaker
hot water

CAUTION Safety goggles must be worn during this activity.

Procedure

1. Using the graduated cylinder, measure 10 mL of cool tap water and pour it into a test tube.
2. Measure the temperature of the water in the test tube. Record the temperature in your notebook.
3. Add a small amount (about 2 mL) of the solute to the test tube.
4. Stopper the test tube and shake it until the solute dissolves.
5. Allow the test tube to stand for a moment. Observe it to check whether there is any undissolved solute.
6. Repeat steps 3, 4, and 5 until, after repeated shaking, there is still some undissolved solute. Record the total number of scoops of potassium nitrate that you added.
7. Prepare a hot water bath by filling the beaker half-full with very hot water.

8. Place the test tube with solution into the hot water bath and allow it to stand for 5 to 10 min (Figure 5.1). Measure the temperature of the solution.
9. Observe the test tube closely. Is there still solute in the bottom of the test tube? Shake the test tube.
10. Add another scoop of the solute. Stopper and shake as before.
11. Continue adding scoops one at a time, and shaking, until some solute remains undissolved even after repeated shaking. Record the number of scoops added in steps 10 and 11.
12. Now hold the test tube under cold running tap water for a minute or two.
13. Measure the temperature of the solution.
14. Make final observations on the contents of the test tube.

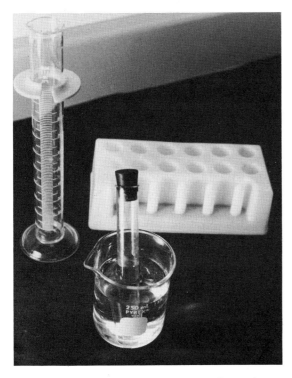

Figure 5.1 *A hot water bath is used to raise the temperature of the solution.*

Observations

1. Describe some properties of the potassium nitrate.
2. Record the following:
   (a) temperature of the cold solution
   (b) number of scoops added
   (c) temperature of the heated solution
   (d) additional number of scoops added
   (e) final temperature of the solution
3. Describe what happened when the solution was cooled.

Questions

1. What effect did an increase in temperature have on the amount of solute that could dissolve in the solvent?
2. What kind of solution did you have when solute remained undissolved in the bottom of the test tube?
3. In this activity, what did you do to make a saturated solution become an unsaturated solution?
4. How does increasing the temperature affect the solubility of potassium nitrate?
5. When the solution was cooled, how did the solute in the test tube differ in appearance from the solute you began with?
6. At the end of the activity, was your solution saturated or unsaturated? Explain your answer.

Challenge

Suppose you have a solution and you wish to recover some of the solute. Think of a method you could use if you had a lot of time to recover the solute. Then think of another, faster way to do this using methods you have now learned.

Solubility Curves

Recall that the solubility of a substance is the number of grams of that substance that will dissolve in 1 L of solvent at a given temperature. The solubility of most substances that dissolve in water is affected by temperature. This effect of temperature on solubility can be shown conveniently with a graph called a **solubility curve**. A solubility curve for potassium nitrate is shown in Figure 5.2. Note, for convenience, the solubility in 100 mL of water is shown.

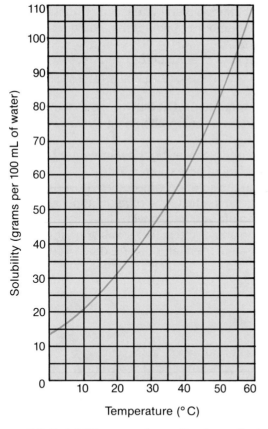

Figure 5.2 *Solubility curve for potassium nitrate*

You can use this graph to find out how much potassium nitrate will dissolve in 100 mL of water at 20°C. Read along the *x*-axis to 20°C and then up to the curve. Then read the amount of solute from the vertical axis. Do you find the mass of the solute to be 33 g? The solubility of potassium nitrate at 20°C is 33 g per 100 mL of water. Note how quickly the solubility of this substance increases as the temperature increases. What is its solubility at 50°C?

Self-check

1. What happens when you cool a saturated solution?
2. Suppose you have a beaker filled with an unsaturated solution of potassium nitrate exposed to the air. If you observed this solution over several days, what do you think you would notice?
3. Use the solubility curve in Figure 5.2 to answer the following questions for potassium nitrate.
   (a) What is its solubility at 15° C?
   (b) To make a saturated solution at 50° C, how much solute would you have to add to 100 mL of water?
   (c) At what temperature will 40 g of potassium nitrate make a saturated solution in 100 mL of water?
   (d) If you had a saturated solution of potassium nitrate in 100 mL of water at 60° C, and you then cooled it to 20° C, how much of the solute would come out of solution?

Ideas and Applications

A crystal is a solid that has a regular shape and smooth, flat surfaces. Crystals form because the particles in a solid arrange themselves in a particular pattern. All the crystals of a substance have the same shape, which is determined by the type of particles and the pattern they form. Crystals of different substances have different shapes. Thus, crystals provide you with another property to help you identify substances.

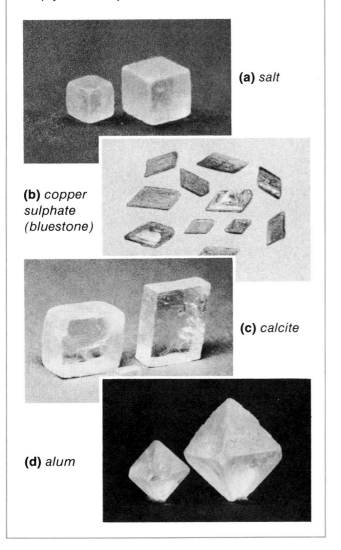

(a) *salt*

(b) *copper sulphate (bluestone)*

(c) *calcite*

(d) *alum*

Supersaturated Solutions

Recall that a saturated solution is a solution in which no more solute can be dissolved in the solute at a given temperature. Under very special conditions, you can make a solution even more concentrated than a saturated solution. Such a solution is called a **supersaturated solution**. A supersaturated solution holds more solute than it would normally hold at a given temperature.

Supersaturated solutions are unstable. A clear supersaturated solution may suddenly form solid crystals of the solute. These crystals settle, leaving behind only a saturated solution.

How can such an unstable situation occur? A saturated solution has to be prepared at a high temperature. At high temperatures, the solubility of the solute is greater. If this hot solution is then cooled to a lower temperature, solubility will be less. The solution will contain more dissolved solute than it normally could hold at the lower temperature. As a result, solid crystals of the solute will settle out of solution.

In Activity 5B, you will prepare your own supersaturated solution to observe its properties.

Problem

How can you make a supersaturated solution?

Materials

test tube	safety goggles
test tube holder	stirring rod
Bunsen burner	stopper
scoopula	water
test tube rack	

15 g of "hypo" (sodium thiosulphate crystals)

CAUTION Safety goggles must be worn during this activity.

Procedure

1. Place about ten drops of water in the test tube.
2. Use the scoopula to place about four of the hypo crystals into the test tube. Stir until they have dissolved.
3. Add crystals, a few at a time, until no more dissolve.
4. Put the test tube in the holder. Heat the solution in the test tube gently, by passing it back and forth slowly through the flame (Figure 5.3). Do not allow the solution to boil.

CAUTION Heat gently. Do not allow the solution to boil.

5. Add crystals until the test tube is about half-full. Continue heating (gently) until all the crystals have dissolved.
6. Remove the test tube from the flame and turn off the Bunsen burner.
7. Avoid shaking the solution, and place it carefully in the test tube rack. Allow it to cool slowly to room temperature. Do not disturb the test tube as it cools.
8. When the solution has cooled to room temperature, drop one hypo crystal into the solution. Observe closely.

Figure 5.3 *Heat gently by passing the test tube back and forth through the flame.*

Observations

1. Describe hypo crystals.
2. What happened when hypo crystals were added to the water?
3. What happened when the solution was heated?
4. Describe the appearance of the solution as it was cooled.
5. What happened when you added the last hypo crystal?

Questions

1. What kind of solution did you have after step 2?
2. What kind of solution did you have after step 3?
3. During step 7, what would you expect to happen as the solution cools?
4. After step 7, what kind of solution did you have? Explain how you know.
5. The crystal that was added in step 8 is called a *seed crystal*. Suggest why.

PART 1

Problem

What is left when a solution evaporates?

Materials

30 g crushed bluestone crystals
hot water (about 60°C) from an electric kettle
100 mL beaker
scoopula
stirring rod
evaporating dish
safety goggles

> **CAUTION** Safety goggles must be worn during this activity.

Procedure

1. Fill the beaker about one-third full with hot water.
2. Using the scoopula, add some bluestone to the water. Stir with the stirring rod.
3. Continue adding bluestone while stirring until no more dissolves.
4. Pour some of the solution into the evaporating dish.
5. Allow the solution to evaporate over the next few days (Figure 5.4).

Observations

1. Each day, describe the contents of the evaporating dish.

Questions

1. Why was a hot bluestone solution, rather than one at room temperature, prepared in this activity?
2. When a solution evaporates, what part of the solution remains as a residue?
3. Why did the solute come out of solution in this activity?

Figure 5.4 *What do you see in the dish as the solution evaporates?*

PART 2

Problem

How can evaporation cause a large crystal to form?

Materials

crystal of bluestone
saturated solution of bluestone
beaker
10 cm thread
stirring rod

Procedure

1. Tie one end of the thread around the crystal and the other end around the stirring rod so that the crystal can be positioned in the solution.
2. Immerse the crystal in the solution with the stirring rod lying across the top of the beaker (Figure 5.5 on the next page). Wind the stirring rod and thread so that the crystal is about halfway down in the solution.

Figure 5.5 *A small bluestone crystal will grow much larger in a saturated bluestone solution.*

3. Set the beaker aside and examine it over a period of several days. You may notice tiny crystals forming on the thread above the seed crystal. These should be removed.

Observations

1. Each day, sketch in your notebook the appearance of the thread and the seed crystal.

Questions

1. Explain why the bluestone crystal grows.

You may now apply what you have learned to a common substance—sugar. In this activity, you will prepare a saturated solution of sugar in water and try to grow a large sugar crystal. Why not try this one at home?

Problem

How can a large sugar crystal be formed?

Materials

small preserving jar or thick-walled glass
saucepan
sugar
water
paper clip
thread
pencil

Procedure

1. Design a table in which to record your observations.
2. Attach the thread to the paper clip, and then wrap the free end around the pencil. Support the pencil over the opening to the jar, as shown in Figure 5.6.

Figure 5.6 *Sugar crystals can be grown in a saturated sugar solution.*

3. Bring about two cups of water to a boil in a saucepan.
4. Add sugar to the boiling water. Constantly stir until no more sugar dissolves. Make sure there is some residue of sugar in the bottom of the pan.
5. Carefully pour the hot solution into the jar.

> **CAUTION** Be careful when pouring the hot solution into the jar.

6. Adjust the paper clip so that it is sitting near the bottom of the jar.
7. Make and record your observations every 15 min as the solution cools to room temperature.
8. Place the jar in an area where it will be undisturbed. Allow the liquid to evaporate to about half its original volume.

Observations

1. Complete the table of observations you designed.
2. Describe the shape of the crystals that formed.

Questions

1. What was the purpose of the paper clip?
2. What happens in a saturated solution of sugar and water as the temperature decreases?

Self-check

1. What is the difference between a saturated solution and a supersaturated solution?
2. What is a seed crystal?
3. Describe how you would grow a crystal using the evaporation technique.

Challenge

Maria began Activity 5C, Part 2, with great care and could hardly wait to look at the crystal. To her disappointment, the next day there was nothing at the end of the thread. Can you suggest why? What did Maria probably not do correctly?

Chapter Objectives

Words to Know

solubility curve
supersaturated solution

Tying It Together

1. State which of the following statements are true and which are false. Correct any false ones and write the corrected statement in your notebook.
   (a) As temperature increases, the solubility of most substances in water increases.
   (b) A saturated solution at 20° C will also be a saturated solution at 40° C.
   (c) A solution that will dissolve more solute at its present temperature is called a saturated solution.
   (d) When a saturated solution is cooled, some of the solute will probably come out of solution.
   (e) If a solution is cooled and no solute comes out of solution, the solution must be supersaturated.
   (f) When a solution evaporates, both the solute and solvent disappear.

2. Define the following, and give an example of each.
   (a) saturated solution
   (b) supersaturated solution
3. What is solubility?
4. Name three factors that determine the solubility of a given solute.
5. The solubility of potassium nitrate at 50° C is 85 g per 100 mL of water.
   (a) How much potassium nitrate would dissolve in 1 L of water at 50° C?
   (b) How much would dissolve in 50 mL of water at 50° C?
6. Suppose you were given a solution of potassium nitrate in water.
   (a) What would you do to find out if it was saturated?
   (b) What would you do to find out if it was supersaturated?
7. Describe how you can "grow" a crystal of bluestone.
8. A student prepared a saturated solution of a substance in water. When the beaker containing the solution and solid was heated, all of the solid disappeared. Explain what happened.

Applying Your Knowledge

1. A jar of liquid honey sat on the kitchen shelf for a long time. The jar became filled with large crystals. What substance were the crystals composed of? How can you make the honey liquid again?
2. Solubility curves for some other substances in water are shown in Figure 5.7.

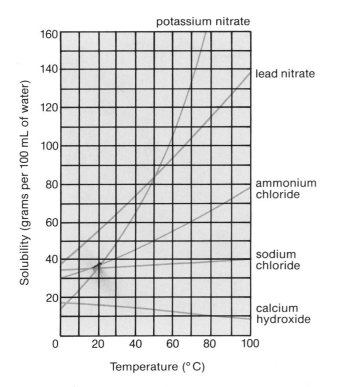

Figure 5.7 *Solubility curves for five substances*

   (a) Which of the substances shown has an almost constant solubility in water regardless of temperature?

   (b) For which substance does the solubility decrease as the temperature increases?

   (c) What is the solubility of ammonium chloride at 50° C?

   (d) At what temperature will potassium nitrate and lead nitrate have the same solubility?

   (e) If you dissolved 50 g of lead nitrate in 100 mL of water at 60° C, how much more would you have to add to it in order to have a saturated solution?

   (f) If you had a saturated solution of lead nitrate in 100 mL of water at 60° C and then cooled the solution to 20° C, how much of the solute would precipitate?

3. Often the soil around a potted house plant will appear white after the surface has dried out. Explain why this happens.

Projects for Investigation

1. The icicle-like structures in the photograph at the beginning of this chapter are called *stalagmites* and *stalactites*. In a library, find out how these are formed. Relate what you find to what you have studied in this chapter.
2. Because of their beauty, crystals are often used in jewellery. For example, you may own or have seen pendants or pins made of purple amethyst, which is a kind of quartz. (Amethyst is Ontario's gemstone. It is found in rocks near Thunder Bay.) Crystals also have many uses in industry. For example, quartz may be used in your watch or in a clock at home. Choose a crystal you are interested in and do research to find out how it forms and how it may be used.

Other Kinds of Mixtures

Key Ideas

- Colloids, suspensions, and emulsions are mixtures.
- Physical properties can be used to distinguish the different kinds of mixtures.
- Filtration can be used to separate the parts of some kinds of mixtures.
- Keeping substances mixed is often more desirable than separating them.

In chapters 4 and 5 in this unit you have learned about mechanical mixtures and solutions. In this chapter you will investigate other kinds of mixtures called colloids, suspensions, and emulsions. These types of mixtures are very common in day-to-day living. You can find them in such places as your medicine cabinet, your refrigerator, your workshop, and your laundry room.

Colloids and Suspensions

The Tyndall Effect

You know that mechanical mixtures are heterogeneous, or non-uniform, whereas solutions are homogeneous, or uniform. A **colloid** is a mixture which may appear to be homogeneous, but the particles are not as small as the particles in a solution. The particles in a colloid are not dissolved, but they are too small to settle. Because they are relatively large, they scatter light. This effect is known as the **Tyndall effect** after John Tyndall, the British physicist (1820–1893) who first reported it (Figure 6.1).

Figure 6.1 *You can see the beam of light as it passes through the colloid. This scattering of light by tiny particles in the colloid is called the Tyndall effect.*

You have probably observed the Tyndall effect when car headlights shine through fog. You can produce the effect yourself when you shine a flashlight on a misty evening. The visible beam produced by a movie projector operating in a smoky or dusty atmosphere is another example of the Tyndall effect.

In dry, dust-free air a beam of light is invisible. If the light is not directed at your eyes, you cannot see it. If, however, there are particles in the path of the light beam, then the particles scatter the light in all directions. The beam becomes visible. This scattering of light occurs only if the particles are of suitable size. Particles of air are not large enough to scatter light. Droplets of water in fog or mist are large enough and do produce this effect. So do particles of smoke or dust. Have you ever noticed that you can see a beam of sunlight in your home because of the dust in the air? This too is an example of the Tyndall effect.

In Activity 6A, you will investigate the properties of colloids. (You will also be able to observe the Tyndall effect for yourself!)

Activity 6A

Problem

What are some properties of colloids?

Materials

4 very clean beakers (250 mL)
ring clamp
stand
funnel
filter paper
flashlight
5 mL laundry starch
15 mL sodium chloride
water
2 stirring rods
masking tape

Procedure

1. Add the laundry starch to one of the beakers, and add the sodium chloride to another beaker.
2. Add water to both beakers so that they are about three-quarters full. Use the masking tape to label them.

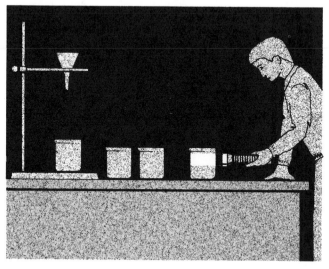

Figure 6.2 *Testing for the Tyndall effect*

3. Stir the contents of each of the beakers thoroughly.
4. To a third beaker, add water to fill it about three-quarters full. Label it with masking tape.
5. Darken the room temporarily and shine the flashlight through the contents of each of the three beakers in turn (Figure 6.2). Observe closely.
6. Prepare the filter paper, filter, ring clamp, and stand for filtration.
7. Place the fourth beaker under the funnel and pour half of the starch mixture into the funnel. Shine the flashlight through the filtrate.
8. Set aside the beaker containing the original starch mixture until next day.
9. Make observations next day on the mixture of starch and water. Once again, darken the room and shine the light through the mixture.

Observations

1. Describe the appearance of the contents of each of the three beakers. Note which ones produced the Tyndall effect.
2. (a) Describe any residue left on the filter paper.
   (b) Describe the filtrate obtained when the starch mixture was filtered. Did the filtrate show the Tyndall effect?
3. Describe the appearance of the starch mixture after it was allowed to stand for a long period of time. Did it show the Tyndall effect?

Questions

1. Which of the three liquids was a colloid? Explain your answer.
2. (a) Why do some particles not pass through filter paper?
   (b) Did any of the particles of starch pass through the filter paper? How do you know?
3. Did the starch mixture show any tendency to settle to the bottom of the beaker?
4. List three properties of colloids.

Colloids in and around the Home

A colloid is a homogeneous mixture in which the particles are large enough to display the Tyndall effect but too small to settle. Some of the larger particles may be separated by filtration, but others are small enough to pass through a filter. The parts of a colloid do not separate when left to stand. Colloids may exist in gases, liquids, or solids (Figure 6.3).

Examples of colloids in a gas include fog and air containing dust. In fog the tiny particles are liquid droplets, and in dust-filled air they are tiny pieces of solids. Aerosols, such as those used in many household sprays, are other examples of colloids in which liquids are mixed with a gas.

Some solid bars of soap are colloids. The soaps which float on water contain tiny bubbles of air. Gelatin desserts are also colloids in solid form. The powdered materials that form the basis of these desserts are mixed in hot water and then refrigerated. The cold temperature causes the particles to clump together, forming a solid which appears homogeneous.

Colloids are common around the kitchen as well. Applesauce is a mixture of tiny pieces of solid apple with spices and sugar, mixed in water. The action of pectin, a chemical occurring naturally in the apple, prevents the apple pieces from breaking down into even smaller pieces. If another chemical, pectinase, is added to the apple pulp, the pectin is broken down. This allows the minute pieces of apple to break down even further and dissolve in the water. The result is apple juice.

Figure 6.3 *What substances make up these colloids?*

Activity 6B

A **suspension** is like a mechanical mixture because it is non-uniform. It is also like a colloid because it contains very small particles as well as the larger ones that will settle.

Problem

What are some properties of a suspension?

Materials

2 beakers (250 mL)
filter paper
funnel
ring clamp
stirring rod
flashlight
30 mL crushed soil
water

Procedure

1. Place the crushed soil in one of the beakers and fill the beaker with water so that it is about three-quarters full. Stir the contents thoroughly.
2. Darken the room temporarily. Shine the flashlight through the mixture.
3. Allow the mixture to stand for several minutes. During this time, prepare the filter paper, funnel, ring clamp, and stand for filtration.
4. After several minutes, observe the mixture once again and note any changes.
5. Stir the soil and water mixture vigorously, and then pour as much as possible into the funnel. Collect the filtrate in the second beaker.
6. Allow the filtrate to stand as long as possible, preferably until next day.
7. Once again, darken the room and shine the flashlight through the filtrate.

Observations

1. (a) Describe the soil and water mixture.
   (b) What effect does stirring have on the mixture?
   (c) Can you observe the Tyndall effect?
2. How does the mixture change when it has been set aside for several minutes?
3. Describe the filtrate and residue.
4. Was there any evidence of further settling within the filtrate?
5. Does the filtrate cause the Tyndall effect?

Questions

1. Describe any differences between the filtrate and the original mixture.
2. What type of particles settled to the bottom of the beaker? What type did not settle?
3. What was the main difference in behaviour between this soil and water mixture and the starch and water mixture of Activity 6A?

Properties and Examples of Suspensions

The main difference between suspensions and colloids is that in a suspension some of the particles are large enough to settle to the bottom due to gravity. When the larger particles in a suspension settle, the portion of the mixture that remains above the settled particles is more like a colloid. A certain amount of settling may continue for a very long period. Like colloids, suspensions display the Tyndall effect.

Suspensions are very common in the home (Figure 6.4). To find one, look in the refrigerator or the medicine cabinet for any container that says "Shake well before using." Many of these mixtures can be combinations of solution, colloid, and suspension, all at the same time.

Figure 6.4 *Tomato juice is a common suspension. The bright red component, which can be separated by filtration, is almost completely odourless and tasteless. The odour and flavour are in the pale liquid.*

Self-check

1. How is the Tyndall effect produced?
2. What is the main difference between a colloid and a suspension?
3. Classify each of the following as solution, colloid, or suspension by the clue given. In some cases, more than one answer is possible.
   (a) The liquid is transparent and cannot be separated by filtering.
   (b) The liquid appears cloudy when a beam of light is passed through it.
   (c) Some particles settle out when the mixture is allowed to stand.
   (d) Some particles are left as a residue when the mixture is filtered.
   (e) When light is passed through the liquid, no beam is seen.
   (f) Some solid particles in the liquid are large and easily visible.
4. A bottle of milk of magnesia displays the instruction, "Shake well before using." What kind of mixture would this be? Explain.

6.2

Emulsions

An **emulsion** is a particular kind of colloid in which tiny liquid droplets are dispersed (finely mixed) in another liquid. An example of an emulsion is homogenized milk. In homogenized milk, tiny droplets of cream are dispersed in skim milk (Figure 6.5). But it is not always easy to get two liquids to mix so completely.

Figure 6.5

(a) *Equipment for homogenizing milk. The milk is sprayed through very small openings. This process, called homogenization, breaks the fat in milk into droplets so tiny that they stay suspended.*

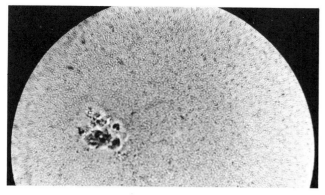

(b) *Microscopic view of homogenized milk*

Activity 6C

Problem

How can you mix water and cooking oil so they stay together?

Materials

100 mL beaker   stirring rod
water   2 mL egg yolk
20 mL corn oil, or other cooking oil

Procedure

1. Put the corn oil into the beaker.
2. Add water to the beaker until it is about half-full. Stir vigorously.
3. Hold the beaker to the light and observe until the swirling comes to rest.
4. Add the egg yolk to the mixture and stir vigorously.
5. Hold the beaker as in step 3 and observe closely.

Observations

1. In your notebook, make a sketch of the mixture after you had added the water but before you had stirred it.
2. How did the mixture change when you stirred it?
3. Describe the appearance after the mixture had come to rest.
4. Describe the appearance of the final mixture which contained egg yolk.

Questions

1. What happened to the oil when you added it to the water?
2. What difference did you notice in the mixture after the egg yolk had been added?

Challenge

Repeat Activity 6C at home, mixing water and cooking oil. Instead of egg yolk, trying using various soaps and detergents as emulsifying agents. Compare their effectiveness.

CAUTION Do not taste these mixtures.

Emulsions in and around the Home

Even when it is stirred, oil mixes only momentarily with water. It comes together into larger and larger drops and finally stays at the surface of the water. Apparently the tiny droplets of oil exert forces of attraction on each other that tend to bring them together and separate them from the water. To mix oil and water permanently a third substance is required.

An **emulsifying agent** is a substance which is added to a mixture in order to keep the parts from separating. For example, egg yolk permits oil and water to form an emulsion (Figure 6.6). The emulsifying agent surrounds the tiny droplets of oil and allows them to mix with the water. This mixture of oil, water, and egg yolk is one that is common to most salad dressings and mayonnaise.

Figure 6.6 *Eggs are used as emulsifiers in salad dressing.*

Emulsions are common not only in your refrigerator but also in the laundry room. The emulsifying agent in the laundry room is soap or detergent. Oils and greases combine with dirt on your skin and clothes. Water alone does not remove the dirt because oils do not mix with water.

Soap or detergent helps to get rid of the dirt because of the nature of the soap particles. One end of a soap particle is attracted to oil or grease, while the other end of the particle is attracted to water. The soap breaks up the oil or grease into small droplets, forming an emulsion. When the water is drained away, the soap particles drag away the droplets. The dirt is dragged away with the droplets. (Figure 6.7).

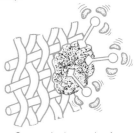
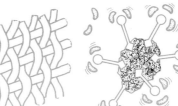

(a) Without soap, grease-laden dirt will not mix with water.

(b) Grease-loving ends of soap molecules attach to dirt.

water-loving end ⟷ grease-loving end
soap molecule

water molecule

(c) Dirt now has water-loving surface. The motion of the water molecules carries dirt away.

Figure 6.7 *How soap works*

Self-check

1. What is an emulsion?
2. What is the purpose of the egg yolk in salad dressing?
3. Name three common emulsions.

Chapter Objectives

NOW THAT YOU HAVE COMPLETED THIS CHAPTER, CAN YOU DO THE FOLLOWING?	FOR REVIEW, TURN TO SECTION
1. Describe the Tyndall effect.	6.1
2. Define colloid and give several examples.	6.1
3. Describe suspension and give several examples.	6.1
4. State some differences between colloids, suspensions, and solutions.	6.1
5. Define emulsion and give several examples.	6.2
6. State the purpose of an emulsifying agent and describe how it works.	6.2

Words to Know

colloid
Tyndall effect
suspension
emulsion
emulsifying agent

Tying It Together

1. In your notebook, make a table similar to Table 6.1. In columns 2, 3, and 5 enter "yes" or "no" for each of the three kinds of mixtures. In column 4, indicate the relative size of the particles (smallest to largest).

2. State which of the following statements are true and which are false. Correct any false statements and write the corrected statement in your notebook.
   (a) The particles in a solution are too small to scatter a beam of light.
   (b) In a suspension, some of the mixture will settle because of gravity.
   (c) The parts of a colloid are easily separated by filtering.
   (d) A colloid is a mixture in which the particles are too small to settle by gravity.
   (e) An emulsion contains two liquids and possibly a third substance.
3. Why do the particles in a colloid pass through a filter?

Table 6.1

TYPE OF MIXTURE	TYNDALL EFFECT	ARE INDIVIDUAL PARTICLES SEEN?	RELATIVE SIZE OF PARTICLES	DO PARTICLES SETTLE?
Solution				
Colloid				
Suspension				

SAMPLE ONLY

4. What is the difference between the particles in a solution and the particles in a suspension?
5. Describe the Tyndall effect.
6. Give four examples of the Tyndall effect from daily life.
7. Give an example of a colloid that consists of the following:
   (a) solid particles in a liquid
   (b) solid particles in a gas
   (c) a liquid in a liquid
   (d) a gas in a liquid
   (e) a liquid in a gas
8. Homogenized milk is not homogeneous and is not a solution. Explain this statement.
9. A colloid can be formed by mixing fine clay with water. Describe the appearance of the colloid after (a) an hour, (b) a week, and (c) a year.

Applying Your Knowledge

1. Look for seven different mixtures in your home. Try to classify them as solution, colloid, suspension, or emulsion. Quite often a mixture falls into two or more of these categories at the same time. Be prepared to give a reason for your classification.
2. Examine the list of ingredients on the labels of bottled spring water found in the supermarket.
   (a) Which ingredients do you think occur naturally in this water?
   (b) Which ingredients may have been added at the bottling plant?

Projects for Investigation

1. In the library, find a book that describes the difference between "hard water" and "soft water." Find out why detergents are more effective than soaps at cleaning clothing in hard water. At the same time, you might try to find out what causes the "ring" that is left around the bathtub after you have a bath. How can you remove it?
2. Design an activity to answer the following questions:
   (a) If milk is filtered, will there be residue?
   (b) Will anything be left behind if milk is allowed to evaporate?
3. (a) Recall that mayonnaise is an emulsion containing oil, water, and egg yolk (with other things such as mustard, lemon juice, or vinegar added for taste). In a cookbook, find a recipe for homemade mayonnaise. Perhaps you would like to try it.
   (b) Design an experiment to find out if it is cheaper to make your own mayonnaise or buy it at the store. Here are a few hints: How many different brands of mayonnaise can you find? Are they all the same price? Do they all have the same ingredients? How do the ingredients in store-bought mayonnaise compare with the ingredients you will use?

Separating the Parts of a Mixture

Key Ideas

- Mechanical mixtures can be separated by filtration, flotation, sedimentation, and centrifugation.
- Solutes and solvents can be separated by evaporation, distillation, and fractional distillation.
- Separation methods are based on differences in the physical properties of the substances.

S ometimes we make mixtures and we want them to stay together. In other situations, we go to great lengths to separate the parts of a mixture. Our natural resources, coal and natural gas, can be separated into many valuable chemicals. Some of these separations require huge industrial plants such as the one shown here.

What methods of separation are used? Besides filtration, there are many other methods. Some are simple and some are more difficult to carry out. In this chapter, you will find out how these methods are used at home, in the community, and in industry.

Separating Mechanical Mixtures

Filtration

Filtration is used to separate parts of a mixture if the particles are of different sizes. The particles to be removed must be too large to pass through the filtering material. Thus, the sand used in the large water tanks shown in Figure 7.1 is able to remove many of the visible particles from the water. But this sand will not remove very small particles or bacteria.

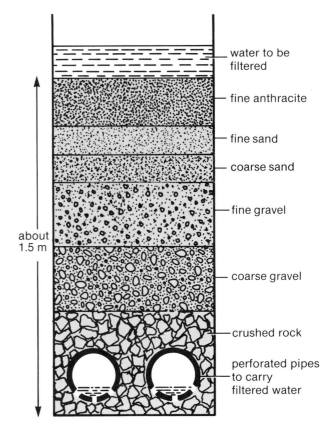

Figure 7.1 *Sand and gravel filter used in a water purification plant*

Sand is the most important filter for purification of water, but other materials can be used as well. Paper is used in laboratories, and other filters may be made from cloth, activated charcoal, metal screens, or plastic screens. Filters are used to remove dust and dirt from air in furnaces and air conditioners. Automobiles also use air filters to prevent dirt from being drawn into the engine. Some of these filters are shown in Figure 7.2.

Figure 7.2 *Three filters used in automobile engines; air filter, fuel filter, and oil filter (removed from its housing)*

Challenge

The person who sprays a rust-preventing substance on the unpainted parts of a car must wear a mask. The mask acts as a filter. There are other jobs which also require people to wear masks. Make a list of as many of these as you can think of.

Flotation

In the production of metals such as copper, zinc, lead, and nickel, the metal must be removed from the natural mixture of mineral and rock. This natural mixture, called **ore** (Figure 7.3), contains valuable mineral and unwanted rock, which can be separated by **flotation**.

The ore is crushed to a fine powder and placed in a tank of liquid that contains a detergent. The tiny particles of nickel are attracted to the detergent, but the particles of rock are not. When air is blown through the liquid the detergent foams. This foam, along with the mineral particles, rises to the top. As the foam is skimmed off, more foam rises and is continually removed. The unwanted particles of rock sink to the bottom and are discarded.

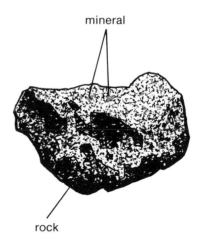

Figure 7.3 *Copper ore is mostly rock.*

Activity 7A

Sedimentation, or settling, can be used to remove solid particles from suspensions. Solid particles, which are more dense than the liquid, will eventually settle to the bottom.

Problem

How can sedimentation be speeded up?

Materials

2 test tubes and stoppers
10 mL garden soil
10 mL alum solution (aluminum sulphate)
test tube stand
water

Procedure

1. Place 5 mL of the soil in each of the test tubes. Add water until the tubes are about three-quarters full. Stopper and shake the tubes to mix the contents.
2. Put one of the test tubes in the test tube stand.
3. Add the alum solution to the other test tube, and place it in the test tube stand.
4. Observe the contents of the two test tubes after 15 min.
5. Put the test tubes in an undisturbed area, and examine them the next day.

Observations

1. Describe the appearance of the two mixtures after they were shaken.
2. How had the contents of the two tubes changed after 15 min?
3. Describe the appearance of each mixture the next day.

Questions

1. What kind of mixture did the test tubes contain after you shook them?
2. What effect did the addition of the alum solution have?

Floc Formation

In some applications, sedimentation may take a long time, especially if the suspended particles are tiny. In the water purification process, sedimentation is speeded up by a process called **floc formation**. Alum is added to the water which contains very fine particles of clay. These clay particles suspended in the water have electric charges on them which tend to keep them apart from each other (Figure 7.4). When the alum is added, the particles of alum attract the clay particles and neutralize the electric charges. The alum and clay particles then fall together to the bottom.

Figure 7.4

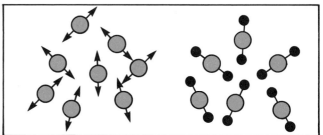

(a) *The particles of a colloid have electric charges that tend to prevent their settling.*

(b) *Alum particles (black spheres) attach themselves to the colloid particles, neutralizing the electric forces. The particles then clump together and settle by gravity.*

Centrifugation

Centrifugation is another method of separating materials which have different densities. The mixture is rotated at high speed. As it whirls around and around, the denser parts of the mixture move to the outer edge of the container.

The device used for this technique is a **centrifuge**. You may have seen a centrifuge used in the kitchen. After lettuce is washed, the water naturally drips off because of gravity.

But, in a lettuce spinner, which is a spinning centrifuge, it is removed much more quickly. Figure 7.5 shows an electrically powered centrifuge for spinning material in test tubes. This type of centrifuge can reach very high speeds. It is an important tool in laboratories.

Figure 7.5 *This electric centrifuge can rotate several thousand times per minute, causing rapid separation.*

Self-check

1. How would you separate the following mixtures:
   (a) fine iron filings and sand
   (b) white sugar and white aquarium sand
   (c) wood chips and sand
   (d) oil and water
   (e) sand and water
2. For each of the mixtures in question 1, what difference in physical properties makes the separation possible?
3. What is floc formation? Name one substance which causes floc formation.
4. What happens when alum is added to a fine suspension of clay in water? Why does this happen?

A Water Purification System

Figure 7.6 shows a water purification plant which supplies a large Canadian city with drinking water. In Figure 7.7 you will be able to trace many of the key features of the purification system.

Water is taken from a lake or other source and pumped through intake screens which prevent coarse material from entering the plant. Further filtering is provided by a microstrainer. The water then enters a flocculator. Here the alum is added and the mixture agitated. Much of the floc will settle in the sedimentation basin. The upper region of this basin feeds into the filter system which contains sand and gravel (Figure 7.1). The final stages consist of chlorination to kill harmful bacteria, and fluoridation to help prevent tooth decay. The purified water is then held in a reservoir, ready for distribution.

Not everything can be removed from a water supply. Filtration removes larger particles and chlorination kills bacteria, making them harmless. But water also contains many dissolved materials. Chemicals such as iron, copper, or calcium may colour the water or change the taste, but they are not normally harmful. Pesticides, herbicides, and waste chemicals from industry can be a major concern. Public water supplies are regularly tested for these substances to ensure that the water is safe for human use. In the long run, the best way to avoid drinking these chemicals will be to prevent them from contaminating lakes and rivers.

Figure 7.6 *The R.C. Harris Water Purification Plant, Scarborough, Ontario*

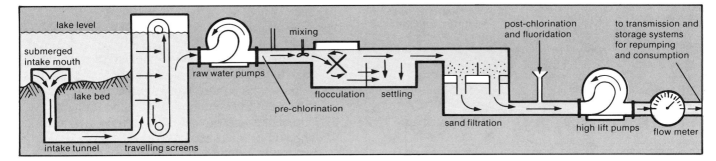

Figure 7.7 *Flowchart for a water purification system*

Self-check

1. How many different separation processes can you identify in Figure 7.7? List them in order.
2. What is the purpose of the alum in the purification process?
3. What other substances are added to water? State the purpose of each.

7.3

Distillation

It is easy to recover the solute from a solution. If you want to recover the bluestone from a solution of bluestone in water, you can simply allow the water to evaporate (Activity 5C, Part 1). Salt, in vast quantities, is obtained by letting salt water evaporate.

But suppose you want to save the solvent as well. Then the simple process of evaporation will not do. You must use a different procedure.

Distillation is a double process involving two changes of state: *vaporization* of a liquid, followed by *condensation* of vapour. It is used when pure solvent is needed. By using distillation you can separate the solvent from the solute (Figure 7.8). The purified solvent which is collected is called the **distillate**. The remaining solution after distillation is called the **residue**.

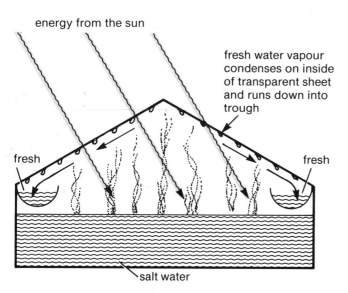

Figure 7.8 *Salt water may be separated into its components by evaporation. Large salt water ponds may be covered by tents of plastic or glass, which keep the sun's heat in. The water evaporates, condenses on the inside of the tent, and drips into the fresh water troughs along the sides. Both the salt and the fresh water may then be used.*

 **Activity 7B**

In distillation, a solution is heated until it boils. The vaporized solvent is not allowed to escape to the open air. Instead, it passes down a tube away from the solution, and then it is cooled. This causes the gas to condense where it can be collected.

Problem

How can distillation be carried out?

Materials

500 mL flask	test tube
stand	boiling chips
clamp	hot plate
rubber stopper (1 hole)	bluestone solution
short length of rubber tubing	cold water
2 glass elbows	safety goggles
500 mL beaker	

> **CAUTION** Safety goggles must be worn during this activity.

Procedure

1. The apparatus will be assembled as shown in Figure 7.9.
2. Your teacher will add bluestone solution to the flask until it is at least one-quarter full. Boiling chips will be added. The purpose of the chips is to prevent the solution from boiling too violently.
3. The beaker will be half-filled with cold water.
4. The end of the rubber tubing will be placed in the test tube, and the test tube will be placed in the beaker of cold water.
5. The solution will be heated on the hot plate until it boils. The heat will be reduced and it will be allowed to continue to boil gently.
6. Before all of the solution has boiled away, the heat source will be turned off.

> **CAUTION** Do not allow the flask to boil dry.

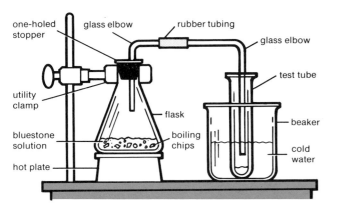

Figure 7.9 *Apparatus for distillation of bluestone solution*

7. Examine the contents of the test tube and the contents of the flask.

Observations

1. Describe the distillate. How is it different from the solution in the flask?
2. Describe the residue. Has the appearance of the solution changed in any way during the distillation?

Questions

1. Has there been separation of solute and solvent? How do you know this?
2. If you continued to boil the contents of the flask until it was dry, what would it consist of?

Challenge

Suppose that the flask in Activity 7B had contained some soil as well as the bluestone solution. Describe what the distillate and the residue would look like.

Fractional Distillation

Suppose you have a mixture of water and alcohol. These two liquids mix completely, forming a solution. You would not be able to separate them by any of the methods mentioned in this chapter—filtration, flotation, sedimentation, centrifugation, floc formation, or ordinary distillation. Even simple evaporation would not separate them as both liquids evaporate. But, because these two liquids have different boiling points, there is a method you could use.

Fractional distillation is a process which is used to separate liquids which have different boiling points. *Controlled* heating is necessary.

In fractional distillation the liquid with the lowest boiling point will boil off first. Once it has boiled away, the temperature will rise. If several liquids are present, each can be collected as its boiling temperature is reached. You would have to change the test tube for each different distillate. Each separate distillate is called a **fraction**.

Figure 7.10 illustrates the apparatus used for fractional distillation. The rubber stopper has two holes, so that a thermometer can be used to measure the temperature of the vapour. During the time that one of the liquids is boiling, the temperature will remain constant. When the temperature begins to rise, this is a signal that a different vapour is present.

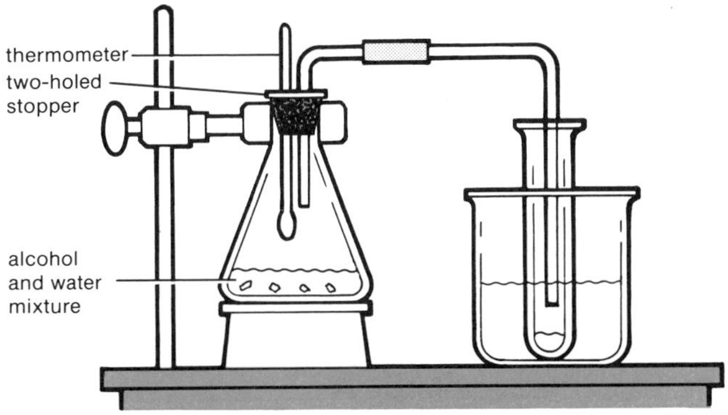

Figure 7.10 *A two-holed rubber stopper allows a thermometer to be inserted into the flask.*

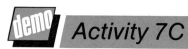

Activity 7C

Problem

How can fractional distillation be carried out?

Materials

500 mL flask
stand
clamp
rubber stopper (2 hole)
short length of rubber tubing
2 glass elbows
thermometer
2 test tubes
test tube rack
boiling chips
safety goggles
clock or watch
hot plate
masking tape
alcohol in water solution
cold water

CAUTION Safety goggles must be worn during this activity.

Procedure

1. Make a table in your notebook similar to Table 7.1. You should allow 40 lines for data.
2. Your teacher will assemble the apparatus as shown in Figure 7.10. The thermometer should be just above the surface of the liquid in the flask.
3. The beaker will be half-filled with cold water.
4. The end of the rubber tubing will be placed in the first test tube, which will be placed in the beaker of cold water.
5. The flask will be heated gently on the hot plate.
6. Record the time and temperature every 30 s while the mixture is heating.
7. When the solution is boiling, continue to record the time and temperature every 30 s.
8. When the temperature of the mixture begins to rise quickly again, the rubber tubing will be transferred to the second test tube.

9. The mixture will be heated until about one-quarter of the original liquid is left in the flask. Then the heat source will be removed.

CAUTION Do not allow the flask to boil dry.

10. Describe the appearance of both fractions. If your teacher asks you to smell the samples, do so carefully, using the proper method.

Observations

Table 7.1

TIME	TEMPERATURE OF VAPOUR
0.5 min	
1.0	
1.5	
2.0	
Etc.	

Questions

1. Draw a graph showing the time and temperature data. Use the horizontal axis for time and the vertical axis for temperature.
2. Describe the shape of the graph.

Challenge

Water and alcohol have different physical properties. With your teacher's guidance, devise several methods you could use to demonstrate these differences. Try your methods, using the two fractions you obtained in Activity 7C.

Fractional Distillation of Crude Oil

The most important use of fractional distillation is in the petroleum industry. Crude oil (petroleum) is a very complex mixture of several liquids, with many solids and gases dissolved in it. To separate these parts, the crude oil is first passed through a furnace, where it reaches a temperature of close to 400°C. Then it enters a tall structure called a fractionating column, or **bubble tower** (Figure 7.11). The bubble tower has several horizontal trays with openings in them, one above the other.

One fraction remains in liquid or solid form, and is removed as the residue. The remaining distillate moves up to the next layer. Here the temperature is slightly lower and one fraction condenses to liquid. The remaining gases pass up to the next level, which is at a lower temperature. This process continues as a different liquid condenses at each level. The fractions with the lowest boiling points condense near the top of the column.

Gasoline has a relatively low boiling point, and condenses near the top of the column. Other important fractions taken off at lower levels are kerosene, heating oil, jet fuel, diesel oil, and heavy lubricating oils and greases. The tall bubble towers of an oil refinery can be seen in Figure 7.12.

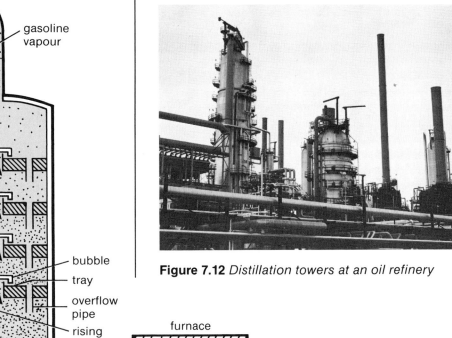

Figure 7.12 *Distillation towers at an oil refinery*

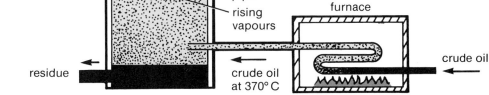

Figure 7.11 *A simplified cross-section of a fractionating column for distillation of crude oil. Although only 4 trays are shown, as many as 30 may be present.*

Self-check

1. What changes of state take place during distillation?
2. How can you obtain drinking water from sea water?
3. How would you separate a mixture of sugar, water, and sand
   (a) if you did not need to save the water?
   (b) if you wanted to save the water?
4. Why is fractional distillation important in the petroleum industry?

Challenge

The temperature of a liquid was measured as it boiled. At first the temperature was 78°C and then it rose to 86°C while it continued to boil. What conclusion can you reach concerning this liquid?

Paper Chromatography

Have you ever spilled some water on your notebook page? You may have noticed that the ink on your page became tinged with colours after the soaking. The various pigments that make up the ink are themselves coloured. When mixed with the water on the page these pigments travel at different rates through the water and the paper and thus become separated.

This method of separation, which depends on differences in solubility, is called **paper chromatography**. Substances that are more soluble in the solvent move faster than substances which are less soluble. Chromatography is an important technique used by technicians in research laboratories. In the next activity, you will carry out paper chromatography.

Science at Work

A team of four scientists at B.C. Research in British Columbia have discovered a method of extracting copper metal from ore without polluting the environment. The method involves the use of rock-eating bacteria. The existence of such bacteria has been known about since the early 1950s, but the separation process could not be controlled because unwanted sulphuric acid was also being produced. (Copper ore contains both copper and sulphur.) In the process discovered by the B.C. Research scientists, the bacteria eat the sulphuric acid and produce valuable sulphur instead.

Problem

How many dyes are there in a sample of ink?

Materials

test tube
strip of filter paper or blank newsprint
a few drops of black ink
water
piece of aluminum foil
test tube rack

Procedure

1. Cut the filter paper so that it will fit inside the test tube from top to bottom. Taper one end of the paper to a tip.
2. Fill the test tube about one-quarter full with water. Do not get any drops on the inside walls of the tube.
3. Streak a few drops of the ink across the paper strip, about one centimetre from the tapered end.
4. Place this end of the filter paper into the test tube, with the tip of the paper in the water. The ink strip should not be in the water (Figure 7.13).

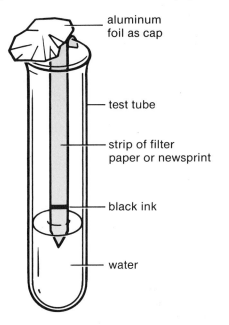

Figure 7.13 *Separating the components of black ink*

5. Secure the top of the paper by folding it over the top of the test tube. Then hold it in place with the aluminum foil as a cover across the top of the tube.
6. When the colours have risen several centimetres, remove the paper and hang it up to dry.

Observations

1. Sketch the paper, showing where the ink was placed.
2. Sketch the paper, showing any changes after chromatography.

Questions

1. How many separate parts of the ink can you detect?
2. Why do some of the parts rise higher in the paper than others?
3. Can you put the separate parts together again? Try it.

Challenge

Try using paper chromatography to separate some of the parts in tomato juice. You will need to filter the juice first and try chromatography with the filtrate. Besides being a solution, what other kind of mixture is tomato juice?

Chapter Objectives

NOW THAT YOU HAVE COMPLETED THIS CHAPTER, CAN YOU DO THE FOLLOWING?	FOR REVIEW, TURN TO SECTION
1. List four methods of separating mechanical mixtures, and tell what property each method depends on.	7.1
2. Describe several examples of filtration.	7.1
3. Tell where flotation is used and how it works.	7.1
4. Explain how sedimentation can be speeded up.	7.1
5. Explain what centrifugation is and why it works.	7.1
6. Describe the separation processes in water purification systems.	7.2
7. Explain how distillation works.	7.3
8. Distinguish between distillation and fractional distillation.	7.3
9. Describe how a bubble tower is used in the fractional distillation of crude oil.	7.3
10. Explain what property is important for separation by paper chromatography.	7.4

Words to Know

filtration
ore
flotation
sedimentation
floc formation
centrifugation
centrifuge

distillation
distillate
residue
fractional distillation
fraction
bubble tower
paper chromatography

Tying It Together

1. State whether each of the following statements is true or false. In your notebook, write the corrected statements.
   (a) Sand in water filtration tanks can remove suspended particles and bacteria from the water.
   (b) Floc formation is used in water treatment to speed up the sedimentation process.
   (c) The liquid that condenses during distillation is called the filtrate.
   (d) Small particles settle more quickly than large particles.
   (e) Fractional distillation can be used to separate two liquids if they have the same boiling point.
2. (a) List four ways of separating parts of a mechanical mixture.
   (b) Which methods depend on substances having particles of different size?
   (c) Which methods depend on particles having different density?
   (d) In which methods is foaming a part of the process?
   (e) Which process is speeded up by the addition of alum?
3. List five uses of filtration in everyday life.
4. (a) What is ore?
   (b) Describe how minerals can be removed from ore.
5. A student mixed clay with water in two cylinders and added alum to one of them. Describe the effect of the alum and explain why it has this effect.

6. What is a centrifuge and how does it work?
7. (a) List the separation methods used in the purification of a public water supply.
   (b) List substances that are added to the water. For each, tell why it is added.
8. (a) In distillation, what two changes of state occur?
   (b) How does the residue after distillation differ from the original solution?
9. (a) What is a bubble tower?
   (b) List several useful substances that are obtained from crude oil.
   (c) Describe how these substances are separated from crude oil.
10. (a) What is paper chromatography?
    (b) On what difference in properties does this method of separation depend?
11. How would you separate each of the following?
    (a) mixture of brass bolts and steel bolts, all painted the same colour
    (b) mixture of sand and water
    (c) mixture of sugar and sand
    (d) mixture of sawdust and sand
    (e) mixture of water, alcohol, and glycerine

Applying Your Knowledge

1. A vacuum cleaner draws up dust in a stream of air. How is the dust removed from the air? Examine several models of vacuum cleaners, and record how different manufacturers have solved the question through different designs.
2. Why do surgeons and nurses wear surgical masks during an operation? What substances would the masks separate?
3. What kinds of impurities do the following filters separate?
   (a) gasoline filter in a car engine
   (b) oil filter in a car engine
   (c) charcoal filter in a kitchen range hood
   (d) charcoal filter in an aquarium

Projects for Investigation

1. How is the water used in swimming pools purified? Investigate the design of different pool purification systems. How are they similar to and how do they differ from the methods of purifying drinking water described in this chapter?
2. You can use paper chromatography to separate the parts of many common substances. An interesting project is to separate the components of coloured leaves, fruits, or vegetables. You will first have to grind the material and dissolve the coloured materials with a solvent. Look up some procedures.
   Caution: Many of the suitable solvents are flammable. Their vapours are poisonous. Avoid breathing the vapours, and do not heat the solvents with open flames.
3. Sewage is a complex mixture of wastes in water. Investigate the sewage treatment system of your municipality. Write a report on the processes involved in separating the wastes from the water. How effective are the processes? How pure is the water that is returned to the source?
4. Do research to find answers to the following questions: What kinds of filters are found in cigarettes? What substances in cigarette smoke can be filtered? Does this mean that cigarettes are safe? What is mainstream smoke? What is sidestream smoke? Taking into account both smokers and non-smokers, how effective are cigarette filters?
5. Find out what electrostatic precipitators and scrubbers are. How do industries use these devices to separate harmful substances from exhaust gases? How effective are they in helping control acid precipitation?
6. In previous studies, you probably investigated the water cycle, which helps maintain the amount of water that circulates in our lakes, rivers, and oceans. Think about what you have learned about the process of distillation. Then apply this to the water cycle. How are they similar? How are they different? You will find it helpful to use illustrations to answer these questions.

Unit Two: Mixtures

MATCH

In your notebook, write the letters (a) to (h). Beside each letter, write the number of the word in the right column that corresponds to each description in the left column.

(a)	non-uniform in composition	1.	solute
(b)	scattering of light	2.	mechanical mixture
(c)	a mixture made up of two liquids	3.	solubility
(d)	what's left on filter paper	4.	residue
(e)	dissolves in the solvent	5.	supersaturated solution
(f)	its parts settle by gravity	6.	emulsion
(g)	amount that dissolves in 1 L of water	7.	Tyndall effect
(h)	has more solute than normal	8.	transparent
		9.	suspension

MULTIPLE CHOICE

In your notebook, write the numbers 1 to 10. Beside each number, write the letter of the best choice.

1. A mixture of pebbles, sand, and water is filtered using filter paper. What will the residue be?
   (a) sand
   (b) water
   (c) sand and water
   (d) sand and pebbles

2. Solutions are different from mechanical mixtures because solutions are
   (a) heterogeneous
   (b) opaque
   (c) homogeneous
   (d) never transparent.

3. Sea water is a solution. What do you call the salt that is dissolved in the water?
   (a) solvent
   (b) solute
   (c) filtrate
   (d) residue

4. Anna placed a single crystal of hypo into a solution of hypo in water. The crystal dissolved very quickly. What can you conclude?
   (a) The solution is unsaturated.
   (b) The solution is saturated.
   (c) The solution is supersaturated.
   (d) Hypo is insoluble in water.

5. A solution of sugar and water is filtered. The filtrate will
   (a) contain water only
   (b) contain water and sugar
   (c) contain sugar only
   (d) be entirely in the residue.

6. When an opaque mixture was allowed to stand, some particles quickly settled to the bottom. What kind of mixture was this?
   (a) solution
   (b) colloid
   (c) emulsion
   (d) suspension

7. Why is alum added to water in a water purification plant?
   (a) to kill bacteria
   (b) to help prevent tooth decay
   (c) to speed up the settling of suspended particles
   (d) to remove dissolved substances from the water

8. A certain mixture did not produce the Tyndall effect. What could this mixture be?
   (a) colloid
   (b) solution
   (c) mechanical mixture
   (d) emulsion

9. Jim wanted to find out which of two plant fertilizers, A or B, was the best for African violets. He divided the plants into two groups of four for the testing. What should he now do?
   (a) Use fertilizer A to fertilize one group and use fertilizer B to fertilize the other.
   (b) Keep all the plants at the same temperature.
   (c) Give all the plants the same amount of light.
   (d) all of the above

10. Some rubbing alcohol and water were mixed, and the rubbing alcohol dissolved in the water. What method could be used to separate the two liquids?
    (a) filtration
    (b) flotation
    (c) distillation
    (d) fractional distillation

TRUE/FALSE

Write the numbers 1 to 10 in your notebook. Beside each number, write T if the statement is true and F if the statement is false. For each false statement, rewrite the statement so it becomes true.

1. A solvent can be separated from a solution by filtration.
2. A suspension is a mechanical mixture.
3. Solutions are heterogeneous mixtures.
4. In order to grow larger, a crystal should be suspended in an unsaturated solution.
5. There is a limit to the amount of solute that will dissolve in a solvent.
6. A colloid will produce the Tyndall effect when light passes through it.
7. In a suspension, the particles are too small to settle by gravity.
8. An emulsifying agent will keep two liquids from mixing together.
9. In an oil refinery distillation tower, the liquids with the higher boiling points separate at the top of the tower.
10. A bleach can remove the colour of a stain without dissolving the stain.

FOR DISCUSSION

Read the paragraphs and answer the questions that follow.

Soft drinks are beverages that are carbonated. In other words, they are solutions of carbon dioxide gas dissolved in water. Carbon dioxide in the water has a refreshing effect, with the ability to quench thirst. The familiar "fizz" of a soft drink is caused by the carbon dioxide gas escaping.

Carbonated waters can be human-made. They also occur in nature in underground springs. These springs contain large amounts of minerals, such as magnesium and calcium, in addition to carbon dioxide. The dissolved materials give a unique taste and aroma to the water. Spring water—known and enjoyed in Roman times—is bottled and sold today in a thriving business.

Many of today's soft drinks contain caffeine, sugar or artificial sweetener, and flavouring extracts. Soft drinks are typically 90% water, 9.5% sugar or other sweetener, and 0.5% flavouring.

In a soft-drink bottling plant, the water used is first treated to remove chlorine that was added at the water purification plant. Any oxygen and dissolved minerals are also removed. Bottles moving along the assembly line are partially filled with flavoured concentrated syrup. They are then diluted with cold carbonated water under high pressure. The high pressure and cool temperature increase the solubility of carbon dioxide in the water.

1. What produces the "fizz" in a soft drink?
2. What solutes are usually found in natural spring water?
3. Give two reasons why bottled spring water is popular today.
4. Name three solutes in a carbonated beverage.
5. What solutes are removed from water early in the bottling process?
6. Suggest why the solutes in question 5 are removed.
7. State two factors that increase the solubility of carbon dioxide in water.
8. If a carbonated soft drink is left open for several hours, it tastes "flat." Why is this so?

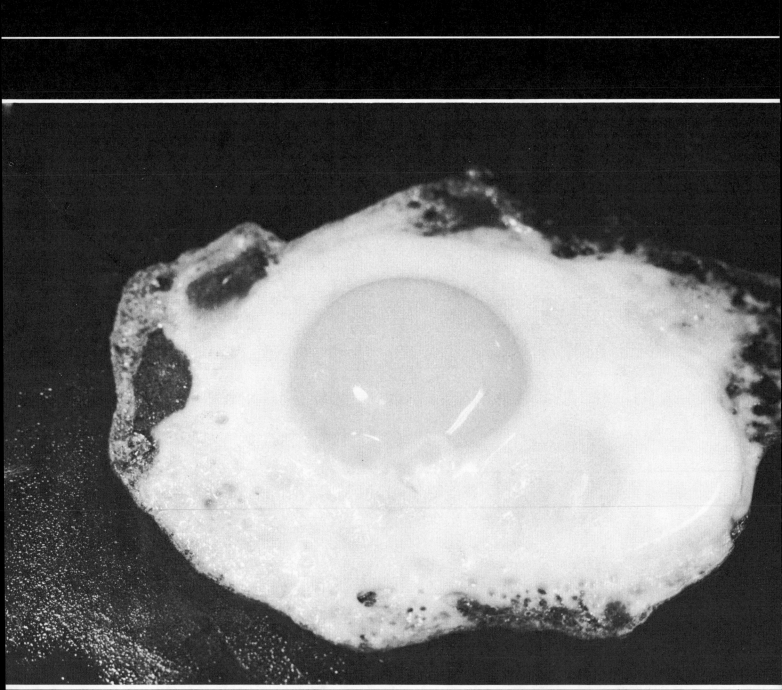

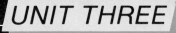

Chemical Change

When you crack open an egg and drop it into a hot frying pan, changes occur in the egg. The transparent part of the egg liquid becomes a white solid. The yellow part of the egg liquid becomes opaque as it solidifies. When you let the egg cool, the white and yellow parts do not return to their original form. You have caused a chemical change.

What is the difference between a chemical change and a physical change? What are chemicals, and how do they affect your life? How are they helpful, and how can they be dangerous? You will find answers to these questions in this unit. You will also learn how to identify some common gases. And you will investigate chemical changes involving one important gas in particular, oxygen.

Chemicals and Chemical Change

Key Ideas

- Pure substances are classified as elements or compounds.
- Chemical change results in new substances with new properties.
- The particle model of matter can help explain chemical change.
- Chemical changes produce substances that can be both useful and dangerous.

Chlorine is a yellowish green gas that is extremely poisonous. When chlorine escaped from a derailed train car in Mississauga, Ontario, in 1979, over 200 000 residents of the area were evacuated for up to six days. Yet controlled amounts of chlorine gas are used to kill harmful bacteria in municipal water supplies—water you can safely drink.

You can see that, even though chlorine is dangerous, it also has properties that make it useful to us. Chlorine is used as a bleach to remove stains. When combined with another substance, sodium, chlorine forms a very familiar and commonly used substance—table salt.

How can some substances be useful and dangerous at the same time? In this chapter, you will investigate changes in matter called *chemical changes* to help you answer this question.

Classifying Matter

All matter may be classified as either pure substances or mixtures. You can see in Figure 8.1 that pure substances can be further classified as elements or compounds. We will define pure substances more precisely later in this section. For now, think of each pure substance as a different kind of matter with its own set of properties. These properties let us tell one pure substance from another. For example, iron can be distinguished from lead because of differences in density, hardness, and resistance to rust.

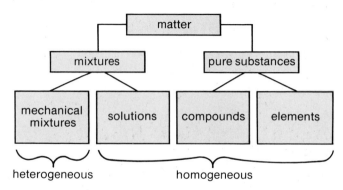

Figure 8.1 *Classification of matter*

The simplest pure substances are **elements**. Elements are pure substances made up of only one kind of particle. For example, gold is an element because it is composed only of gold particles. Compounds are pure substances made up of two or more elements. Pure substances are always homogeneous. But each pure substance is a different kind of matter with its own set of properties.

There are only 92 naturally occurring elements on earth. There are also a few more elements which scientists have been able to make using nuclear reactions. Table 8.1 on the next page shows the names of some elements. You probably have heard of the elements in the left-hand column, but the ones in the right-hand column may be less familiar to you.

Table 8.1 *Different Elements*

SOME ELEMENTS YOU MIGHT KNOW	SOME ELEMENTS YOU MIGHT NOT KNOW
Aluminum	Actinium
Calcium	Californium
Chlorine	Gallium
Copper	Krypton
Hydrogen	Plutonium
Mercury	Strontium
Oxygen	Thorium
Sulphur	Uranium
Zinc	Yttrium

Different elements have different properties. They may be solids, liquids, or gases. Some are totally harmless, and others are very dangerous. You, the earth, and the entire universe are made up of elements (Figure 8.2).

Figure 8.2

(a) *Distribution of elements in the earth's crust*

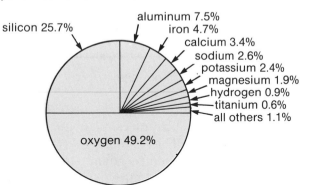

silicon 25.7%
aluminum 7.5%
iron 4.7%
calcium 3.4%
sodium 2.6%
potassium 2.4%
magnesium 1.9%
hydrogen 0.9%
titanium 0.6%
all others 1.1%
oxygen 49.2%

(b) *Elements in the human body*

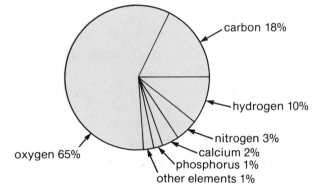

carbon 18%
hydrogen 10%
nitrogen 3%
calcium 2%
phosphorus 1%
other elements 1%
oxygen 65%

Problem

What properties of elements can you observe (Figure 8.3)?

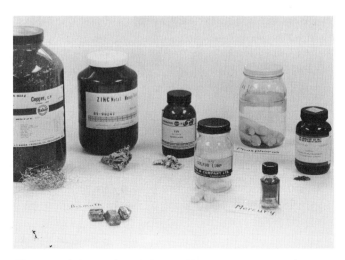

Figure 8.3 *Note the visible differences among these common elements.*

Materials

10 samples from the following list of elements:

aluminum	lead
bismuth	magnesium
cadmium	mercury
calcium	nickel
carbon	nitrogen
chromium	silicon
cobalt	sulphur
copper	tin
helium	tungsten
iodine	zinc
iron	

CAUTION Many elements are poisonous and dangerous. Follow very carefully your teacher's instructions for observing the elements.

Procedure

1. In your notebook, make a table similar to Table 8.2. List the names of the elements that you will examine.
2. Carefully observe the elements. In the Properties column, record what you observe. Use the following questions as a guide.
   (a) What colour is the element? (If the element is not visible, say so.)
   (b) What is the state of the element?
   (c) Is the element shiny, bendable, etc.? You may have to scrape the surface with a sharp object to see the underlying surface.

> **CAUTION** Be careful when using sharp tools. Make any scraping motions away from yourself.

   (d) Is the element heavy or light for its size compared to the others? In other words, how would you rate its density?

Observations

Table 8.2

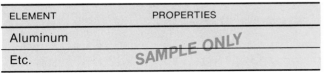

ELEMENT	PROPERTIES
Aluminum	
Etc.	

Questions

1. Which of your elements are gases? Which are liquids?
2. (a) Which of your elements do you think are metals?
   (b) What general properties do metals have?
   (c) Are there any elements which had only some properties usually associated with metals? Name them.

Elements and Compounds

How can there be so many different chemicals if there are only 92 naturally occurring elements? If you think of the English alphabet this may become clear. The 26 letters of the alphabet are used to make thousands of words—long words, short words, words with repeated letters, and words with letters in different order. In the same way, the 92 elements can be combined to make millions of different **compounds**.

Atoms and Molecules

According to the particle model for matter, all matter is made up of tiny particles. The smallest particle of an element that has the properties of the element is called an **atom**. In an element, every atom is exactly the same as every other atom. For example, the element hydrogen is composed only of hydrogen atoms. The element oxygen is composed only of oxygen atoms.

Atoms join together to form larger particles called **molecules**. Compounds are made up of molecules containing two or more different kinds of atoms. For example, water is a common compound. Water consists of molecules made up of two atoms of hydrogen and one atom of oxygen. See Figure 8.4 (a).

Some molecules of elements are made up of atoms of the same element. For example, the element oxygen consists of molecules containing two oxygen atoms. See Figure 8.4 (b).

Figure 8.4

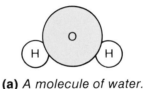
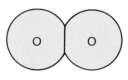

(a) *A molecule of water. In a compound there are two or more kinds of atoms.*

(b) *A molecule of oxygen. In a molecule of an element there is only one kind of atom.*

When compounds form, a certain number of atoms of one type combine with a certain number of atoms of another type. Molecules always have the same composition. For example, *one* atom of magnesium always combines with exactly *one* atom of oxygen to form the compound magnesium oxide.

It is possible to form more than one compound from the same elements. For example, carbon dioxide and carbon monoxide both contain only carbon atoms and oxygen atoms. But carbon dioxide and carbon monoxide have very different properties. Similarly, water and hydrogen peroxide are different compounds which are both made up only of hydrogen atoms and oxygen atoms. But the properties of water and hydrogen peroxide are quite different (Figure 8.5).

Figure 8.5

(a)

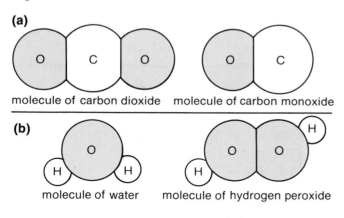

molecule of carbon dioxide   molecule of carbon monoxide

(b)

molecule of water   molecule of hydrogen peroxide

Symbols for Elements and Compounds

Words are not needed to explain the meanings of the symbols circled in Figure 8.6. These symbols can be found on many household products. Symbols are used because they present important information quickly and directly.

Figure 8.6 *These symbols can be found on various products. They warn of potential dangers to the user.*

They are also understandable no matter what language you speak. Scientists use symbols for a similar reason. Some of the symbols they use to represent elements are shown in Table 8.3. Why do you think some elements must have two-letter symbols?

Table 8.3 *Some Elements and Their Symbols*

ELEMENT	SYMBOL
Aluminum	Al
Calcium	Ca
Carbon	C
Cobalt	Co
Copper	Cu
Hydrogen	H
Iodine	I
Magnesium	Mg
Nitrogen	N
Oxygen	O
Phosphorus	P
Zinc	Zn

Just as atoms combine to form molecules, the symbols of atoms can be combined to form symbols of molecules. Some examples of these symbols, called **chemical formulas**, are shown in Table 8.4. Chemical formulas are more than just short forms of names. They represent a lot of chemical information. They show the kind and the number of atoms present in a molecule of a substance. Since each symbol represents the exact mass of one atom, the formula represents the exact mass of a molecule. Because of this, chemical formulas show the exact proportion of each element in the compound.

Self-check

1. Why is chlorine dangerous and why is it useful?
2. (a) What is an element?
   (b) What is a compound?
3. Distinguish between an atom and a molecule.
4. What is the difference between a solution and a pure substance?
5. Why are symbols for elements used?
6. What is represented by these chemical formulas?
   (a) H_2O
   (b) CO_2

Table 8.4 *Some Chemical Formulas of Substances*

SUBSTANCE	FORMULA	ATOMS IN MOLECULE	HOW SUBSTANCE IS USED
Oxygen	O_2	2 atoms of oxygen	You breathe this to stay alive.
Water	H_2O	2 atoms of hydrogen 1 atom of oxygen	It makes up 70% of your body.
Glucose	$C_6H_{12}O_6$	6 atoms of carbon 12 atoms of hydrogen 6 atoms of oxygen	A sugar used for quick energy.
Bluestone (copper sulphate)	$CuSO_4$	1 atom of copper 1 atom of sulphur 4 atoms of oxygen	Kills fungi on fruit crops; source of copper in electroplating.
Alum	$Al_2(SO_4)_3$	2 atoms of aluminum 3 atoms of sulphur 12 atoms of oxygen	Hastens settling when added to suspensions.
Carbon dioxide	CO_2	1 atom of carbon 2 atoms of oxygen	Plants use this to produce food.
Acetylsalicylic acid (ASA)	$C_9H_8O_4$	9 atoms of carbon 8 atoms of hydrogen 4 atoms of oxygen	Reduces headaches and inflammations.

Physical Change and Chemical Change

Reviewing Physical Change

When sugar dissolves in water to form a solution, the solution tastes sweet. The solution shares the properties of both the sugar molecules and the water molecules, and the molecules are not changed. If the parts of the solution are separated, the sugar still has its original properties and so does the water. When water is heated to 100°C, it becomes a gas. But gaseous water is still the same substance as liquid water. If heat is removed, the gas returns to the liquid state (Figure 8.7). The change of state does not change the water molecules themselves. Their speed and spacing may change, but no new type of particle is formed.

Both dissolving and changing state are examples of **physical change**. During a physical change, the molecules of the starting material or materials remain the same. No new kind of matter is formed.

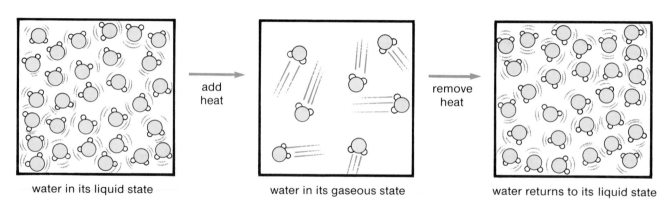

water in its liquid state   add heat   water in its gaseous state   remove heat   water returns to its liquid state

Figure 8.7 *Adding or removing heat causes the physical change of water from one state to another. The substance itself, water, remains the same.*

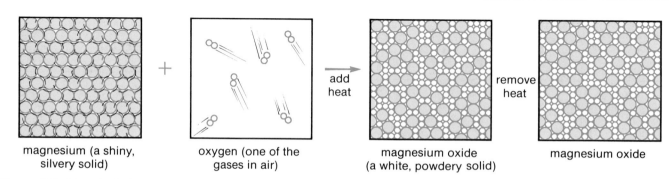

magnesium (a shiny, silvery solid)   +   oxygen (one of the gases in air)   add heat   magnesium oxide (a white, powdery solid)   remove heat   magnesium oxide

Figure 8.8 *Chemical change occurs when heat is added to magnesium. The magnesium combines with oxygen in the air to produce a new substance with new properties, magnesium oxide. If the heat is removed, the magnesium oxide compound will not change back to the individual elements magnesium and oxygen.*

Chemical Change

In a **chemical change**, a new substance with new properties is formed. When iron is exposed to oxygen in the air, a crumbly, reddish brown solid, which we call rust, is formed. The properties of rust are different from the properties of iron. For example, you know that one of the physical properties of iron is that it can be attracted to a magnet. But rust is not attracted to a magnet. The iron has changed into a different substance with different properties.

During a chemical change, the particles of the starting materials are broken down or rearranged to form new particles with new properties (Figure 8.8). They do not simply arrange themselves differently or move differently. They actually change identity.

When you describe how a substance behaves in a chemical change, you are describing its **chemical properties**. For example, one of the chemical properties of iron is that it combines with oxygen to form rust. One of the chemical properties of solid magnesium metal is that it combines with oxygen to form a white powdery solid, called magnesium oxide.

PART 1

Problem

How can you distinguish physical from chemical change?

Materials

2 birthday candles
test tube
beaker
hot plate
safety goggles
Plasticine or other stand for candle
match

> **CAUTION** Safety goggles must be worn for this activity.

Procedure

1. Place one of the candles into the test tube. Put this test tube in water in the beaker, and heat it.
2. Observe and record any changes which happen to the candle.
3. Set the other candle securely in its stand, and light it.
4. Let the candle burn for a few minutes. While it is burning observe and record any changes.

Observations

1. Describe the physical properties of candle wax.
2. Describe what happened when the birthday candle was heated in the test tube.
3. Describe the burning candle.
4. How was the candle different after it had been burned?

Questions

1. (a) When the candle was heated in the test tube, did you observe a physical or a chemical change? How do you know?
   (b) Is it possible to put the candle back the way it was before it was heated?
2. (a) When the candle was burned, was this a physical or a chemical change? How do you know?
   (b) What is one chemical property of candle wax?
   (c) Is it possible to put the candle back the way it was before it was burned?
3. When a candle burns, are any new substances made? If so, can you see them? Explain.

PART 2

Problem

What other physical and chemical changes can you observe?

Materials

test tube
evaporating dish
flask
scoopula
potassium iodide
lead nitrate
distilled water

CAUTION The chemicals used in this activity contain lead which is a poison. Do not pour this material down the drain. Carefully pour the material into a large filter provided by your teacher. Your teacher will then properly dispose of the residue.

Procedure

1. Observe and record the physical properties of the potassium iodide and lead nitrate.
2. Fill a test tube half-full with distilled water. Add a small amount of potassium iodide and mix. Describe the solution.
3. Pour some of the potassium iodide solution into an evaporating dish and set this aside until the next class. If your teacher directs you to, you could speed up evaporation by warming the evaporating dish on a hot plate.
4. Place 25 mL of distilled water into the flask. With a scoopula, add a small amount of lead nitrate. Swirl the flask until the lead nitrate dissolves. Describe the solution.
5. Add the potassium iodide solution to the lead nitrate solution. Describe what happens.
6. Following your teacher's directions, carefully dispose of the material in the flask.

Observations

1. (a) Describe potassium iodide and lead nitrate.
   (b) Describe the two solutions.
2. After the potassium iodide solution was left for some time in the evaporating dish, what remained?
3. What happened when the two solutions were mixed?

Questions

1. What kind of change occurred to the potassium iodide solution in the evaporating dish? Explain your answer.
2. What kind of change occurred when the two solutions were mixed? Explain how you know this.

Challenge

In a chemical change, is there any matter created or destroyed? Repeat Part 2 of Activity 8B, using the apparatus shown here. The test tube must be small enough to fit inside the stoppered flask. Find the mass of the flask and everything in it. Then, keeping the stopper in, tip the flask so that the two solutions mix. Determine the mass of the flask and the new substances. Has there been any change?

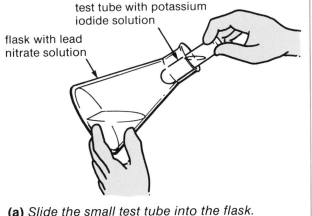

test tube with potassium iodide solution

flask with lead nitrate solution

(a) *Slide the small test tube into the flask. Do not let the solutions mix.*

(b) *Put a stopper in the flask.*

(c) *Determine the mass of the flask and everything in it.*

Reviewing Chemical Change

Chemical changes, often called **chemical reactions**, are going on all around you. When any substance burns, it undergoes a chemical reaction. When you fry an egg, or bake a cake, or allow glue to harden, you are carrying out chemical reactions.

All life on earth depends on a chemical reaction called *photosynthesis*. Plants use photosynthesis to produce sugar from water, carbon dioxide, and energy from the Sun. When you eat that sugar you use it in another chemical reaction that occurs inside your body. The sugar combines with oxygen from the air you breathe to produce the energy you use to run, study, and stay alive.

Some chemical reactions, like the explosion or burning of flammable materials, may be spectacular and happen very fast (Figure 8.9). Others, such as the rusting of iron, may be slow and hard to detect. For some reactions, such as photosynthesis, energy must be supplied to make the reaction happen. For other reactions, such as the chemical reactions in your body, energy is released. But in every chemical reaction a new substance with new properties is formed.

Figure 8.9

121

Word Equations for Chemical Reactions

In a chemical reaction the starting materials are called the **reactants**. The substances produced are called the **products**. Thus, for example, iron and oxygen (reactants) react to form rust (product). A chemical reaction can be written as a word equation. A **word equation** shows at a glance the reactants and products of a reaction.

$$reactant(s) \longrightarrow product(s)$$

In this statement, the arrow ($\longrightarrow$) means "produce(s)."

The rusting of iron can be represented as

$$iron + oxygen \longrightarrow rust$$

You would read this as "Iron plus oxygen produces rust." Iron and oxygen are the reactants, and rust is the product. The reaction in Activity 8B, Part 2 can now be expressed in words as

$$lead\ nitrate + potassium\ iodide \longrightarrow$$
$$potassium\ nitrate + lead\ iodide$$

Try reading this equation aloud. In this reaction, lead nitrate and potassium iodide are the reactants, and potassium nitrate and lead iodide are the products.

Some reactions involve the presence of energy. For example, in order to cook an egg, energy must be added. Energy is not a substance, so it is not a reactant. But it can still be expressed in an equation as

$$raw\ egg + energy \longrightarrow cooked\ egg$$

You would read this as "raw egg plus energy produces cooked egg." The energy term can also be added to the photosynthesis equation:

$$carbon\ dioxide + water + energy \longrightarrow$$
$$sugar + oxygen$$

Energy can also appear as a product in a reaction. One of the reactions that occurs in your body produces energy. Thus, in this case, energy appears with the products:

$$sugar + oxygen \longrightarrow$$
$$carbon\ dioxide + water + energy$$

Challenge

In activity 4F you carried out chemical reactions using two different kinds of bleach. Design an experiment in which you test these two kinds of bleach on samples of hair. After the chemical change, examine the hair carefully. Which type of bleach does less damage to the hair?

CAUTION Use samples of hair for this experiment. Never use any laundry bleach on any part of your body!

Challenge

In a beaker, make a solution of bluestone (copper sulphate). Place an iron nail in the solution, and watch what happens over the next 30 min. What evidence can you see that a chemical reaction is occurring? What are the reactants? Can you identify one of the products? Try writing a word equation for this reaction.

Activity 8C

PART 1

Problem

What are some characteristics of chemical changes involving elements that combine to form compounds?

Materials

magnesium ribbon
steel wool (iron)
chlorine gas (cylinder)
charcoal (carbon)
oxygen (cylinder)
sulphur
crucible tongs
Bunsen burner
gas bottles
deflagrating spoon
glass plates
safety goggles

> **CAUTION** All parts of this activity are teacher demonstrations. Wear safety goggles during these demonstrations.

Procedure

1. In your notebook, make a large table like Table 8.5.
2. For each of the pairs of elements do the following:
   (a) Observe and record the physical properties of the elements before the chemical change.
   (b) Observe and record the chemical change.
   (c) Observe and record the physical properties of the new substance formed.
3. Your teacher will clean a 4 cm strip of magnesium ribbon with emery paper. Then the ribbon will be held with crucible tongs and heated in the flame of a Bunsen burner (Figure 8.10).

Figure 8.10 *Do not look directly at the burning magnesium.*

> **CAUTION** Do not look directly at the burning magnesium.

4. Your teacher will hold a piece of steel wool in crucible tongs. The steel wool will be heated in the flame of the Bunsen burner and then immediately placed in a bottle of chlorine gas. The bottle will be covered immediately.

> **CAUTION** Chlorine is a very poisonous gas. Only perform this demonstration in a fume hood.

5. Your teacher will hold a few pieces of charcoal in the crucible tongs and heat them until they begin to glow. Then they will be placed in a bottle of oxygen.

6. Your teacher will place a small amount of sulphur in a deflagrating spoon (Figure 8.11). The sulphur will be heated until it begins to burn and then it will be placed in a bottle of oxygen. The bottle will be covered immediately.

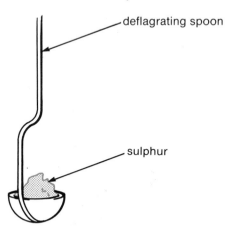

Figure 8.11 *A deflagrating spoon*

CAUTION Perform this demonstration in a fume hood. The product of the reaction is a poisonous gas.

Observations

(See Table 8.5.)

Questions

1. What happened to the elements in each of the chemical changes?
2. In each case, what was needed to get the reaction started?
3. In which of these reactions was energy released? Explain how you know this.
4. Using the information supplied by your teacher, complete the word equations for the chemical changes in this activity.

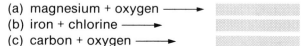

(a) magnesium + oxygen ⟶
(b) iron + chlorine ⟶
(c) carbon + oxygen ⟶
(d) sulphur + oxygen ⟶

PART 2

Problem

How can a compound be broken down into its elements?

Materials

electrolysis apparatus
water
wooden splints
dilute acid

Table 8.5

PAIR OF ELEMENTS	PHYSICAL PROPERTIES OF ELEMENTS	CHEMICAL CHANGE	PHYSICAL PROPERTIES OF NEW SUBSTANCES
Magnesium Oxygen			
Steel wool (iron) Chlorine			
Charcoal (carbon) Oxygen			
Sulphur Oxygen			

SAMPLE ONLY

Procedure

Your teacher will set up the apparatus shown in Figure 8.12. Observe what happens.

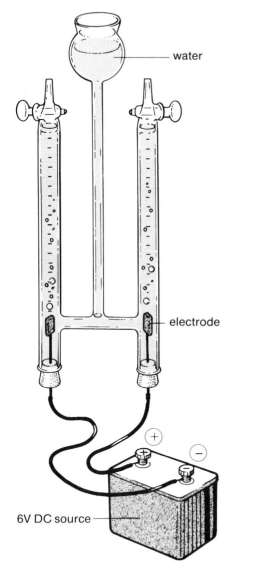

Figure 8.12 *Electrolysis apparatus for Activity 8C, Part 2*

Observations

1. What happens when the electricity is turned on?
2. What happens when the electricity is turned off for 2 min?

3. Describe the physical properties of the substances that are collected above the liquid.
4. Compare the amounts of each new substance.
5. (a) Describe what your teacher does to test the substances. Observe closely. You will learn more about these tests—and perform them yourself—in Chapter 9.
   (b) Tell what happens in the tests.

Questions

1. (a) Did you observe any physical changes? Explain your answer.
   (b) Did you observe any chemical changes? Explain your answer.
2. (a) What happens when the energy supply (the electricity) is turned on?
   (b) What happens when it is turned off?
   (c) What happens when it is turned on again?
3. (a) What properties do the two gases have in common?
   (b) What difference in properties makes it possible to tell that the two gases are different?
4. Write a word equation for the reaction.

Ideas and Applications

After performing many experiments, scientists have found that in a chemical reaction the total mass of the reactants is always equal to the total mass of the products. To demonstrate this, all reactants and products must be included in any calculations. It is especially easy to neglect gases that may be reactants or products. Gases have mass, even if they are not visible, and this mass cannot be ignored. For example, when iron rusts, the mass of rust produced will be greater than the starting mass of iron. This is because oxygen gas combined with iron to produce the rust. If you forgot about the oxygen, it would appear that matter was gained in the reaction. And, of course, it wasn't.

Self-check

1. What is the difference between a physical change and a chemical change?
2. For each of the following, tell whether it is a physical or chemical change. Give reasons for each answer.
   (a) boiling some asparagus
   (b) burying garbage
   (c) baking a cake
   (d) dicing carrots
   (e) running a car motor
   (f) shredding paper
   (g) drying your hair
   (h) filtering tomato juice
   (i) chopping wood
   (j) burning wood
3. Give one example of a chemical reaction that requires energy.
4. Give one example of a chemical reaction that releases energy.
5. (a) Write a word equation for the reaction of iron and oxygen. heat → iron oxide
   (b) Explain what the word equation means.

The Salt of the Earth

Over 400 000 000 years ago, shallow seas covered most of North America. Coral reefs grew in the Great Lakes, dividing the seas into shallow basins. As the water evaporated, salt beds formed. Between Windsor and Goderich in Ontario, the salt beds average more than 200 m deep. These deposits are covered by 300 m to 600 m of rock and earth. Special equipment is needed to mine them.

Salt lowers the freezing point of water. Therefore, salt is used on our roads and highways to melt ice in winter.

▲ Large amounts of salt are used in the meat-packing industry. Salt preserves food by preventing the growth of some bacteria and moulds.

If your home has a water-softener, you know that you must d salt to it regularly. The sodium in the salt trades places with ie unwanted calcium and magnesium particles in hard water. "Soft" water (water without calcium and magnesium particles) does not leave deposits in kettles and hot water pipes.

An electric current may be passed through a salt solution to obtain other useful substances. This is done industrially on a much larger scale than you see here. The substances produced are chlorine gas, hydrogen gas, and caustic soda (sodium hydroxide).

You may have observed your teacher demonstrating the chemical reactivity of sodium metal in water. Sodium is made from salt.

The hydrochloric acid used in your school lab can be made from salt.

Chemicals and Society

Many of the substances which affect you either directly or indirectly—sandwich wrap, latex paint, plastics, polyester fabrics, many pesticides and fertilizers—were unknown only a century ago. Chemical reactions are responsible for the astounding number of new substances that are produced daily around the world. Today it is often easy to take these and other substances for granted. But it is important to realize that many of these useful substances may also be dangerous.

In the introduction to this chapter, you read how chlorine can be both useful and dangerous. In this section, you will consider other chemicals that have both beneficial and dangerous uses.

Chemicals in Common Products

Certain products are useful in the home because they do not *react* (combine chemically) with other substances (Figure 8.13). For example, you would not want plastic bowls to react with the food stored in them, or plastic wrap that leaves a taste or odour on the food wrapped in it.

Nor would you want clothes made from synthetic (human-made) fabrics such as polyester if the fabrics disintegrated in normal use. For some products, the less reactive they are, the better.

On the other hand, people find some other products useful because they do react with other substances (Figure 8.14). Reactive products may help with cleaning, polishing, painting, and many other activities. Glues, detergents, solvents, polishes, and paints are typical examples. They work because they contain reactive chemicals.

Products that contain reactive chemicals are much more dangerous than those which do not. Special care must be taken in storing and using them. They must be kept safely away from children and pets. In many cases they must be kept away from moisture and extreme (high or low) temperatures. The reactive chemicals contained in some common products are listed in Table 8.6.

Figure 8.13 *These products usually do not react with other materials during normal home use.*

Figure 8.14 *This person is relying on the reactive chemicals in the polish to make the copper pan shiny again.*

Table 8.6 *Chemicals Found in Some Common Products*

CHEMICAL	FOUND IN	CHEMICAL FORMULA	HAZARD
Caustic soda (sodium hydroxide)	Oven cleaners; drain cleaners	$NaOH$	Poisonous; burns skin; corrosive
Caustic potash (potassium hydroxide)	Drain cleaners	KOH	Poisonous; burns skin; corrosive
Muriatic acid (hydrochloric acid)	Iron cleaner; concrete etching fluid	HCl	Poisonous; burns skin; corrosive
Toluene	Contact cement	C_7H_8	Poisonous; flammable
Acetone	Glues	C_3H_6O	Poisonous; flammable
Ketones	Glues	Various related compounds	Poisonous; flammable
Lead (tetraethyl lead)	Gasoline	$Pb(C_2H_5)_4$	Poisonous; accumulates in the body
Petroleum distillates; mineral spirits*	Spray paint; furniture polish; paint solvents; thinners	Mixtures of compounds called hydrocarbons	Poisonous; flammable
Pentachlorophenol	Wood preservatives	C_6Cl_5OH	Poisonous
Methanol (wood alcohol)	Paint solvent	CH_3OH	Poisonous; flammable

*These are a series of compounds related to gasoline and natural gas.

The chemicals listed in Table 8.6 are always named on product labels because they are a danger to human health and/or the environment. For this reason, they are called **hazardous chemicals**. Products that contain hazardous chemicals will give some warning on the container. A suggested first aid treatment may also be described, or an *antidote* (a substance that counteracts the effect of a hazardous chemical) may be named. Before you use products such as those listed in Table 8.6, be sure you are aware of the hazardous chemicals they contain. Be sure you know the appropriate first aid treatment or the antidotes for the chemicals.

In 1962, chemistry textbooks claimed that the gases helium, neon, argon, krypton, xenon, and radon do not react with other substances. By year's end, new textbooks were being written because of experiments done at the University of British Columbia by Dr. Neil Bartlett. Bartlett was working with a very reactive compound called platinum hexafluoride. It was so reactive that he wondered whether it might react with xenon gas. Xenon gas is very unreactive, but it is the most reactive of the gases named above. Bartlett designed an experiment, performed it, and the reaction occurred. A new substance, a yellow solid called xenon platinum hexafluoride, had been formed. Since then, other xenon compounds have been formed. Compounds involving krypton and radon have also been formed. So far, though, no compounds involving helium, neon, and radon have been produced.

Neil Bartlett

Why Are Some Chemicals Hazardous?

Chemicals may be hazardous for a number of reasons. The most common cause of danger is that the chemicals are poisonous, carcinogenic (cancer causing), flammable, corrosive, or very reactive.

Hazardous chemicals may show their effects quickly, or they may be slow acting. It is generally easier to recognize the effect of a fast-acting chemical, since its effect will show up within 96 h. Gasoline petroleum products, and sodium hydroxide are examples of fast-acting hazardous chemicals.

The effect of a slow-acting hazardous chemical is harder to identify precisely because such a chemical is so slow in its action. For example, leaded gasoline is a fast-acting hazardous chemical, because it is so flammable. But it also contains a lead compound that is a slow-acting poison. Many people work with leaded gasoline and breathe its fumes for many years without noticing the effect of the lead compound. Eventually, in some cases, changes in the red blood cells, general weakness, or paralysis may result. Because these effects do not appear immediately and because not everyone experiences them, people are often less careful with leaded gasoline than they should be.

Chemicals, the Environment, and You

Recently, we have become increasingly aware of the chemicals discharged into the air, water, and soil. Some of these chemicals are hazardous. As such, they can affect the environment severely. Other human-made substances, such as plastics, are not considered hazardous. But because these substances are so unreactive, they do not easily and quickly decompose (break down) in the environment as many other substances do.

Large-scale industrial waste chemicals are concentrated and come in large quantities from identified locations. Thus, they are in many ways easier to deal with than the wastes produced by individual people. The exhaust from the motor vehicles we use and the chemical products we unthinkingly pour down the drain are other examples of pollution produced by people (Figure 8.15).

You are a member of a society that uses many chemicals and chemical products. These can be very helpful if they are understood and used properly. But used improperly, they can be dangerous to humans and the environment. During your lifetime, you will have to make many decisions about the acceptable use of chemicals and chemical products. These chemical products might be in your home, your community, your country, and in the industries that support your community and country. Being able to apply your knowledge of chemical change (and science in general) to everyday life can help you make informed decisions that may benefit you and your community.

Figure 8.15 *How long do you think it would take for these products to decompose in the environment if they were simply discarded after use?*

Ideas and Applications

In the 1970s, a new chemical substance was developed which worked as a flame-retardant when it was used to treat fabrics. Cloth treated with this chemical did not burn easily. As a safety precaution, children's sleepwear was regularly treated with the chemical. The result was a decrease in the number of accidents in which children were badly burned. But researchers later discovered that this useful chemical was also cancer-causing. Thus, its use was discontinued. Soon there were further advances in chemistry, and manufacturers developed new fabrics which did not burn easily, even without the chemical treatment.

Self-check

1. (a) List three elements that can react with oxygen.
   (b) Name the compounds produced.
2. (a) Describe how water can be broken down into its elements.
   (b) Write a word equation for this reaction.
   (c) Is this reaction a physical change or a chemical change? Explain your answer.
3. In general, why are products that depend on reactions for their usefulness more dangerous than products that are normally unreactive?
4. Why is it harder to identify slow-acting chemicals than fast-acting chemicals?
5. Why is it easier to control wastes from an industrial plant than wastes from homes and individuals?

Chapter Objectives

NOW THAT YOU HAVE COMPLETED THIS CHAPTER, CAN YOU DO THE FOLLOWING?	FOR REVIEW, TURN TO SECTION
1. Explain the difference between a pure substance and a mixture.	8.1
2. Distinguish between an element and a compound.	8.1
3. Define the terms "atom" and "molecule."	8.1
4. Explain the difference between physical and chemical change, and give examples of each.	8.2
5. Identify the reactants and the products in a chemical reaction.	8.2
6. Explain why the total mass of reactants always equals the total mass of products.	8.2
7. Write a word equation for a chemical reaction.	8.2
8. Give examples of chemical reactions involving elements.	8.2
9. Explain why chemicals can be both useful and dangerous.	8.3
10. Describe how you can be a more responsible user of chemical and chemical products.	8.3

Words to Know

element
compound
atom
molecule
chemical formula
physical change
chemical change
chemical property

chemical reaction
reactant
product
word equation
hazardous chemical

Tying It Together

1. Explain how all matter can be classified into two main categories.
2. Which of the following types of matter are homogeneous: elements, compounds, solutions, colloids, mechanical mixtures? Give reasons for your answer.
3. (a) What is the name for the smallest possible particle of an element?
   (b) What is the name for the smallest possible particle of a compound?

4. (a) What is the minimum number of elements needed to make a compound?
   (b) Name three compounds that contain only this number of elements.
   (c) Name a compound which contains more than the minimum number of elements.
5. What information is given by the chemical formula H_2O?
6. For each of the following, tell whether it is a physical change or a chemical change. Give a reason for each answer.
   (a) cutting the lawn   (d) denting a fender
   (b) barbecuing a steak   (e) making tea
   (c) bending a paper clip   (f) bleaching a stain
7. What is a chemical property?
8. Give an example of a chemical property of
   (a) candle wax   (c) water
   (b) chlorine   (d) sulphur.
9. Complete the following word equations:
   (a) iron + ▨▨▨ ⟶ rust
   (b) energy + water ⟶ ▨▨▨ + oxygen.
10. Distinguish between
    (a) an atom of an element and a molecule of an element

(b) a molecule of an element and a molecule of a compound.

11. Give three examples of unreactive products that you use in your home.
12. (a) What are hazardous chemicals?
    (b) Give three examples of hazardous chemicals.
13. How should hazardous chemicals be stored in the home?
14. (a) Name a hazardous chemical you have used.
    (b) Explain how you should discard it when you have finished using it.
15. Describe some benefits of the use of chemicals.
16. Name some harmful effects of chemicals.

Applying Your Knowledge

1. (a) Give two examples of physical changes which are not mentioned in this chapter.
   (b) Give two examples of chemical changes which are not mentioned in this chapter.
2. A black material is heated. A gas is released and a white substance remains. Was the original material an element or a compound? Explain your answer.
3. Tell what one molecule of each of the following consists of:
   (a) sulphuric acid (H_2SO_4)
   (b) nitric acid (HNO_3)
   (c) aluminum chloride ($AlCl_3$)
4. When you burn a large piece of firewood, the ashes have a mass much less than the mass of the original wood. Yet scientists say that matter is not destroyed in a chemical change. What has happened to the matter in the wood? Explain this in terms of the particle model of matter.
5. Write the word equation for the burning of magnesium in oxygen. Which has the greater mass, the magnesium metal or the white ash? Explain your answer. Design an experiment which would confirm your answer.
6. Write word equations for the following reactions:
   (a) zinc and sulphuric acid react, producing zinc sulphate and hydrogen gas
   (b) sodium metal reacts with chlorine gas, producing sodium chloride

Projects for Investigation

1. Chemical products and the trucks and railway cars carrying chemicals are marked with hazard symbols such as the ones shown in Figure 8.16. Find out the exact meaning of each of these symbols. In your own home, find examples of as many of these symbols as you can.
2. Mercury is an element that is both useful and very poisonous. Research and report on the uses of mercury. Include in your report information about mercury poisoning.
3. All organisms (plants and animals) obtain the substances necessary for life, such as clean air, water, and food, from the environment. Likewise, all organisms put substances back into the environment through their various functions. Although some substances that we discard as waste are useful to the environment, far more of them are harmful pollutants. **Waste management** refers to the methods we can use to reduce and contain waste in order to prevent pollution. Find out how efforts at waste management are affecting the transportation, use, and elimination of chemicals (hazardous and otherwise) in your province. You might want to present your findings in the form of a poster.

Figure 8.16 *What do these symbols mean? Which of them can you find in your home?*

9

Investigating Gases

Key Ideas

- Gases have many physical properties in common.
- It is possible to identify gases by their chemical properties.
- Standard tests let us identify gases.
- The different chemical properties of gases make them useful in many ways.

Helium gas may be used in balloons because it has two particularly useful properties. Helium has the physical property of being less dense than air. Thus, a helium-filled balloon floats in air. Helium also has the chemical property of not burning or reacting with chemicals. Thus, balloons filled with helium gas will not explode if contact is made with a spark or flame.

Helium, like many other gases, is clear and colourless. How can you tell helium from other clear, colourless gases such as hydrogen or oxygen if all three are invisible? In this chapter, you will first review the physical properties of gases. Then you will find out how to identify gases by their chemical properties.

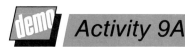

Most gases are clear and colourless. But the gas produced in this activity is *visible*, so you will be able to observe it more easily.

Problem

What are some physical properties of gases?

Materials

copper	syringe
concentrated nitric acid	rubber stopper
large, stoppered container	safety goggles
large flask	

CAUTION Do not allow nitric acid to come into contact with skin or clothing. The gas produced is very poisonous. This activity is a teacher demonstration and should be carried out under the fume hood.

Procedure

1. Observe and describe the physical properties of the copper and the nitric acid.
2. Your teacher will place the copper into the large container and add a very small amount of nitric acid.
3. Observe the changes that occur in the container.
4. Describe the shape that the gas takes as it is formed.
5. When the container is full, the mouth of the flask will be placed over the mouth of the container. The flask and container will be tipped so that some of the gas enters the flask. Describe what happens to the volume of the gas as it enters the flask.
6. Your teacher will remove some of the gas with the syringe and cover the end of the syringe by pressing it into a rubber stopper. The plunger of the syringe will be pushed as in Figure 9.1. What happens to the volume of the gas inside the syringe?

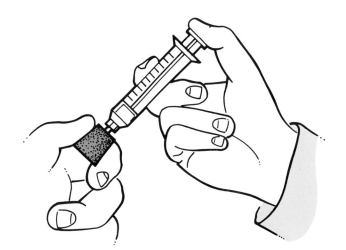

Figure 9.1 *What happens to the volume when the plunger is pushed in?*

Observations

1. Describe the copper and the nitric acid.
2. Describe what happened when the nitric acid was added to the copper.
3. How did the volume of the gas change when it was allowed into the flask?
4. How did the volume of the gas in the syringe change when the plunger was pushed in?

Questions

1. What indicates that a chemical reaction occurred in the container?
2. Why are the words "variable" or "indefinite" good words to describe the shape and volume of a gas?
3. Gases are fluids. (*Fluids* are substances that are not rigid and can flow.) What did you see that indicated this property of gases?
4. Gases can be compressed. What did you see that showed this property of gases?

Physical Properties of Gases

1. Gases have no definite shape. They take the shape of their containers such as different balloons.
2. Gases expand to fill the available volume. Air freshener sprayed into one corner of a room, for example, will spread to fill the room.
3. Gases can be compressed. Propane tanks, for example, can contain gas for many hours of cooking (Figure 9.2).
4. Gases are fluids. Natural gas, for example, may be transported through pipelines (Figure 9.3).

Figure 9.2 *Propane gas canister*

Figure 9.3 *A pipeline under construction*

Chemical Change and Gases

Because many gases are invisible, it is difficult to know when they take part in chemical reactions. If the reaction takes place in a liquid, you may see bubbles (Figure 9.4). The bubbles are evidence that a gas is produced. Some gases also have an odour that allows you to identify them. For example, the dangerous gas hydrogen sulphide smells like rotten eggs. Because of the difficulty of identifying them, scientists have devised **standard tests** for gases. You can use these standard tests to identify particular gases.

Figure 9.4 *Bubbles can be seen, although the gas itself is invisible.*

Activity 9B

Water vapour is a very common clear and colourless gas. In this activity you will use a special test paper which changes colour in the presence of water vapour.

Problem

How can you test for the presence of water vapour?

Materials

hot plate
beaker
water
cobalt chloride paper

Procedure

1. Heat the water in the beaker until it is near boiling.
2. While the water is heating, hold the test paper near the hot plate to ensure that it is thoroughly dry.
3. Hold the piece of dry test paper over the water in the beaker.

Observations

1. Describe the cobalt chloride paper. Was there any colour change as you dried it?
2. What happened when you held the paper over the beaker of water?

Questions

1. How can you test for the presence of water vapour?

Self-check

1. Why is helium used in balloons?
2. List four physical properties of gases.
3. When an antacid tablet reacts with water, explain how you know that one of the products is a gas.

Oxygen

You know that oxygen is one of the components of the air we breathe. How is pure oxygen obtained? Purified air is first converted into a liquid. This liquid consists of liquefied nitrogen, argon, and oxygen. Each of these substances has a different boiling point. Thus, industries that manufacture oxygen can use this property to separate oxygen from the liquefied mixture.

When the liquefied air is boiled, nitrogen and argon boil off first. Nitrogen has a boiling point of -196°C. Argon boils ten degrees higher at -186°C. Oxygen, with a boiling point of -183°C, has the highest boiling point of the three. Thus, it remains after the nitrogen and argon have boiled off.

About 89% of this manufactured oxygen is piped directly to the industry using it. Another 10% is liquefied for shipment in special insulated tanks. The remaining 1% is compressed at high pressure for transportation in steel cylinders.

Oxygen is a very reactive element. Oxygen atoms combine readily with other atoms and molecules. Most of the oxygen on earth is in compounds such as water, sand, and the ores of many metals. Oxygen in compounds makes up about 50% of the mass of the earth's crust, and about 65% of your body's mass.

How much oxygen do you think there is in the air? You will find out in Activity 9C.

Steel consists mainly of iron, and iron reacts with oxygen in the air. In this activity, you will see what is produced in this reaction. Then you will be able to infer how much of the air is oxygen.

Problem

What is the proportion of oxygen in the air?

Materials

steel wool (cleaned in alcohol to remove any protective coating)
2 graduated cylinders (100 mL)
2 large beakers
water

Procedure

1. Make a table similar to Table 9.1 in your notebook.
2. Fill both large beakers half-full with water.
3. Push the piece of steel wool to the bottom of one of the cylinders. Then place the cylinder upside down in a beaker of water.
4. Place the second cylinder upside down in the other beaker of water (Figure 9.5).
5. Record the level of water in the two cylinders. This level tells you the volume of air which is trapped inside.
6. Store the apparatus until the next period.

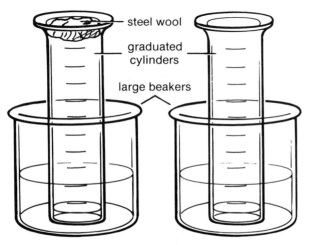

Figure 9.5 *Apparatus for Activity 9C*

7. After at least one day, record the level of water in the two cylinders.
8. Leave the apparatus set up for several days. On the final day, record the final levels of water.

Observations

Table 9.1 *Data for Activity 9C*

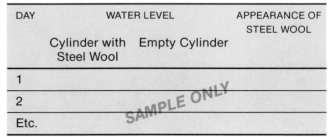

DAY	WATER LEVEL		APPEARANCE OF STEEL WOOL
	Cylinder with Steel Wool	Empty Cylinder	
1			
2			
Etc.			

Questions

1. Why was the empty cylinder used?
2. Write a word equation for the chemical change that occurred.
3. Why did the reaction stop when it did?
4. (a) How many millilitres of air were used up in this reaction?
   (b) What fraction of the air was used up?
   (c) What fraction of the air is oxygen? How did you infer this?
5. Calculate the percentage of oxygen in the air using the formula

$$\% \text{ oxygen in the air} = \frac{[(\text{original volume of air}) - (\text{amount of air left})]}{\text{original volume of air}} \times 100$$

Challenge

You can use a set-up like that in Activity 9C to test ways of preventing rusting. Before putting the steel wool in the cylinder, dip it in motor oil. Compare the rusting of the steel wool in the motor oil with the rusting of the untreated sample. Does any rust form? Does it form more slowly? You could try this experiment using other protective coatings, such as the ones used to spray the underside of automobiles.

In this activity, you will produce oxygen. To speed up the chemical reaction you will need to add a catalyst. A **catalyst** is a substance that speeds up a chemical reaction without itself being used up in the reaction.

Problem

How can you make and test for oxygen?

Materials

test tube	dilute hydrogen peroxide
test tube stand	manganese dioxide
wooden splint	safety goggles
scoopula	

CAUTION Wear safety goggles for this activity.

Procedure

1. Pour hydrogen peroxide into a clean test tube until it is one-quarter full. Observe it carefully for several seconds and record your observations.

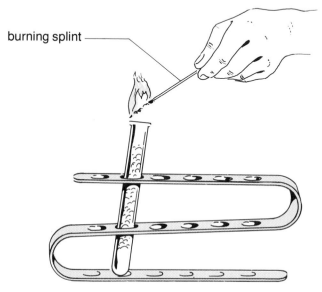
burning splint

Figure 9.6 *What happens when a glowing splint is held to the mouth of the test tube?*

2. Use the scoopula to add a small amount of manganese dioxide to the test tube. Observe and record your observations.
3. Light the wooden splint. Blow out the flame so that it just glows.
4. Lower the glowing splint into the mouth of the test tube (Figure 9.6).
5. When the reaction has stopped, add a little more hydrogen peroxide to the test tube. Observe and record your results.
6. When the reaction in step 5 stops, add a little more manganese dioxide. Observe and record your results.

Observations

1. Describe hydrogen peroxide and manganese dioxide.
2. Could you see any signs of a chemical reaction in the hydrogen peroxide alone?
3. What happened when manganese dioxide was added to the hydrogen peroxide?
4. What happened when the glowing splint was placed in the mouth of the test tube?
5. When more hydrogen peroxide was added did the reaction start up again?
6. When more manganese dioxide was added did the reaction start up again?

Questions

1. Identify the gas which was produced.
2. (a) What evidence do you have that the hydrogen peroxide was used up in the reaction?
   (b) Was the manganese dioxide used up? Explain how you know this.
   (c) Which substance is the catalyst in this reaction?
3. What do you think would happen if you put a *burning* splint into oxygen? If you have your teacher's permission, try this.
4. In this chemical change there are two products. You already identified one. The other product, which stays in the test tube, is water. Write a word equation for the chemical change that occurred in this activity.

Ideas and Applications

Catalysts are used for many different industrial processes. By speeding up the rate of reaction, they can make a process more efficient. Catalysts are also used to control the reactions in automobile engines. A car engine gives off dangerous gases through its exhaust system. To reduce the amounts of these gases, all modern cars have a mixture of catalysts in the exhaust system in a device called a *catalytic converter*. This device causes exhaust gases to react more completely than they would without the catalytic converter. As a result, smaller amounts of dangerous gases are emitted.

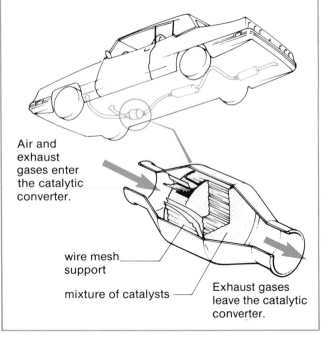

Air and exhaust gases enter the catalytic converter.

wire mesh support

mixture of catalysts

Exhaust gases leave the catalytic converter.

Self-check

1. List three compounds which contain oxygen.
2. What is a catalyst?
3. What common compound can break down, forming oxygen?
4. What is the purpose of a catalytic converter in an automobile?

Carbon Dioxide

Carbon dioxide is another colourless, odourless gas. It is not poisonous, but if it is present in high concentrations, it can cause suffocation.

Carbon dioxide plays several different roles:

1. It can be used as a raw material for industrial processes such as the manufacture of sodium bicarbonate and sodium carbonate. (Sodium bicarbonate, also known as baking soda, is found in many antacid products. Sodium carbonate, also known as soda ash, is used in the manufacture of glass. Sodium carbonate is also used in water softeners and in making other kinds of chemicals, paper, and detergents.)
2. It is one of the reactants in the important process of photosynthesis. (You will find out more about this important process in Unit Five.)
3. It is a by-product of many processes, such as the burning of fossil fuels (coal, natural gas, and petroleum).
4. It can be found in such varied items as soft drinks (it produces the "fizz"), in fire extinguishers, and in railroad refrigeration cars (in the form of dry ice).
5. As a pollutant, it is responsible for the greenhouse effect. (You will find out about the greenhouse effect in Chapter 10.)

In the next activity, you will make carbon dioxide. You will also observe how it can be distinguished from other colourless, odourless gases.

Problem

How can you make and test for carbon dioxide?

Materials

retort stand	baking soda
3 test tubes	vinegar
test tube clamp	limewater
one-holed stopper	wooden splint
glass elbow	

Procedure

1. Prepare the apparatus as shown in Figure 9.7. Put baking soda in one test tube and limewater in the other.
2. Add vinegar to the baking soda, and immediately insert the stopper into the test tube.
3. Observe any reaction.
4. Put baking soda into one test tube, but extend the glass elbow into an empty test tube. After about half a minute, lower a burning splint into the mouth of the test tube.

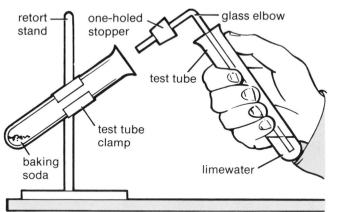

Figure 9.7 *Apparatus for Activity 9E*

Observations

1. Describe the physical properties of vinegar, baking soda, and limewater.
2. List all the changes that occurred when the vinegar was added.
3. What happened to the limewater?
4. What happened to the burning splint?

Questions

1. Describe the reaction of carbon dioxide and limewater.
2. Scientists say that "carbon dioxide does not support combustion." What did you observe that supports this statement?
3. Is carbon dioxide more dense or less dense than air? Explain how you know.
4. (a) What were the reactants in the reaction that produced carbon dioxide?
   (b) Besides carbon dioxide, the other products (which stay in the test tube) are sodium acetate and water. Write a word equation for this reaction.

Challenge

You can do "magic" with carbon dioxide at home. Set up the apparatus shown. Light the candle and allow your audience to see that it continues to burn. Now add vinegar to the baking soda, and hold the large jar slightly tipped over the candle flame. They won't believe what they see! Be prepared to repeat this several times with fresh baking soda and vinegar. Can you explain to your audience what has happened?

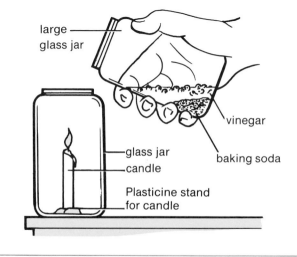

Ideas and Applications

In baking, the reaction between baking soda and acid causes the lightness of cakes and muffins. There are many acidic substances that are used in cooking. Some examples are vinegar, buttermilk, and fruit juices. Carbon dioxide is formed in the batter and trapped as small bubbles when the batter is cooked. If there is no acidic substance in the recipe, then *baking powder* is used instead of baking soda. Baking powder contains baking soda already mixed with a dry acid. During baking, the same reaction occurs, and bubbles of carbon dioxide are formed.

Hydrogen

Hydrogen is the least dense of all the gases. This physical property was used back in the 1930s to make large airships such as the Hindenberg float in air. Hydrogen is also a very reactive gas. This chemical property makes it useful as a fuel for propelling more modern airships such as the space shuttles. However, as you can see in Figure 9.8, hydrogen's useful properties—physical and chemical—have resulted in disastrous accidents. When accidents such as these occur, scientists and technologists must search for safer ways to use the properties of elements such as hydrogen.

Figure 9.8

(a) *In May of 1937, the hydrogen-filled Hindenberg caught fire while landing at Lakehurst, New Jersey. This fire was probably due to a discharge of lightning near a hydrogen leak. Thirty-six passengers and crew died in that inferno. The Goodyear blimp, which in contrast is filled with helium gas, can't catch fire in this way.*

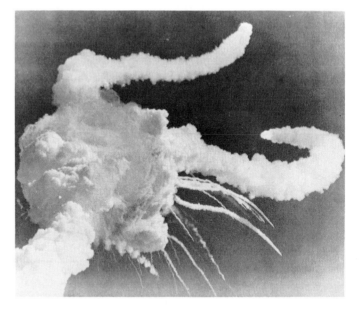

(b) *The United States space shuttle Challenger was destroyed during a launch on January 28, 1986 when a seal in its right booster rocket burned through and caused the rupture of the main liquid hydrogen fuel tank. Challenger's seven crew members were killed in the explosive fireball formed by this hydrogen-oxygen reaction.*

Activity 9F

Because hydrogen is less dense than air, it must be collected in a special way.

Problem

How can you make and test for hydrogen?

Materials

250 mL beaker   water
Bunsen burner   paper towelling
test tube   safety goggles
wooden splint
lump of freshly cut calcium metal

CAUTION Wear safety goggles for this activity. Do not touch calcium with wet fingers. It may burn your skin.

Procedure

1. Fill a 250 mL beaker half-full with water.
2. Follow the directions in Figure 9.9 in order to fill a test tube with water and turn it upside down.
3. Obtain a lump of freshly cut calcium on a piece of paper. Observe the physical properties of the metal.
4. Place the calcium in the beaker of water. Place the test tube over the calcium and let the gas collect until the test tube is full. Do not remove the test tube from the beaker yet.
5. Light a wooden splint.
6. Remove the test tube of gas from the beaker, keeping it upside down. Quickly and carefully place the burning splint into the mouth of the test tube. Record your observations.

Observations

1. Describe calcium.
2. What happened when the calcium was added to the water?
3. What happened when the burning splint was placed in the mouth of the test tube?

Questions

1. Why is the water in the test tube displaced by the gas?
2. Why is the test tube kept upside down until the gas is tested?
3. (a) What are the reactants in this reaction?
   (b) Besides hydrogen, the other product is *calcium hydroxide*. Write a word equation for this reaction.
4. (a) Can you guess what compound is produced when hydrogen burns? The reactants are hydrogen and oxygen.
   (b) Write a word equation for the burning of hydrogen.

Figure 9.9

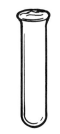 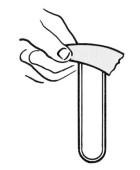 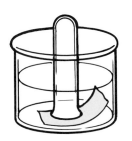 

(a) *Fill a test tube with water, as full as possible.*   (b) *Put a paper towel on the surface of the water.*   (c) *While holding the towel, turn the test tube upside down and put it in the beaker of water.*   (d) *Remove the paper towel.*

Ideas and Applications

You have seen in Activity 8C, Part 2 that electrical energy can be used to split a compound into its elements. (Water was broken down into hydrogen and oxygen.) You also know from Figure 9.8 that when elements combine together, a great amount of energy may be produced. In a world in which energy is so important, scientists are working to develop new sources of energy. One of the most exciting of these new sources is hydrogen power. This involves harnessing the simple reaction of combining hydrogen and oxygen to produce not only water but also controlled energy. You can read more about hydrogen as a source of energy in Chapter 10.

Self-check

1. What two common household substances can be used to produce carbon dioxide?
2. How is carbon dioxide involved in baking cakes?
3. (a) In a drawing, show how to collect hydrogen gas which is produced as bubbles.
    (b) Why must the gas be collected in this way?

Summary: Standard Tests for Gases

TEST FOR WATER VAPOUR

Blue cobalt chloride test paper turns pink in the presence of water vapour.

TEST FOR OXYGEN

When a glowing splint is put into pure oxygen, the splint bursts into flame.

TEST FOR CARBON DIOXIDE

Carbon dioxide is the only clear colourless gas which turns limewater milky. Any gas that does this must be carbon dioxide.

TEST FOR HYDROGEN

Hydrogen is the only clear colourless gas which burns with a "pop" when a burning splint is put into it.

Chapter Objectives

NOW THAT YOU HAVE COMPLETED THIS CHAPTER, CAN YOU DO THE FOLLOWING?	FOR REVIEW, TURN TO SECTION
1. Describe four physical properties of gases.	9.1
2. Explain how the physical properties of gases make them useful.	9.1
3. Tell why standard tests are necessary to distinguish one gas from another.	9.2
4. Describe how to test for the presence of water vapour.	9.2
5. Define "catalyst," and give an example of a reaction which is affected by a catalyst.	9.3
6. Describe the standard test for oxygen.	9.3
7. Explain how to determine the percentage of oxygen in air.	9.3
8. Explain how to produce and test for carbon dioxide.	9.4
9. Describe the standard test for hydrogen.	9.5

Words to Know

standard test
catalyst

Tying It Together

1. (a) List two properties of helium which make it useful for balloons.
   (b) Which of these is a physical property and which is a chemical property?
2. List four physical properties of gases.
3. Why is it sometimes difficult to know when a gas is involved in a chemical change?
4. (a) What is a catalyst?
   (b) Give an example of a reaction that is affected by a catalyst.
5. Using diagrams, describe how you can make and test for oxygen.
6. (a) What is baking powder?
   (b) Describe how baking powder reacts in cake batter.
7. How is limewater used to identify carbon dioxide?
8. What property of hydrogen makes it dangerous in balloons?

Applying Your Knowledge

1. Divers using SCUBA (Self Contained Underwater Breathing Apparatus) equipment can stay underwater for a long time (Figure 9.10). What physical property of gases makes this activity possible? Explain this property in terms of the particle model of matter.

Figure 9.10

145

2. What is the source of the hydrogen gas which is produced when calcium reacts with water? Explain the reaction in terms of the particle theory.

3. In the splint tests for gases you use either a burning splint or a glowing splint. Explain how you could use a splint to find out whether a mystery gas is oxygen, hydrogen, or carbon dioxide.

4. In our bodies there are catalysts called *enzymes*. What do you think they do?

Projects for Investigation

1. Find out how carbon dioxide is useful in fire extinguishers. Draw a diagram showing what is inside the extinguisher, and explain it to your class. If possible, demonstrate the action of one of these fire extinguishers.

2. In the library, find out when SCUBA equipment was first used. What are the limits of exploration using SCUBA gear? Are there any dangers in a normal dive?

Science in Your Life

The ideas and applications of science are everywhere.
Find as many of them as you can in these photos.

10

Reactions Involving Oxygen

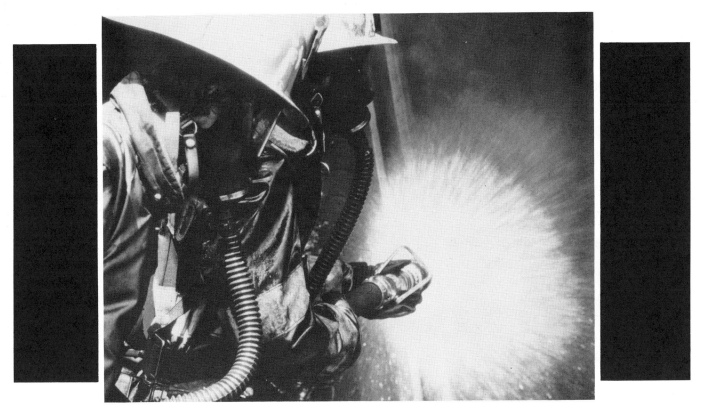

Key Ideas

- There are many fuels which are burned to release energy.
- The products of combustion can be identified.
- Metals can be separated from ore.
- Chemical reactions can cause air pollution.

You could live for several weeks without food and for nearly a week without water. But without oxygen, you could survive for no more than a few minutes. We need oxygen in the air we breathe, but we also use it in our cars, in our furnaces, and in many industrial processes. In this chapter you will learn about some chemical reactions involving oxygen. You will also find out why these reactions can cause air pollution and how air pollution may be changing the climate of the earth.

Combustion of Fuels

Combustion means burning. In combustion, a substance joins with oxygen in a chemical reaction that gives off light and heat. The discovery of a way to start fire was an important early example of a combustion reaction (Figure 10.1).

Figure 10.1

From the earliest times, combustion of fuel has produced the warmth that humans need in order to survive. Until recently, it also provided the only lighting available at night (Figure 10.2). For many centuries the most commonly used fuels came from vegetation. People used reeds, dried leaves, peat, and wood for light and heat. Today in industrialized countries such as Canada, fossil fuels (coal, oil, and natural gas) are used to supply energy.

In modern society, we use fuels for many purposes. The heat from combustion may be used to turn turbines that generate electricity (Figure 10.3). In internal combustion engines such as those in automobiles, the energy from combustion drives pistons in cylinders. The moving pistons cause the car to move.

Some of our most common fuels are separated from crude oil by *fractional distillation*. (Recall what you learned about fractional distillation in Chapter 7.) These fuels are all hydrocarbons. **Hydrocarbons** are compounds that contain only the elements hydrogen and carbon. The natural gas burned in your Bunsen burner contains hydrocarbons (mainly methane).

Figure 10.2 *Before electricity was widely available, many people used kerosene lamps as a source of light.*

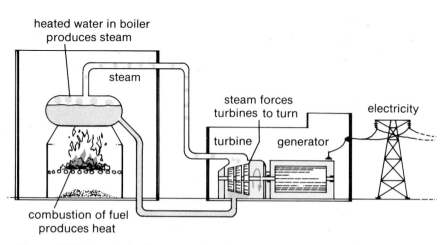

Figure 10.3 *How heat from combustion can be used to produce electricity*

How many of the hydrocarbons in Table 10.1 do you recognize?

Table 10.1 *Different Hydrocarbons*

NAME	FORMULA	STATE	USE
Methane	CH_4	Gas	Makes up a main part of natural gas
Ethane	C_2H_6	Gas	Used in making alcohol and acetic acid
Propane	C_3H_8	Gas	Burned in portable stoves and heaters
Butane	C_4H_{10}	Gas	Burned in cigarette lighters
Pentane	C_5H_{12}	Liquid	Used as a solvent
Hexane	C_6H_{14}	Liquid	Makes up a main part of certain motor fuels and cleaning solvents
Heptane	C_7H_{16}	Liquid	Makes up a main part of turpentine
Octane	C_8H_{18}	Liquid	Makes up a main part of gasoline

Ideas and Applications

Natural gas has very little odour. You might not notice the smell if pure natural gas were leaking in your home. Because leakage of the gas is dangerous, substances with odours, such as *tert butyl mercaptan* and *dimethyl sulphide* are added. You can detect any leak because of the smell.

Activity 10A

In this activity you will burn a hydrocarbon, either methane or candle wax, and identify the products of combustion.

Problem

What are the products of combustion?

Materials

Bunsen burner
candle and stand
beaker tongs
limewater
250 mL beaker
large test tube
test tube holder
ice water
cobalt chloride paper
safety goggles

CAUTION Wear safety goggles for this activity.

Procedure

1. Light the candle or Bunsen burner.
2. Fill the test tube with ice water. Wipe the outside of the test tube to be sure it is dry.
3. Using the test tube holder, hold the large test tube at an angle about 5 cm to 10 cm above the flame. Keep it there for a few seconds.
4. Observe the outside of the test tube. Touch a strip of cobalt chloride paper against the outside of the test tube. Record your observations.
5. Pour some limewater into the beaker, and pour it out again. A few drops of limewater should remain in the beaker. Using beaker tongs, hold the beaker upside down over the flame for a few seconds. Record your observations.

Observations

1. (a) What did you observe on the outside of the test tube?
   (b) Was there a change in the cobalt chloride paper?
2. What did you observe in the beaker when it was held over the flame?

Questions

1. What two products of combustion were you able to identify in this activity? Explain how you identified them.
2. Write word equations for the combustion of natural gas and candle wax.
3. Was there any difference between the products of combustion of natural gas and the products of combustion of candle wax? If there was, describe the difference.

By-products of Combustion

If fuels were absolutely pure hydrocarbons and if they were always burned in ideal conditions, then energy, water, and carbon dioxide would be the only products. This is not the case, however. Combustion also produces **by-products**, which are products other than the desired ones. An example of a by-product is *soot*, which consists of the element carbon (Figure 10.4). In automobile engines, other products of combustion include nitrogen oxides and carbon monoxide.

Nitrogen oxides are gases formed when fuels are burned at high temperatures in engines and furnaces. Nitrogen oxides react with sunlight to produce ozone, a very reactive form of oxygen. (There is a layer of ozone in earth's upper atmosphere. It prevents most of the sun's ultraviolet radiation from reaching earth's surface. Small amounts of ozone make the air smell fresh. In large amounts and concentrations, ozone is dangerous to breathe.)

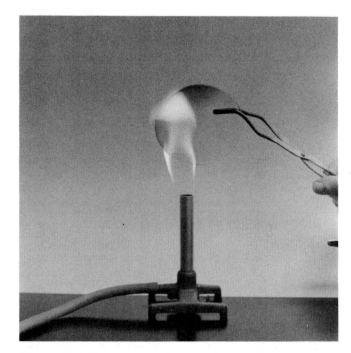

Figure 10.4 *Soot from a Bunsen burner flame*

Hydrogen as a Fuel

In a world where energy is so important, scientists are always trying to find new fuels to supplement or replace fossil fuels, which are limited in supply. Hydrogen would seem to be the perfect fuel. It combines with oxygen to form only water. It gives off light and heat, but no soot, smoke, or other kinds of pollution. Hydrogen could be burned as a fuel in homes, cars, ships, and planes, and could be used to produce electricity. It has already been used to propel rockets and to put spacecraft into orbit.

There is a plentiful supply of hydrogen. Since most hydrogen occurs combined with oxygen in water, it is easy to produce. But at present the process of converting hydrogen into fuel is too expensive to be economical. The storage of the hydrogen also presents a problem because the gas is so reactive. In the future, however, hydrogen may become a common fuel for the average consumer (Figure 10.5).

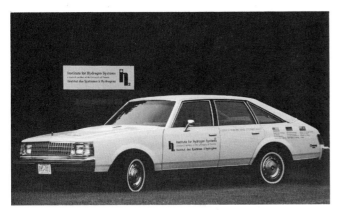

Figure 10.5 *This car looks no different from any other. But it is actually powered by hydrogen, rather than gasoline. The large tank needed for the hydrogen is hidden in the trunk.*

Self-check

1. (a) What is combustion?
   (b) How has combustion been important through human history?
2. (a) What is a hydrocarbon?
   (b) Name three common hydrocarbons and tell how each is used.
3. In your notebook, write the following word equation. Complete it by filling in the blanks.

   hydrocarbon + oxygen ⟶ ___ + ___ + energy

4. Name three by-products of combustion.
5. Give two reasons why hydrogen is not commonly used as a fuel.

Ideas and Applications

Spacecraft require electrical power to run their computers and other electrical systems. Ordinary batteries are too heavy to be practical. A different source of electricity is needed. Hydrogen and oxygen can be used in a *fuel cell*. As the two elements react together in the cell they produce energy which is used to make electricity, and water which is used for drinking.

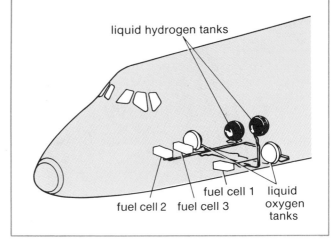

liquid hydrogen tanks

fuel cell 1
fuel cell 2   fuel cell 3
liquid oxygen tanks

Oxygen and the Production of Metal from Ore

Metals are essential in the production of skyscrapers, houses, cars, appliances, computers, knives, microchips, and thousands of other things that we take for granted in our lives. Canada is one of the world's largest suppliers of metals. In this country we produce nickel, zinc, silver, uranium, gold, lead, and copper, among others. The mineral industry provides jobs for many people (Figure 10.6).

Gold is one of the few metals that occurs uncombined. In other words, gold can be found in its pure elemental form, and not as a compound. Most naturally occurring metals, however, exist as compounds with sulphur or oxygen. For example, copper ore usually contains copper together with sulphur in a compound. Iron ore contains iron together with oxygen in a compound. In Chapter 7, you learned how ore can be treated so that the rock is separated mechanically from the metal-containing compound. After that separation, chemical changes are necessary to produce the pure metal.

In the case of metals that exist as compounds with sulphur, the chemical change usually involves reacting the compound with oxygen. The oxygen atoms attract the sulphur atoms, leaving the desired metal as well as sulphur dioxide gas as a by-product of the reaction. In the case of metals that exist as compounds with oxygen, the chemical change usually involves reacting the compound with carbon. What by-product will this result in?

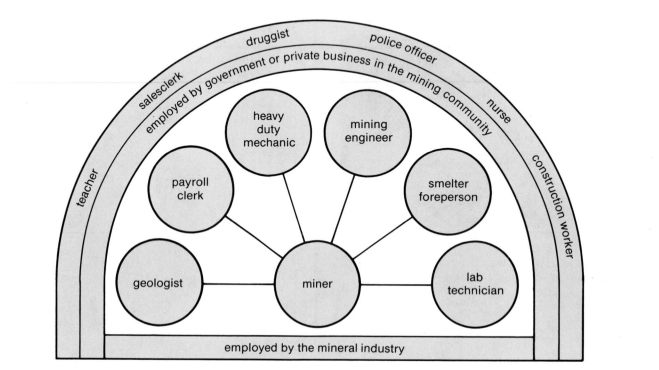

Figure 10.6 *Jobs in the mineral industry*

Problem

How can you separate a metal from a compound, and what products are produced?

Materials

Bunsen burner	glass elbow
2 test tubes	powdered charcoal
retort stand	limewater
test tube clamp	copper oxide
one-holed stopper	safety goggles

> **CAUTION** Wear safety goggles for this activity. Do not allow any liquid to enter the hot test tube. Make sure your test tubes are dry and not cracked or scratched.

Procedure

1. Mix some copper oxide with at least twice its mass of powdered charcoal in a test tube. The test tube should be less than one-quarter full.
2. Fill the other test tube half-full with limewater.
3. Insert the stopper and glass elbow as shown in Figure 10.7. Put the end of the glass elbow in the limewater.
4. Heat the mixture by moving the Bunsen burner back and forth under the test tube.

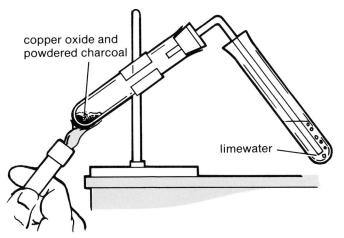

copper oxide and powdered charcoal

limewater

Figure 10.7 *Apparatus for Activity 10B*

5. After some gas has bubbled through the liquid, remove the tube from the liquid. Then turn off the burner.

> **CAUTION** Remove the test tube with limewater before turning off the Bunsen burner.

6. Allow the test tube to cool and observe the contents of the test tube. Rinse away any remaining charcoal, and try to identify the product you can see.

Observations

1. Describe the two starting substances.
2. What happened to the limewater as gas bubbled through it?
3. Describe the substance that remained in the test tube after it was cooled.

Questions

1. What were the reactants in this reaction?
2. What observation told you that a chemical reaction was taking place?
3. What two substances were produced? Explain how you know this.
4. (a) Write an equation for this reaction.
   (b) Would "energy" be on the left side or the right side of this equation? Explain your answer.

Self-check

1. In what form do most metals occur in nature?
2. Why is gold an unusual metal?
3. How can some metals be separated from compounds?
4. Name several jobs associated with the mineral industry.
5. Several of the jobs shown in Figure 10.6 in the outer ring may seem at first glance to have little connection with the mineral industry. Pick two of these jobs and explain how they are connected with the mineral industry.

The Combustion of Fuels and Air Pollution

Acid Precipitation

Earlier, you read about two of the by-products formed from the combustion of fossil fuels: nitrogen oxides and carbon monoxide. These gases are also produced in the industrial processes that separate useful fuels from crude oil, and metals from their compounds. Another by-product of these processes is a clear, colourless gas with a suffocating odour detectable only in large concentrations. This gas is sulphur dioxide. Sulphur dioxide is highly corrosive. It is one of the major causes of acid precipitation.

Acid precipitation results when sulphur dioxide and nitrogen oxides combine with water vapour in the atmosphere to produce sulphurous acid and nitric acid. These acids fall to the ground mixed with precipitation (rain or snow).

Acid precipitation causes chemical changes in soil and bodies of water (Figure 10.8). Over many years, it can reduce soil fertility, retard tree growth, and kill fish and plant life in rivers and lakes. In the province of Ontario alone, 4000 lakes have already died due to acid precipitation. In other words, fish and plant life can no longer live in these lakes. By the year 2000, environmental specialists estimate that 48 000 lakes could die if preventive measures are not taken.

Must the activities of human industrial society result in increased air pollution? We pollute when we use energy. And we pollute

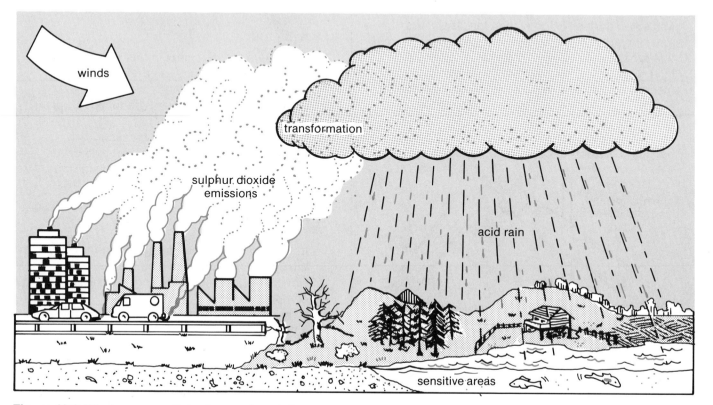

Figure 10.8 *Winds carry pollutants over long distances—hundreds and even thousands of kilometres.*

when we refine metals. How much does it cost to install pollution controls on automobiles? Should industrial plants spend more money to reduce sulphur dioxide emissions and other pollutants?

Reducing pollution is possible. The two pictures in Figure 10.9 show that industries have taken steps to ensure that air pollution caused by their plants is reduced. Pollution control is expensive. Ultimately, we, as consumers, may help pay the price of stronger pollution control measures in the form of higher-priced goods. But we will all benefit from pollution control because we will get cleaner air and water.

Figure 10.9

(a) *The plant life around this metals-separation complex was killed by sulphur dioxide.*

(b) *Because new methods were used to remove sulphur dioxide from the exhaust gases, plants were able to grow in the area again.*

Too Much Carbon Dioxide: The Greenhouse Effect

In Activity 10B, one of the products in the reaction of copper oxide and carbon was carbon dioxide. When ores that contain oxygen, such as iron ore, are purified, carbon dioxide is one of the by-products. This gas is also another of the by-products of the combustion of fossil fuels.

Excess carbon dioxide in the atmosphere can act like a blanket, trapping heat that would usually escape from the earth's atmosphere. This trapping of heat occurs in the same way as a greenhouse traps heat (Figure 10.10). Because of the similarity, this warming effect is called the **greenhouse effect**.

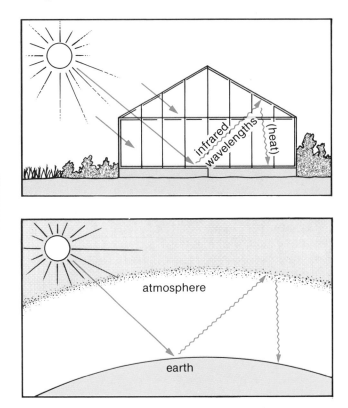

Figure 10.10 *The greenhouse effect. The glass panes of a greenhouse allow sunlight to pass through but prevent heat from escaping. Excess carbon dioxide in the atmosphere can act in the same way.*

If we continue to burn large amounts of fossil fuels, the proportion of carbon dioxide in the atmosphere may increase. This may increase the greenhouse effect. Some weather experts think that the earth's climate will gradually become warmer. In January, you might find this an appealing idea, but it could cause serious problems. If the ice sheets at the North and South Poles were to melt, the ocean level would rise, and coastal areas around the world would be flooded.

Air Pollution: Everyone's Responsibility

Each province in Canada has inspectors who make sure that the level of harmful gases in the air stays within safe limits. The limits are set by the governments we elect. Our governments are advised by doctors and scientists. In the end, it is up to all of us, as members of society, to decide how much pollution we will tolerate. It is also up to us to decide how much we are prepared to pay or can afford to pay for cleaner, safer air.

Self-check

1. List four substances that pollute air.
2. What is acid precipitation?
3. List two sources of compounds that cause acid precipitation.
4. What is the greenhouse effect?
5. Why is air pollution everyone's responsibility?

Chapter Objectives

NOW THAT YOU HAVE COMPLETED THIS CHAPTER, CAN YOU DO THE FOLLOWING?	FOR REVIEW, TURN TO SECTION
1. List several fuels and explain how they are used.	10.1
2. Explain the term "hydrocarbon," and describe the products of combustion of hydrocarbons.	10.1
3. List several by-products of combustion.	10.1
4. Describe the advantages and disadvantages of using hydrogen as a fuel.	10.1
5. Describe how a metal can be produced from its ore.	10.2
6. Describe the causes and effects of two kinds of air pollution.	10.3

Words to Know

combustion
hydrocarbon
by-product
acid precipitation
greenhouse effect

Tying It Together

1. (a) What are the products of combustion of natural gas?
   (b) What are some possible by-products of combustion?
2. The chemical formula of propane is C_3H_8. What type of compound is propane? What further information is given in the chemical formula?

3. (a) Describe two ways that hydrogen might be useful as a fuel.
   (b) Why is hydrogen not widely used as a fuel now?
4. Describe the reaction that occurs in a fuel cell on a spacecraft.
5. (a) List five metals which are found in Canada.
   (b) Name one metal which is mined as an element.
6. What reaction occurs when powdered charcoal is heated with copper oxide? Explain how you can identify each of the products of this reaction.
7. List two major sources of air pollution.
8. Describe some of the effects of air pollution.
9. In what way is carbon dioxide in the atmosphere like the glass in a greenhouse?

Applying Your Knowledge

1. The burning of wood produces by-products in addition to the major products, light and heat. What very noticeable by-product can you think of that is produced during the combustion of hydrocarbons?
2. In Chapter 8 you observed the combustion of several elements. Describe any characteristics which are common to all the combustion reactions you have seen. Write word equations for all these reactions.
3. Scientists are interested in a source of "clean" energy. What does this mean, and why would they be interested in this?
4. How is rusting similar to combustion? How is it different?

Projects for Investigation

1. Canada is a world leader in the production and exporting of metals. Choose one metal mined and processed in Canada. Find out where it is obtained and the chemical processes used to extract the metal from its location/ore. Summarize your findings in the form of a poster.
2. Sometimes a combustion reaction can occur suddenly, with no warning. Find out what this kind of reaction is called, how it occurs, and how you can guard against its occurrence. How do industries keep this reaction from occurring?
3. Corrosion of metals costs society large amounts of money. Several methods are used to protect metals from corrosion. These include painting, galvanizing, and electroplating. Find out how each of these methods can protect metals from corrosion.
4. Sometimes lakes that have been killed by acid rain can be saved. Certain chemical reactions can be used to make lakes less acidic. Find out how this is done. How can plant and animal life be returned to the lake?
5. There are several methods of reducing air pollution from industries. Research and report on the method of your choice. How effective is it? How expensive is it? Are any other countries using this (or other) methods?

Unit Three: Chemical Change

MATCH

In your notebook, write the letters (a) to (j). Beside each letter, write the number of the word in the right column that corresponds to each description in the left column.

(a)	a poisonous, carcinogenic, or corrosive chemical	1. physical change
(b)	shows the reactants and products	2. element
(c)	a pure substance made up of more than one kind of element	3. molecule
(d)	a new substance is formed	4. hazardous
(e)	formed from atoms	5. word equation
(f)	the simplest pure substance	6. chemical reaction
(g)	molecules remain the same	7. standard test
(h)	oxides of sulphur and nitrogen mixed with rain or snow	8. ore
(i)	a metal compound usually containing sulphur or oxygen	9. catalyst
(j)	a way of identifying a gas	10. acid precipitation
		11. compound

MULTIPLE CHOICE

In your notebook, write the numbers 1 to 10. Beside each number, write the letter of the best choice.

1. Which of the following types of matter is *not* homogeneous?
   (a) element
   (b) pure substance
   (c) mechanical mixture
   (d) solution

2. Which of the following is true for a chemical change?
   (a) No new particle is produced.
   (b) A new substance with new properties is produced.
   (c) The starting material and the product are the same.
   (d) The product can easily be changed back to its original form.

3. Compounds are
   (a) not pure substances
   (b) made up of more than one element
   (c) the same as elements
   (d) rarely found on earth.

4. All gases have many of the same properties. Which of the following is *not* a property of gases?
   (a) All gases are compressible.
   (b) All gases are flammable.
   (c) All gases are fluids.
   (d) All gases have no definite shape.

5. Which material is used as a standard test for oxygen?
   (a) limewater
   (b) a burning splint
   (c) steel wool
   (d) a glowing splint

6. Catalysts are substances that
   (a) are not used up in a reaction
   (b) test for carbon dioxide
   (c) react vigorously to form oxygen
   (d) are of no useful purpose.

7. What can cobalt chloride paper be used to test for?
   (a) water vapour
   (b) oxygen
   (c) carbon dioxide
   (d) hydrogen

8. Which of the following is *not* true about combustion?
   (a) Combustion means burning.
   (b) In combustion, a substance joins with oxygen.
   (c) Combustion uses substances called fuels.
   (d) Combustion uses up light and heat.

9. Which of the following is *not* a product of combustion?
   (a) water
   (b) hydrocarbons
   (c) energy
   (d) carbon dioxide
10. Which of the following is a disadvantage of using hydrogen as a fuel?
   (a) It is reactive and difficult to store.
   (b) It combines with oxygen to form only water.
   (c) It is in plentiful supply.
   (d) It does not pollute the environment.

TRUE/FALSE

Write the numbers 1 to 10 in your notebook. Beside each number, write T if the statement is true and F if the statement is false. For each false statement, rewrite the statement so it becomes true.

1. All matter may be classified as either pure substances or mixtures.
2. Different elements have identical properties.
3. The smallest particle of an element that has the properties of the element is called a molecule.
4. Changing state is an example of a physical change.
5. A burning splint "pops" when put into hydrogen.
6. Oxygen is not a very reactive element.
7. Baking soda and baking powder are used in baking because they form carbon dioxide gas.
8. Hydrocarbons contain carbon, hydrogen, and oxygen.
9. By-products of combustion are undesired products such as soot.
10. Canada is one of the world's largest suppliers of metals.

FOR DISCUSSION

Read the paragraph and answer the questions that follow.

Imagine that a spill of chlorine gas has occurred on a road 1 km from your school. All people within 5 km are to be evacuated. You are in charge of the evacuation procedures.

1. Draw a rough map of your school area. Mark on it where the spill has occurred. Then draw a circle with a radius of 5 km around that point.
2. Explain how you would communicate with people in the area. What would you tell them?
3. What arrangements would you make to move the people from the area?
4. How would you care for those people who had no friends or relatives to help them?
5. Which groups or organizations would you call in to assist you? What responsibilities would you give to each?
6. After the danger was over, how would you let people know they could return to the area?

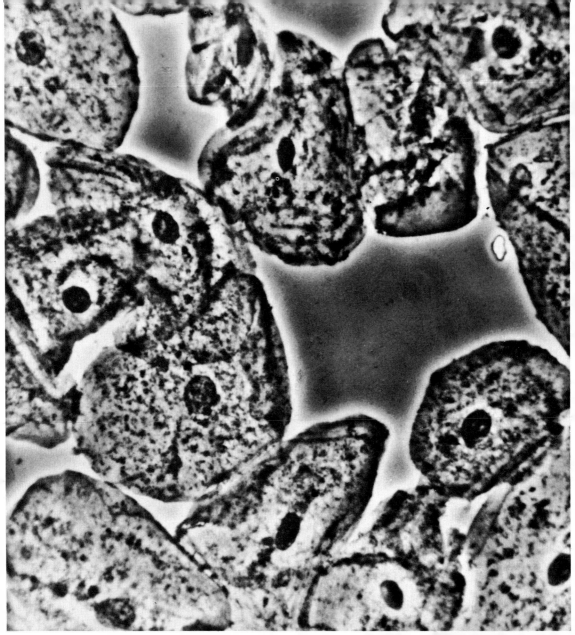

The Cell and Its Systems

The cell is the basic building block of life. It is the smallest unit that can continue to function on its own. Cells make up all living things, from the smallest and simplest to the largest and most complex. It is amazing to think that each of us, with more than 60 000 000 000 000 (60 million million) cells in our bodies, came from one tiny cell, the fertilized egg.

Scientific curiosity has led to the discovery of and knowledge about cells. Scientists and technicians have devised and used instruments that help them see and understand things that can't be seen with the unaided eye. In this unit, you will follow in their footsteps by using the microscope to observe and examine cells and their structure. You will learn how cells function.

11

Using the Microscope

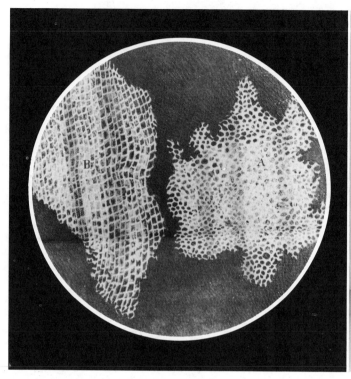

Key Ideas

- Microscopes help us see things too small to be seen with the unaided eye.
- Microscopes are sensitive instruments that must be cared for properly.
- A wide variety of objects can be viewed under low, medium, and high power of a microscope.
- For very high magnification or for three-dimensional images, electron microscopes are used.

S o far as we know, the English scientist Robert Hooke (1635–1703) was the first person to observe and describe the appearance of cork cells. Using a microscope similar to the one shown here, the fascinated Hooke drew these strange, air-filled, honeycomb-like structures. This enabled him to explain the properties of cork—its low density, its water-resistance, and its insulating abilities. Can you explain these properties using Hooke's drawings?

In this course, you will develop skills in using a microscope. (The microscope you will be using is basically similar to Hooke's microscope.) You will also find out how technology has provided us with microscopes that are millions of times more powerful than Hooke's.

What Is a Microscope?

Much of the living world is too small for human eyes to see. Our eyes can only see objects that are larger than 0.1 mm. Look at the blocks of dots in Figure 11.1. In the first block, you can probably see individual dots. In which block or blocks does the colour appear solid to you? How might you determine whether these blocks of colour are also made up of many dots which are very close together? The dots must be farther apart than 0.1 mm in order for most of us to see separate dots.

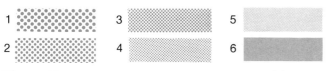

Figure 11.1 *Which block or blocks appear solid to you?*

The microscope is an instrument used for viewing objects normally too small for the eye to see. The first microscopes were single lenses held in place by a clamp (Figure 11.2). These magnifiers are called **simple microscopes** because only one lens was used.

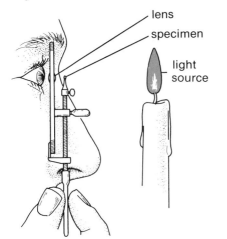

Figure 11.2 *An early microscope. The lens was clamped in a space between the two brass plates. The object was mounted on a point and brought into focus by means of the screws.*

The Compound Light Microscope

One of the microscopes in Figure 11.3 is similar to the microscope you will use in your class. These microscopes are called **compound light microscopes**. They use a series of lenses, and like the simple microscope, they use light to view an object. Often they are just called compound microscopes.

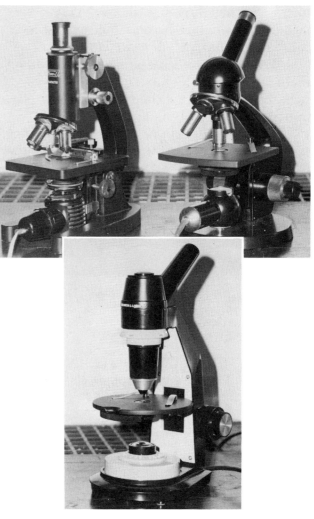

Figure 11.3 *You will be using a compound microscope like one of these.*

Parts of the Compound Light Microscope

THE OPTICAL SYSTEM

In modern microscopes, a series of lenses is supported in a tube called the *body tube* (Figure 11.4). The lens closest to the eye is called the *eyepiece*. The lens closest to the object you want to view is called the objective lens. There may be two, three, or four sets of *objective lenses*. The shortest is the low-power objective lens. If the microscope has three sets, the next is the medium-power objective lens. The longest is called the high-power objective lens.

All the objective lenses are attached to the *revolving nosepiece* (Figure 11.5). This lets you turn the lenses so that they will come into place over the object and be in line with the eyepiece.

Together, the eyepiece, body tube, revolving nosepiece, and the objective lenses make up the **optical system** of the microscope.

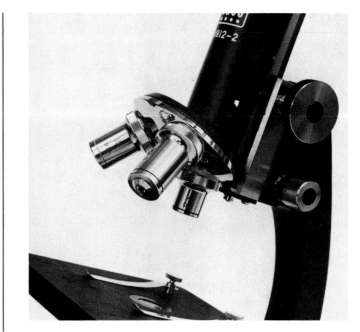

Figure 11.5 *The objective lenses are held in a revolving nosepiece. This allows you to change the lenses from low to medium to high power.*

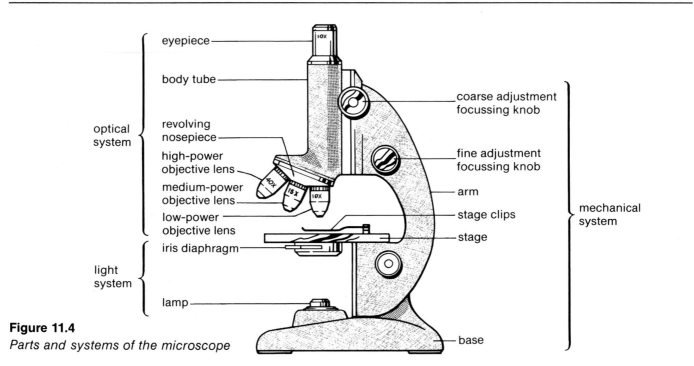

Figure 11.4
Parts and systems of the microscope

THE MECHANICAL SYSTEM

The **mechanical system** allows the parts of the compound microscope to move. It consists of the frame and the adjusting knobs. The *base* supports the frame of the microscope. The C-shaped upright structure, called the *arm*, is attached to the base. The arm is used to carry the microscope. The *stage* is the platform on which you place the slide. When you are to view an object on a slide, you secure it on the stage by the two *stage clips*. The slide should now be directly over the central opening. This is a hole in the stage which permits light to pass through to the eyepiece. The *focussing knobs* are located in different places on different types of microscopes. (To *focus* means to make something sharp or clear.) The *coarse adjustment knob* alters the distance between the lenses and the stage. This is called the working distance. The *fine adjustment knob* brings the object into sharper focus. You will need to use this knob for medium- and high-power focussing.

THE LIGHT SYSTEM

The **light system** of the microscope provides the correct amount of light so you can view the object properly. Light is needed to make the images on the slide more visible. A *mirror* or a *lamp* is located on the base of the microscope. This is the main source of light for the object. The *iris diaphragm* is located under the stage. It works rather like the iris of your eye in that it regulates the amount of light that comes up from the lamp or mirror to the object through the opening. It can be opened or closed by a small lever under the stage. See Figure 11.6 (a). Some microscopes have a *disc diaphragm* with holes of different sizes. See Figure 11.6 (b).

Figure 11.6 *The diaphragm regulates the amount of light that comes from the lamp or mirror to the object.*

(a) *The iris diaphragm opens like the eye.*

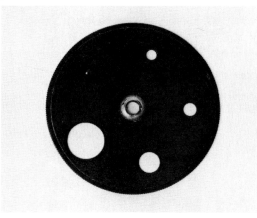

(b) *The disc diaphragm has openings of different sizes.*

Before using a microscope, you should be familiar with its parts (Figure 11.4) and what the parts are used for. You should also know how to care properly for a microscope. Carefully read the ten hints for handling a microscope on the next page.

Hints for Handling a Microscope

1. Use both hands to carry the microscope. Carry it by the arm, and use your other hand for additional support under the base. Always place the microscope on the desk or table carefully and gently.
2. Keep the microscope in an upright position when using liquids or when it is not in use.
3. Keep the stage clean and dry. If any liquids are spilled on the microscope, wipe them up immediately with a piece of tissue.
4. Focus with the low-power lens first, before using any of the other lenses.
5. Focus by moving the lens away from the slide. In other words, focus by increasing the working distance (the distance between stage and objective lens).
6. Call your teacher if the lenses are dirty.
7. Call your teacher if the adjustments do not work freely.
8. Remove the slide from the microscope stage when it is not being used.
9. When putting the microscope away, place the stage clips parallel to the arm, and place the low-power objective lens over the opening.
10. Keep your microscope covered when it is not in use, and keep your work area clean and tidy.

Self-check

1. What is the main difference between a simple microscope and a compound microscope?
2. Explain the term *focus*.
3. Your teacher will give you a diagram of a microscope. Label it carefully.
4. Why is it important to care properly for a microscope?
5. Make a table that lists all the parts of the microscope and their functions. Group the parts according to the three systems of the microscope: optical system, mechanical system, and light system.

Using a Compound Light Microscope

Determining Magnification

To determine the *magnifying power* of a microscope, you need to know two things:

1. The magnifying power of each objective lens. This is engraved on the side of the objective. Refer again to Figure 11.4.
2. The magnifying power of the eyepiece. This is engraved beside the lens. Again, see Figure 11.4.

Magnification is the product of the magnifying power of the objective and the eyepiece lenses. For example, if the low-power objective's magnifying power is 5x and the eyepiece's magnifying power is 10x, the total magnification at low power is as follows:

5 x 10 = 50x

In other words, the total magnification at low power is 50x.

In the next two activities, you will have a chance to practise using and observing objects through a microscope.

A slide prepared for viewing with water and a coverslip is called a **wet mount**. In this activity, you will have a chance to gain skill in preparing a wet mount and using a microscope. Then you will be ready to find out how to measure the **field of view**— the area that can be observed through the eyepiece.

PART 1

Problem

How can you prepare and view a wet mount of a newspaper letter?

Materials

letter "a" or "e" cut from a newspaper
beaker of water
dropper
coverslip
microscope slide
compound microscope

Procedure

1. Using the dropper, place a drop of water on a clean glass slide.
2. Place the prepared letter carefully on top of the drop of water (Figure 11.7). Be careful not to smudge the slide with your fingers.

prepared letter on drop of water

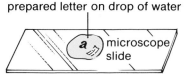

Figure 11.7 *Place the prepared letter carefully on top of the drop of water.*

3. Take a clean coverslip. Hold it between your thumb and forefinger, and touch the edge of the coverslip to the slide. (The water will spread along the edge.) Hold the coverslip to the slide as shown in Figure 11.8. Lower the coverslip onto the slide so that it covers the object. If you have done this correctly, there will be no air bubbles between the slide and coverslip. The slide you now have is called a *wet mount*.

Figure 11.8

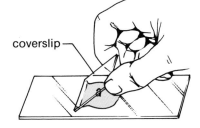

(a) *Holding the slide between thumb and forefinger, touch the edge of the coverslip to the drop of water.*

(b) *There should be no air bubbles between the slide and the coverslip.*

4. Turn on or adjust the light source on the microscope.
5. Make sure the low-power objective lens is "clicked" into position.
6. Place your prepared slide on the centre of the stage under the stage clips. The letter should be over the centre of the opening. Make sure the coverslip is on the top side of the slide. The letter on the slide should be facing you.
7. With your eye at stage level, use the coarse adjustment to bring the object and the objective lens as near to each other as possible without touching the coverslip. See Figure 11.9 on the next page.

Figure 11.9 *With your eye at stage level, use the coarse adjustment to decrease the working distance.*

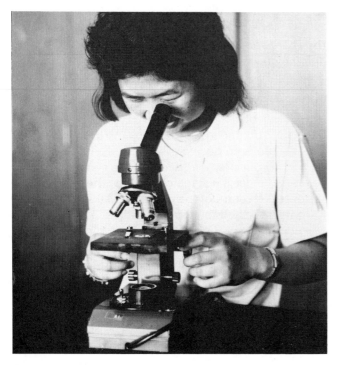

Figure 11.10 *Slowly move the coarse adjustment to increase the working distance and focus the object.*

8. Now, with your eye to the eyepiece, slowly move the coarse adjustment to increase the distance between the letter and the lens (Figure 11.10). Continue this process until the image is focussed clearly. If the image is not clear, repeat steps 7 and 8. Make sure the letter is centered over the opening.
9. Adjust the diaphragm so that the letter can be seen as clearly as possible.
10. To observe the letter under medium and high power, follow the steps listed in Table 11.1. Move the revolving nosepiece to bring the next highest lens into position. Make sure that you hear the "click" sound to ensure that the objective lens is in place. *When using the medium- and high-power objectives, you should only use the fine adjustment.*

CAUTION If your microscope has an oil immersion lens, get instructions for its use from your teacher.

Table 11.1 *Focussing Your Microscope— The Basic Rules*

- To find the object, always begin your examination with the low-power objective lens.
- To bring the object into focus, increase the working distance between object and objective lens.
- Never allow the objective lens to come in contact with the coverslip. If you have difficulty focussing under a higher power, start over again with the lower power.
- Always centre the object in the field of view before turning to a higher objective lens.

Observations

1. Observe the slide without the microscope. Draw what you see.
2. Draw a diagram of what you see in the field of view under
   (a) low power
   (b) medium power
   (c) high power.

3. View the letter under the low-power objective lens. Describe what happens to it in each case if you move the slide
   (a) to the right
   (b) to the left
   (c) toward you
   (d) away from you.
4. What effect does closing the diaphragm have on the field of view with
   (a) low power?
   (b) medium power?
   (c) high power?

Questions

1. Under which objective lens could you view the entire letter?
2. Calculate the magnifications for each objective lens. Write the magnifications under your three diagrams from Observations, question 2.
3. Describe the difference between the letter as it appears to the unaided eye and as it appears under the high-power objective.
4. Why must you always focus on an object first with the low-power objective, before moving to the medium or high power?

PART 2

Problem

How can you determine the diameter of the field of view?

Materials

compound microscope
clear plastic ruler

Procedure

1. Place a clear plastic ruler on the stage of the microscope.

2. Using the low-power objective, focus and view the ruler through the eyepiece.
3. Move the ruler so that one of the markings is at the left edge of the field of view (Figure 11.11). Measure the field of view across the diameter.

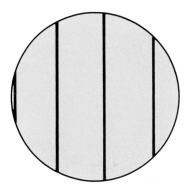

Figure 11.11 *The diameter of the field of view shown here is 3.5 mm.*

4. Calculate the diameter of the field of view for the medium- and high-power objectives using the following formula:

$$\text{diameter of medium-power or high-power field} = \text{diameter of low-power field} \times \frac{\text{magnification at low power}}{\text{magnification at medium or high power}}$$

Observations

1. What is the diameter of the field of view for
   (a) the low-power objective?
   (b) the medium-power objective?
   (c) the high-power objective?

Questions

1. Which objective lens allows you to see the largest field of view?
2. Why do you think that the diameter of the field of view is calculated, rather than measured, for the medium- and high-power objectives?

Challenge

You can find the actual size of objects under the microscope if you know the diameter of the field of view. Line up the letter in the centre of the field of view. Count the number of times you think it would fit across the diameter. For example, in this illustration the letter fits across the field of view seven times.

To determine the actual size of an object, use the formula below. This will give you the actual size of the object, whether it is seen under low, medium or high power.

$$\frac{\text{actual size}}{\text{of object}} = \frac{\text{diameter of field of view (mm)}}{\text{number of times object fits across the field of view}}$$

Problem

How can objects be viewed using the compound microscope?

Materials

2 strands of hair
newspaper picture
sugar
cornstarch
4 microscope slides
compound microscope
iodine solution
beaker of water
dropper
coverslips

Procedure

1. Prepare a wet mount of the two strands of hair (crossed) and a wet mount of a small piece of the newspaper picture. Follow directions given in Activity 11A.

CAUTION Iodine is a corrosive liquid that will stain the skin, clothing, desk tops, and floors. Take care to avoid spilling. Quickly wipe up any accidental spills, and rinse with water.

2. Prepare a wet mount of cornstarch using iodine solution instead of water.
3. Place some sugar crystals on a slide, but do not use a coverslip or water.
4. For each slide, follow these steps:
   (a) Focus the slide under the low-power objective lens of your compound microscope using the procedure given in Activity 11A.
   (b) In your notebook make a diagram of what you see under the low-power objective lens.
   (c) Focus the slide under the medium- or high-power objective lens using the procedure given in Activity 11A.
   (d) Make a diagram of what you now see through the microscope.

Observations

1. Was it possible to see where the two strands of hair crossed? Explain.
2. Which magnification allowed you to see the whole of the sugar crystal?

Questions

1. Describe the differences between what you saw under the low- and medium- or high-power objectives.
2. Why is it not possible to see the whole object in the field of view under medium or high power?
3. Why is it not possible to see the whole depth of an object?

Self-check

1. Describe, simply, how you would prepare a wet mount of the letter "y."
2. What are the four most important rules to remember when focussing any object using the compound microscope?
3. What are the steps needed to find the actual size of an object as seen under the microscope?

Other Types of Microscopes

In 1938, two scientists at the University of Toronto made the first microscope in North America to use electron beams instead of light rays. They were James Hillier and Albert Prebus of the Physics Department. It was called the **transmission electron microscope (TEM)**. See Figure 11.12. The fact that the TEM uses electron beams instead of light is very important. The best light microscope can only magnify objects about 1600 times. This is because there is a limit in our ability to see two separate points which are close together. By using electron beams, however, we can distinguish between two objects which are very close together.

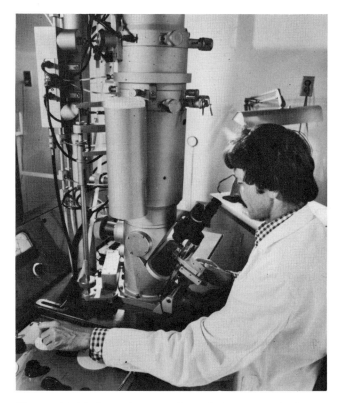

Figure 11.12 *A modern version of the transmission electron microscope (TEM)*

171

The two types of microscopes also differ in the way in which they focus objects. Light microscopes use glass lenses, but the TEM uses electromagnets as lenses (Figure 11.13). A glass slide is used to mount the specimen for examination with a light microscope. With a TEM, the glass slide is replaced by a thin copper grid (Figure 11.14). The TEM has revealed many fine details of the structures of the cells of living things.

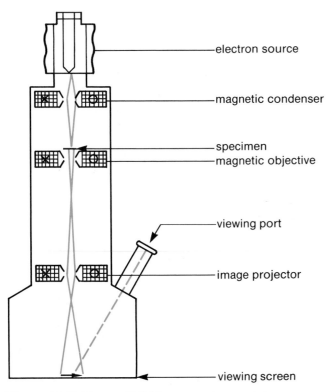

Figure 11.13 *How a TEM forms an image. The electrons are emitted from the source to pass through a series of electromagnetic lenses. (These focus the beam of electrons just as the objective and eyepiece would in a compound light microscope.) The image is formed on the screen at the bottom and is viewed through a viewing port on the side.*

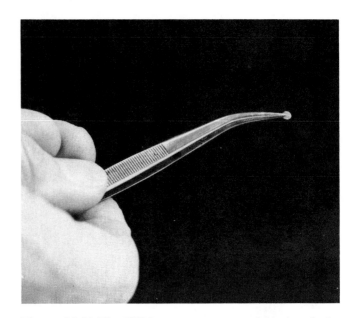

Figure 11.14 *The TEM uses a copper grid instead of a glass slide to hold the object.*

Another type of microscope that uses electron beams instead of light is the **scanning electron microscope (SEM)**. The SEM reveals objects, not as flat surfaces, but in three dimensions. The SEM was developed in 1938 but was not available commercially until 1965. The SEM works rather like a television picture tube. A special lens in the microscope focusses the beam of electrons into a fine ray that scans the surface of the specimen. The rays are then picked up and transmitted onto a monitor. Specimens are harder to prepare for SEM than for the TEM. SEM specimens can take up to three days to prepare. Figure 11.15 shows the same object viewed with the SEM under two different magnifications.

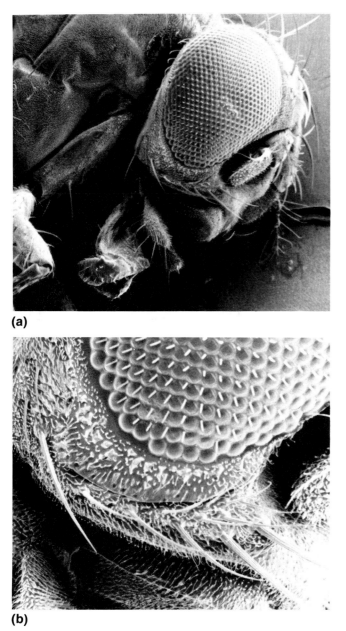

(a)

(b)

Figure 11.15 *A fruit fly as magnified through the SEM (a) 150 times and (b) 600 times*

Chapter Objectives

NOW THAT YOU HAVE COMPLETED THIS CHAPTER, CAN YOU DO THE FOLLOWING?	FOR REVIEW, TURN TO SECTION
1. Distinguish between a simple and compound light microscope.	11.1 and 11.2
2. Identify all the parts of a microscope.	11.2
3. Handle a microscope properly.	11.2
4. Prepare a wet mount.	11.3
5. Focus using low, medium, and high power.	11.3
6. Calculate the magnification of a microscope.	11.3
7. Determine the size of the field of view.	11.3
8. Observe a variety of objects using a microscope.	11.3
9. Distinguish between a transmission electron microscope and a scanning electron microscope.	11.4

Words to Know

simple microscope
compound light microscope
optical system
mechanical system
light system
magnification
wet mount
field of view
transmission electron microscope (TEM)
scanning electron microscope (SEM)

Tying It Together

1. What are two differences between a simple microscope and a compound microscope?
2. List four points you should follow to care properly for your microscope.
3. What are the four basic rules to be followed in focussing a microscope?
4. List the parts of the microscope in
   (a) the mechanical system
   (b) the optical system
   (c) the light system.
5. Describe how to prepare a wet-mount slide.
6. Describe what steps you would take to focus an object under the medium-power objective lens.
7. List two differences between the following:
   (a) the compound light microscope and the transmission electron microscope
   (b) the compound light microscope and the scanning electron microscope
   (c) the scanning electron microscope and the transmission electron microscope
8. How would you determine
   (a) the magnification of a compound microscope?
   (b) the diameter of the field of view?
9. The following objects are viewed using a specific magnification. State whether the object is likely being viewed with a compound light microscope or an electron microscope:
   (a) a leaf magnified 15x
   (b) a computer chip magnified 200x
   (c) a sugar crystal magnified 10 000x
   (d) a salt crystal magnified 5x
10. In your notebook, tell whether the following statements are true or false. If false, correct the statement so that it is true.

(a) Our eyes see only objects that are larger than 0.01 mm.

(b) Three systems—optical, electrical, and light—combine to form a functioning microscope.

(c) The transmission electron microscope can produce three-dimensional images.

(d) When bringing an object into focus, always begin with the high-power objective lens.

(e) When you move a slide to the right under the microscope, the slide appears to move away from you.

(f) Electron microscopes can give better magnifications of objects than compound microscopes.

Applying Your Knowledge

1. Draw each of the following letters and numbers as they would look when viewed through the low-power objective lens of a compound microscope: 7, 8 C, B, S, 9, f, h.

2. A sugar crystal is viewed through a compound microscope. If the objective lens is 10x and the eyepiece is 15x, how much larger does the crystal appear to be?

3. Determine the actual size of the fruit fly shown in Figure 11.16 if the diameter of the field of view is 6 mm and if it fits three times across the diameter of the field of view.

Figure 11.16

4. Give three possible reasons for seeing only darkness when you first look in a microscope.

5. What happens to the size of the field and the brightness of illumination as you move to higher magnifications?

Projects for Investigation

1. Lenses are used to make telescopes as well as microscopes. How are the lens arrangements in these two instruments different? Do research to find out how the arrangement of lenses differs in a compound microscope and a refracting telescope.

2. You can construct a microscope using two hand lenses and a cardboard tube about 6 cm long. Attach the two lenses in slots near each end of the tube. Use your "homemade" microscope to view different objects.

Challenge

A simple lens can be made by filling a flask with water. Explain how the effect in this photograph occurs.

The Basic Unit of Life: The Cell

Key Ideas

- The cell is the basic unit of life.
- Cells contain many structures called organelles.
- Each organelle has a specific function in a cell.
- The structure of many kinds of cells can be seen under the microscope.

Trees show many colours during the year. In the spring they are a soft green but as the summer ends the leaves change to deep reds and even purple. What is there within the structure of the leaf that allows the tree to change its colours from green to yellow or red? You will learn in this chapter how this occurs, but here is a clue: the yellow colours were present even in the spring.

In this chapter, you will also learn that a cell is made up of parts which have definite structures and functions.

Cell Size and Shape

The **cell** is the basic unit of life for all living things. Cells vary greatly in size. A pneumonia bacterium is one of the smallest known living things that exists as a single cell (0.0001 mm). The ostrich egg, however, is probably the largest single cell (75 mm). The longest cells in the human body are nerve cells, which can measure up to 1000 mm. A human egg cell measures about 0.2 mm, while a sperm cell is only about one-third this size. In fact, 100 000 sperm cells tightly packed together would only just be visible to your eyes. Blood cells are also quite small. Leaf cells vary in size but are about the size of a human egg cell.

Cells have so many different sizes! And, as you can see in Figure 12.1, different cells also have very different shapes. The shape of a cell is normally determined by the *function* (job) the cell will perform. For example, the function of muscle cells is to stretch and contract. Thus, they are usually long and thin. Fat cells store fat. Thus, they are usually round and "fattish" in shape.

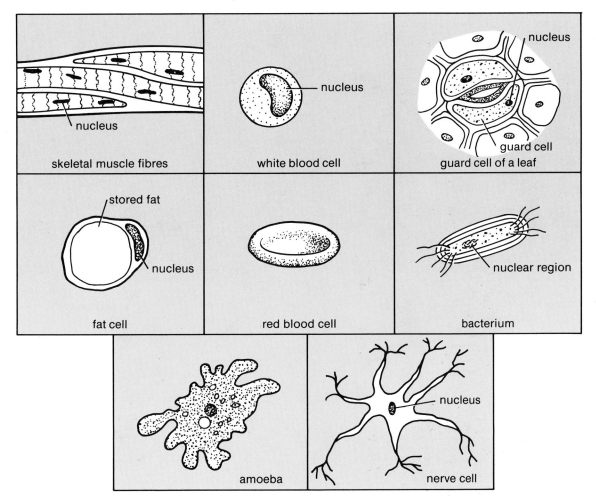

Figure 12.1 *Different kinds of cells (not drawn to scale)*

Cell Theory

In the nineteenth century, two scientists, Matthias Schleiden and Theodor Schwann, developed a theory to summarize ideas about the cell. The following three statements form what is known as the **cell theory**.

1. Cells are the basis of structure of all living things.
2. Cells are the basis of function for all living things.
3. All cells are formed from pre-existing cells.

Schleiden and Schwann also noted that some cells, such as bacteria, exist on their own. Other cells, such as nerve cells, are part of larger living things.

Cell Structure and Function

You now know that a cell has a size and a shape (its *structure*). But what is a cell? Each cell can perform all the required functions of life. If parts of a cell were removed, it would not be able to live. Each cell contains smaller structures called **organelles**, which carry out the different processes necessary for the cell to live. An organelle cannot exist on its own because it is not a living thing. However, it can perform its functions within the cell together with other organelles so that the cell can live.

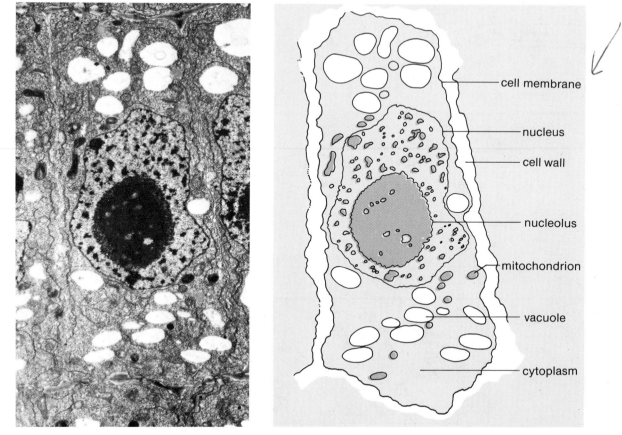

Figure 12.2 (a) *A plant cell as shown by the TEM. The sketch shows the main parts of the plant cell.*

178

Each cell, regardless of its shape or size, has three parts that are responsible for the basic structure and function of the cell. These three parts are the **cell membrane**, the **cytoplasm**, and the **nucleus** (Figure 12.2).

Plant cells also have an additional structure found outside the cell membrane, called the **cell wall**. It can be seen more easily than the cell membrane in most plant cells. The cell wall is non-living.

Ideas and Applications

Bacteria are the simplest and smallest living things. Like plant cells, they have a cell wall which holds their shape. Penicillin is an antibiotic—a substance which can kill or prevent the growth of bacteria. Penicillin interferes with the production of a substance needed for bacterial cell walls. This helps prevent the growth of bacteria.

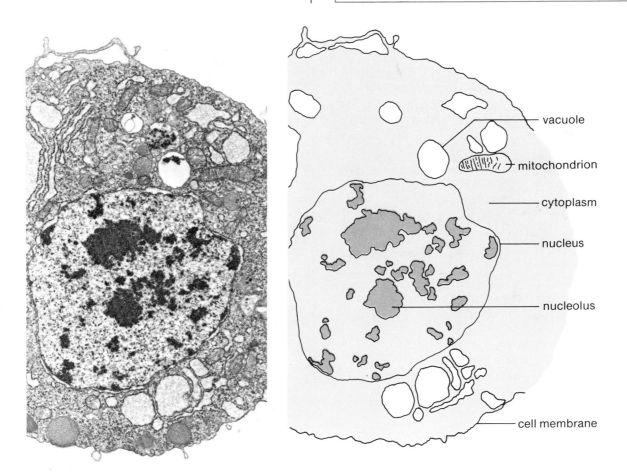

Figure 12.2 (b) *An animal cell shown by the TEM. How many cell parts can you identify?*

Problem

How can you view a living plant cell with a compound microscope?

Materials

forceps	compound microscope
microscope slide	paper towel or tissue
coverslip	iodine solution
dropper	
water	

small section of white onion
  (with outer dry skin removed)

Procedure

1. Strip a small section of skin (membrane) from your onion. It will come off easily if the inner curved surface of the onion is folded, as shown in Figure 12.3.

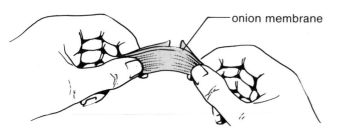

Figure 12.3 *Fold back the onion section to see the skin (membrane) you will need.*

2. Put a drop of water on a slide. Using forceps, place the membrane on the drop of water. Be careful not to let the membrane fold over (Figure 12.4).

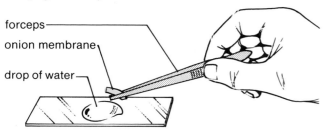

Figure 12.4 *Using the forceps, place the skin carefully on the slide.*

3. Place another drop of water on top of the onion membrane.
4. Place the coverslip on the slide as shown in Figure 12.5.

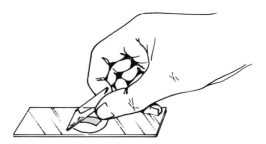

Figure 12.5 *Carefully place the coverslip on the slide.*

5. View your prepared slide with the compound microscope using the low-power objective lens.
6. Make a diagram of at least three cells and label any parts you can see.
7. Remove the slide from the microscope and carefully draw off the water by placing a piece of paper towel or tissue at one side of the coverslip (Figure 12.6).

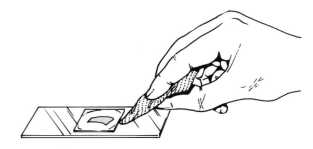

Figure 12.6 *Place a piece of paper towel or tissue at the edge of the coverslip to draw off the water.*

CAUTION Iodine is a corrosive liquid that will stain the skin, clothing, desk tops, and floors. Take care to avoid spilling. Quickly wipe up any accidental spills, and rise with water.

8. Using the dropper, place a drop of iodine solution at one edge of the coverslip (Figure 12.7). It will spread across underneath the coverslip.

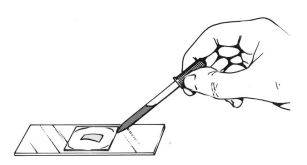

Figure 12.7 *Place the drop of iodine at one edge of the coverslip. The iodine will spread under the coverslip.*

9. View the slide again using first the low-power then the medium-power and finally, the high-power objective lens.
10. Make a diagram of one cell as it appears under the high-power lens. Label the nucleus, cell wall, and cytoplasm. Show the magnification of your diagram.

Observations

1. What structures could you see without the iodine?
2. What structures could you see more easily after adding iodine?

Questions

1. Does the addition of the iodine solution make it easier or more difficult to locate structures within the cell? Explain.
2. Could you see the cell membrane? Why or why not?

Challenge

Using forceps, peel a piece of skin from a section of either a tomato, green pepper, or *Elodea*. Then carefully use a scalpel to scrape the inner surface of the skin to remove the flesh. (Note: The scalpel is extremely sharp! Be very careful.) The skin should be clear. Place the cleaned skin carefully on a slide so that the outer surface is uppermost. Add two drops of iodine solution and place the coverslip in the usual manner. Observe your slide under all magnifications, starting with low power. Make a diagram of two cells that are beside each other, and label all visible parts. Based on your observations here, and in Activities 12A and 12B (following), describe the differences you see between plant cells and animal cells.

Scrape the inner surface of the tomato to remove the flesh. The skin you are going to use should be clear (transparent).

Problem

How can you view a living animal cell with the microscope?

Materials

microscope slide
coverslip
dropper
water
iodine solution
compound microscope
wooden toothpick (clean and sterile)

CAUTION Make sure you wash your hands carefully before and after putting the clean toothpick in your mouth. Throw away the toothpick after it has been used once. Remember the warning about the iodine solution in Activity 12A.

Procedure

1. Add two drops of iodine solution to the slide.
2. Using the *blunt* end of the toothpick, gently scrape the inside of your cheek and remove some of the saliva from this area (Figure 12.8).

Figure 12.8 *Using the blunt end of the toothpick, gently scrape the inside of your cheek.*

3. Spread this scraping onto the slide.
4. Place the coverslip on the slide in the usual manner for a wet mount.
5. Observe the slide with the compound microscope under all magnifications. Remember to start with the low-power objective lens.

Observations

1. Make a diagram of two cells under high power.
2. Label any structures you recognize.

Questions

1. What is the general shape of the cells?
2. Can you see all structures clearly within the cell? Why or why not?

Self-check

1. What is a cell?
2. List the three parts of the cell theory.
3. Define the terms *structure* and *function* as they apply to a cell.
4. What are the three basic parts of a cell?
5. (a) What is the cell wall?
   (b) Where is it found?

The Internal Structure of Cells

The Cell Membrane

The outer living surface of a cell is called the **cell membrane**. All cells possess cell membranes. The cell membrane regulates the passage of certain substances, such as gases and chemicals in solutions, into or out of the cell. It also separates one cell from other surrounding cells. With the compound microscope it looks like a very thin line.

The Nucleus

The nucleus is the central "control centre" of a cell. It organizes and directs the functions of the particular cell in which it is found. It is also responsible for the production of new cells. The nucleus itself looks like a golf ball and floats within the cytoplasm of the cell (Figure 12.9).

The nucleus is surrounded by two membranes called the **nuclear membrane**. There are many holes in the surface of the nucleus. These are the *nuclear pores*. They permit substances to pass in and out of the nucleus.

Within the nucleus is a fluid, similar to the fluid in the cytoplasm. This is called the *nuclear fluid*. The grainy substance in the nuclear fluid is the **chromosomes**. The chromosomes are made up of a nucleic acid called **DNA (deoxyribonucleic acid)**. DNA is the material that is passed on from one generation to the next when the cell divides.

When the cell is not dividing, the chromosomes are not visible as individual threads, but as dark masses called **chromatin**. DNA also provides the information for the production of the proteins in cells. It is these proteins that make the cell work.

The nucleus also contains one or more **nucleoli** (singular: **nucleolus**). They are round but do not have a membrane around them. They store some proteins and also make an acid called **RNA (ribonucleic acid)**. RNA is very similar in structure to DNA. However, RNA can pass out of the nucleus but DNA cannot. RNA takes information from DNA to make proteins in the cytoplasm. In Activity 12A, the nucleolus may have been visible as a darkly stained dot within the nucleus of the onion cell.

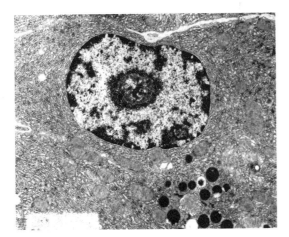

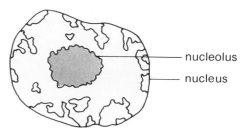

Figure 12.9 *The nucleus floats in the cytoplasm.*

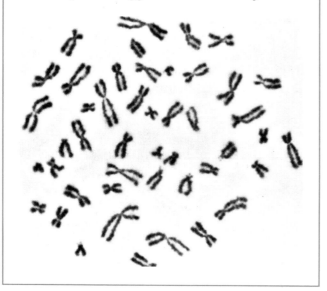
The Cytoplasm

The **cytoplasm** is the fluid within which all the cell organelles are found. It is where most of the reactions of the cell occur. It is like a factory. It can receive materials, break down material to form useful substances, manufacture new substances to be delivered to other cells, and excrete waste materials.

Ribosomes are small dot-like structures found throughout the cytoplasm. Ribosomes are made of RNA. They are the places where proteins are produced.

The Cell's Powerhouse

The **mitochondrion** (plural: **mitochondria**) is an organelle responsible for releasing energy from food. It is often called the power house of the cell. The number of mitochondria each cell has depends upon the amount of energy the cell needs. For instance, a muscle cell needs more energy than a fat cell. Therefore, the muscle cell will have more mitochondria than a fat cell. A mitochondrion looks like a bag with another bag inside it. The inner, larger bag becomes folded to increase the surface area (Figure 12.10).

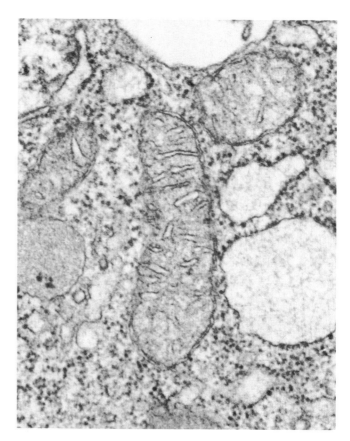

Figure 12.10 *The mitochondrion sometimes looks like a bean. It has folded membranes inside. It is on these membranes that energy is released from sugar. This process, called respiration, uses up oxygen and produces carbon dioxide.*

Chloroplasts

Plant cells are green because they possess organelles called **chloroplasts** (Figure 12.11). Chloroplasts contain a green pigment called **chlorophyll**. It enables a green plant to convert the energy of sunlight into the chemical energy of food. You will learn about this in Chapter 15.

Vacuoles

Another type of organelle within the cytoplasm is the **vacuole**. Vacuoles are surrounded by a membrane and are filled with watery substances and pigments (Figure 12.12). They are large in plant cells but usually small in animal cells.

Vacuoles also serve to store waste materials for the cell. It is these vacuoles that store the fat in body cells and yellow and red pigments seen in the fall on tree leaves.

The yellow and red colours of fall leaves are contained in the vacuoles of the leaf cells. The leaves change colour because there is no more chlorophyll. The green colouring of the leaves disappears. The other colours now become visible.

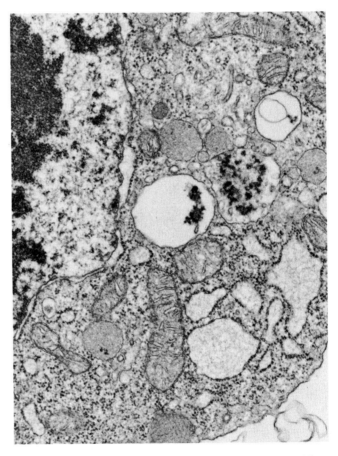

Figure 12.12 *Animal vacuoles are usually small. What other organelles can you identify?*

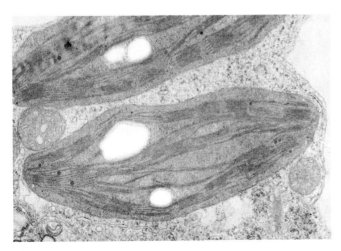

Figure 12.11 *The chloroplast makes the cell green. It is where photosynthesis takes place.*

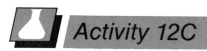
Problem

How can you view vacuoles in plant and animal cells?

Materials

outer skin of red onion (not the dry part)
microscope slide
coverslip
dropper
water
prepared slide of human skin
compound microscope

Procedure

1. Prepare a wet mount of the outer red skin of the red onion.
2. Observe the prepared slide under the low-, medium-, and high-power objectives. The clear areas in the cytoplasm contain a pink material. These are vacuoles. They contain a red pigment.
3. Draw and label at least two onion skin cells.

4. Observe the prepared slide of human skin under the low-power objective lens. The layers of cells directly beneath the skin look white and empty. Figure 12.13 will help you locate the fat cells. They are very large because the vacuoles within the cells are filled with fat. The nucleus of the cell is just visible at the corner of the cell.
5. Focus under the high-power objective lens. Draw and label at least two human skin cells.

Observations

1. On your drawing of onion skin cells, label the cytoplasm, vacuole, nucleus, cell membrane, nucleolus, chromatin, and cell wall.
2. On your drawing of human skin cells, label the cytoplasm, vacuole, nucleus, and cell membrane.

Questions

1. How are the vacuoles in an animal fat cell different from the vacuoles in most animal cells?
2. Why would you see the vacuoles in these cells, but not in the onion skin cells in Activity 12A?

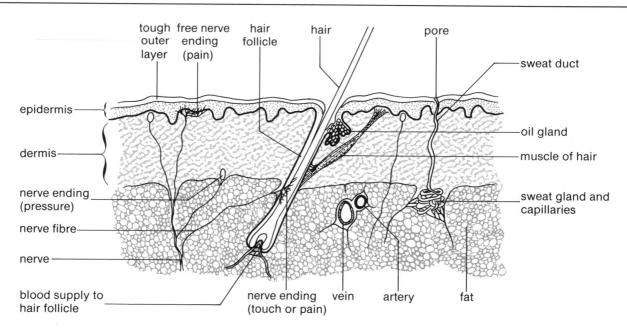

Figure 12.13 *Skin has many layers. Each structure within the skin has a specific function.*

186

Challenge

Examine Figure 12.13. Each structure within each layer in the skin has a special function. Which cells protect you against cold? Which cells need replacing regularly? Name two additional functions of the skin. Which cells are associated with these functions?

Other Cell Structures

Some structures within the cell are made of tiny tubes. Two of these important structures are called **flagella** (singular: **flagellum**) and **cilia**. (The singular, cilium, is hardly ever used.) A flagellum is a long, hair-like whip that swings back and forth to move the cell through a liquid environment. See Figure 12.14. Cilia are numerous tiny hairs, usually shorter in length compared with flagella. You can observe cilia on the single-celled organism called paramecium (plural: paramecia). See Figure 12.15.

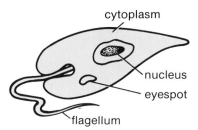

Figure 12.14 Euglena *is a single-celled organism which uses a flagellum to pull itself through its water environment.*

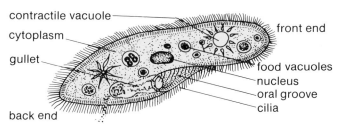

Figure 12.15 *Paramecium is a free-living, one-celled organism.*

Problem

How can you observe cilia in paramecia?

Materials

compound microscope
dropper and water
microscope slide
coverslip
live culture of paramecium
paper tissue

Procedure

1. Place a drop of water on a microscope slide.
2. Place a small piece of paper tissue on the drop of water.
3. Put one drop of paramecium culture on top of the paper tissue. The tissue prevents the paramecia from swimming too quickly.
4. Add the coverslip to the slide using the usual method for a wet mount.
5. Observe your paramecia closely. Sketch one.

Observations

1. Use Figure 12.15 to help you label the sketch you made.
2. Describe how a paramecium swims.

Questions

1. What causes the turning movement of the paramecium?

Challenge

Add some stained yeast cells to your paramecia slide. Describe the role of the cilia in the feeding process.

Self-check

1. Describe in your own words the structure and function of the following:
   (a) cell membrane
   (b) nucleus
2. (a) What is an organelle?
   (b) Name three organelles and state the function of each.
   (c) Where are these three organelles found in the cell?
3. List two functions of the cytoplasm.
4. What is the function of the mitochondrion?
5. What is one difference between cilia and flagella?
6. (a) Is the cell shown in Figure 12.16 a plant cell or an animal cell? Give reasons for your answer.
   (b) Name at least five cell parts and their functions.

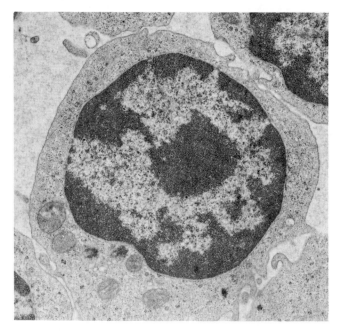

Figure 12.16 *Is this a plant or animal cell?*

Chapter Objectives

NOW THAT YOU HAVE COMPLETED THIS CHAPTER, CAN YOU DO THE FOLLOWING?	FOR REVIEW, TURN TO SECTION
1. Describe the importance of cell size and shape.	12.1
2. State the three parts of the cell theory.	12.2
3. Name the three main parts of all cells and an additional structure that occurs only in plant cells.	12.3
4. Prepare wet mount slides of living plant and animal cells, and draw and label the main parts.	12.3
5. Describe the structure and function of the following organelles: cell membrane, cell wall, nucleus, cytoplasm, mitochondrion, chloroplast, vacuole, cilia, and flagella.	12.4
6. Observe, describe, and sketch cell structures in plant and animal cells and in single-celled organisms.	12.3, 12.4

Words to Know

cell
cell theory
organelle
cell membrane
cytoplasm
nucleus
cell wall
nuclear membrane
chromatin
chromosome
DNA (deoxyribonucleic acid)
nucleoli (singular: nucleolus)
RNA (ribonucleic acid)
ribosome

mitochondrion (plural: mitochondria)
chloroplast
chlorophyll
vacuole
flagella (singular: flagellum)
cilia

Tying It Together

1. You have looked at many plant and animal cells in this chapter. Copy Table 12.1 into your notebook, and put a check mark beside the features that are present in plant and animal cells. This will help you see better how plant and animal cells are different and how they are alike. Do not write in this textbook.

Table 12.1

FEATURE	PLANT CELL	ANIMAL CELL
Cell membrane		
Cell wall		
Nucleus		
Cytoplasm		
Chloroplast		
Regular shape		
Large vacuoles		
Mitochondria		

SAMPLE ONLY

2. What influences the shape of a cell?
3. (a) What are the three statements of the cell theory?
   (b) Is it possible to prove the first two statements? Explain.
4. Make labelled diagrams to show the structure of the following:
   (a) a cell of the skin of an onion
   (b) a human cheek cell
5. Explain the importance of the following:
   (a) the cell wall
   (b) mitochondria
   (c) cilia
   (d) ribosome
   (e) nucleus
   (f) chloroplast
6. What is the difference between the following:
   (a) cell wall and cell membrane
   (b) cytoplasm and vacuole
   (c) nucleus and nucleolus
   (d) RNA and DNA

Applying Your Knowledge

1. *Elodea* is a plant that lives under water in the shallows of lakes and ponds. Suggest why its leaf cells may be more difficult to study through the microscope than the cells of an onion.
2. In previous studies, you learned about the characteristics of living things. What characteristics apply to life for a single cell?

Projects for Investigation

1. Here is a project to do at home. (Make sure an adult knows what you are doing before you start.) Using only edible materials and a mixing bowl, construct a cell in as much detail as possible. Of course, include the organelles. Try to represent the approximate shape and proportions of the cell.

Cell Activities

Key Ideas

- Single cells divide, to produce identical pairs of cells.
- In mitosis, each daughter cell receives an identical set of chromosomes.
- In larger organisms, cells are organized into tissues, organs, and systems.
- Cell membranes regulate the passage of substances into and out of cells.
- Cells are affected by their environment.

In 1981, a young man who had lost his leg because of cancer attempted to run across Canada in his Marathon of Hope. He wanted to alert people to the increasing need for continued cancer research. This man was Terry Fox. He died before he could complete his run, but many others, such as Steve Fonyo, used his idea for collecting funds in aid of cancer research. They have been running, walking, and even wheelchairing across Canada and other countries.

What is cancer? It is a disease caused when cell activities have gone wild. It is the second leading cause of death in Canada. A **cancer** may be formed when cells in the body start to divide in an uncontrolled manner. What makes cells divide in an orderly way at some times but divide too fast at other times, thus producing a cancer?

Cell Division

All cells have to divide at some time. Some divide when they are worn out or need repair. Skin cells, for example, are constantly being rubbed off the body. (You don't notice this constant removal unless you have a sunburn. Then great sheets of these flat cells "peel" off all at once.)

Normal body cells need to divide when they get too large. The exchange of substances such as oxygen with the environment can only occur at the surface of the cell. As shown in Figure 13.1, the area of the surface does not increase as the volume of the cell increases. A large cell, therefore, is slow to get supplies and get rid of wastes. Thus, there is a limit to the size of single-celled organisms. If an organism is to become much larger, it must do so by increasing the *number* of cells, not by enlarging its cells.

If a new cell is to be formed, it must be able to carry out all the activities necessary for life. The instructions for these processes are contained in the nucleus. Therefore, the most important stage in cell division is the duplication of the nucleus. This process is called **mitosis**.

(a)   **(b)**

Figure 13.1 *The volume of B is twice the volume of A. Is the surface area of B twice the surface area of A? Can B, which is twice the volume of A, take in twice as much oxygen as A? Can B take in the oxygen as quickly as A? Would B be better off if it divided in half?*

Mitosis

Mitosis ensures that the *chromosomes* (hereditary material) in each new nucleus contain a complete set of instructions for life. Chromosomes are visible only during cell division. The word *chromatin* is used to describe the chromosomes at other times when this material is dispersed throughout the nucleus.

Cell division does not actually start until mitosis nears completion. Mitosis is a continuous process, but it is easier to think of it occurring in four stages.

Figure 13.2 shows only the parts of the cell that are important in mitosis. Before mitosis begins, the cell assembles all the molecules needed to duplicate the nucleus. Then, during the first stage of mitosis, each chromosome assembles materials to form an exact duplicate of itself. The chromosome and its duplicate remain joined at one point so that they move as a pair. Next, the nucleolus and the nuclear membrane start to disappear. The *centrioles* move to opposite ends of the nucleus. See Figure 13.3(a) on the next page. Fibres called *rays* begin to form between them. (Note: Most plant cells do not have centrioles.) Eventually the rays meet to form the *spindle*.

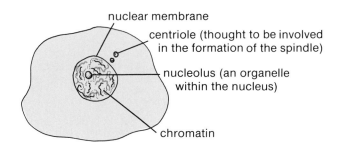

nuclear membrane

centriole (thought to be involved in the formation of the spindle)

nucleolus (an organelle within the nucleus)

chromatin

Figure 13.2 *In this simplified diagram of an animal cell, only the parts involved in mitosis are shown.*

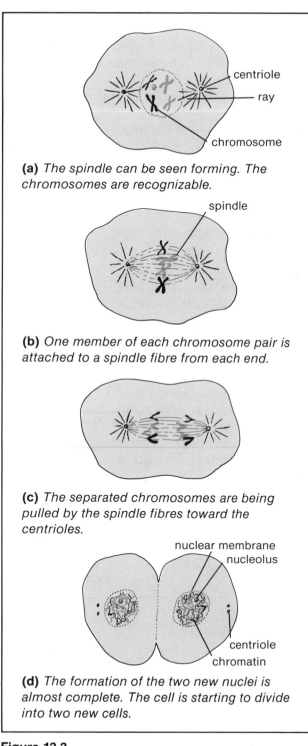

(a) *The spindle can be seen forming. The chromosomes are recognizable.*

(b) *One member of each chromosome pair is attached to a spindle fibre from each end.*

(c) *The separated chromosomes are being pulled by the spindle fibres toward the centrioles.*

(d) *The formation of the two new nuclei is almost complete. The cell is starting to divide into two new cells.*

Figure 13.3

During the second stage, the spindle forms. The chromosome pairs drift toward the middle of the cell until they form a definite group there. The connection between each pair of chromosomes now breaks. The two halves of each pair become attached to spindle fibres from opposite ends. See Figure 13.3(b).

During the third stage, two new nuclei begin to form. The spindle fibres gradually contract. This pulls the halves of each chromosome pair to opposite ends of the cell. See Figure 13.3(c).

During the fourth and last stage, the nuclear membrane and nucleolus reappear. The chromosomes gradually become a network of chromatin. The spindle fibres disappear. The centrioles duplicate. As this process finishes, cell division begins. The cell membrane begins to separate the cytoplasm into two new cells. In animal cells, the cell membrane divides by "pinching off" the membrane between each new nucleus. See Figure 13.4(a). Plant cells form a *cell plate* of cellulose between the new cells. See Figure 13.4(b).

The original cell has divided into two identical cells. These cells enter the first stage and continue to grow until they are ready to divide again.

Figure 13.4

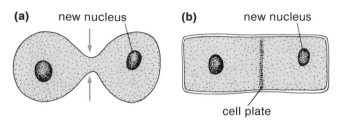

(a) *Animal cells divide by pinching off the membrane between the two new nuclei.* (b) *Plant cells form a cell plate of cellulose between the new nuclei.*

You can observe different stages in the life cycle of a cell. When the slide you'll be using in this activity was prepared, the cells were caught at different stages in the continuous process of cell division and growth.

Problem

What stages of plant cell division can you recognize?

Materials

prepared slide of onion root tip
compound microscope

Procedure

1. Place the slide of the onion root tip cells on the microscope. Examine the cells under the low-power objective lens. The most actively dividing cells are those near the root tip (Figure 13.5). Move your slide so that these cells are in the centre of view.

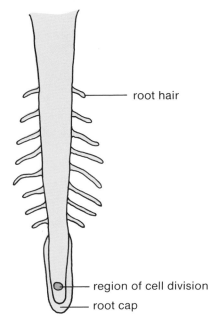

Figure 13.5 *An onion root tip. Only the cells just behind the root cap are actively carrying out mitosis.*

2. Change to the medium-power objective and then to the high-power objective. Observe the cells.
3. Move the fine adjustment slightly to bring different structures into view.
4. Very slowly, move the slide along the length of the root tip. Find cells at different stages of cell division.

Observations

1. Sketch and label cells at the following different stages of cell division:
   (a) when chromosomes are not visible
   (b) where doubled chromosomes are visible
   (c) when two cells are splitting

Questions

1. Why do so few cells show stages of mitosis even in the root tip?
2. Name at least two other places that you might look to find cells dividing.

Challenge

Use a prepared sheet from your teacher or refer to Figures 13.2, 13.3, and 13.4(a) which show mitosis in animal cells. Compare the stages of cell division in animal cells with those you saw in plant cells.

Cells in Groups

After cell division, cells continue to grow and develop until they reach their most effective size. In many-celled organisms (such as you), the total number of cells also increases during growth. During development in many-celled organisms, cells may become specialized to perform certain jobs. A group of cells that performs a specific job is called a **tissue**. In larger organisms, specialized tissues that are grouped to perform a specific job are called an **organ**. Organs, in turn, may be grouped together into **organ systems** to perform a major function.

Let's look at an example. In a human, the heart is an organ of the circulatory system. The job of the circulatory system (made up of the heart, arteries, and veins) is to distribute blood throughout the body. The job of the heart, which is an organ, is to pump the blood. The heart is composed of muscle tissue as well as other types of tissue. The muscle tissue of the heart is composed of specialized cells that are able to contract and relax so that the heart can function (Figure 13.6).

Self-check

1. Why do cells need to divide?
2. What is the importance of mitosis?
3. (a) What is a chromosome?
   (b) Why are chromosomes important to organisms?
4. Describe, in order, the changes that take place in a cell during cell division.
5. When cells are not undergoing cell division, what are they doing?
6. How do plant cells and animal cells differ in the process of cell division?
7. What is the difference between a cell, a tissue, an organ, and an organ system? Give an example of each type.

circulatory system

heart

The heart is an organ in the system.

heart tissue

a heart muscle cell

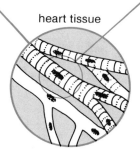

Figure 13.6 *Specialized cells perform certain functions within tissues, organs, and organ systems in the human body.*

How Materials Enter and Leave Cells

Recall that the cell membrane makes up the outer surface of the cell. The cell membrane regulates the passage of substances, such as gases and chemicals in solution, into and out of the cell. The cell membrane is said to be a **differentially permeable membrane**. This means that it allows only certain substances to enter and leave the cell.

Recall that **diffusion** refers to the movement of substances from an area of high concentration to an area of low concentration. Diffusion occurs throughout our body as cells take in or remove substances. When cells store food, such as sugars, they can convert the sugars into larger particles so that they cannot diffuse through the membrane.

Diffusion also occurs in the lungs when gas exchange takes place. Oxygen and carbon dioxide are small molecules that can diffuse through the membranes in both directions. Oxygen is usually at higher concentrations outside the cell, so more oxygen will diffuse into the cell than diffuses out (Figure 13.7).

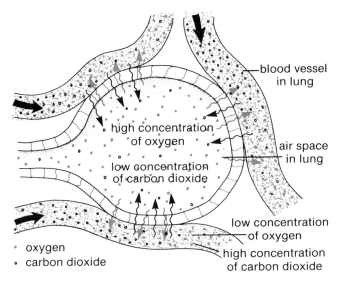

Figure 13.7 *Diffusion between the blood and the air spaces in the lungs*

Challenge

In Chapter 3, you used the particle theory to explain how substances spread through liquids and gases. Review the particle theory, and use it to explain the following:
(a) how the molecules from an open bottle of perfume are able to move about in air
(b) that diffusion involves movement from an area of high concentration to an area of low concentration (as in a cup of instant coffee)
(c) that differentially permeable membranes allow only certain substances to pass through (as in a tea bag)

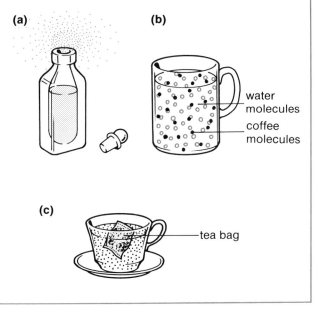

In this activity, you can demonstrate diffusion through a differentially permeable membrane. In this case, the cell is represented by a piece of dialysis tubing and a piece of glass tubing.

Problem

How can a model represent the differentially permeable membrane of a cell?

Materials

10 cm dialysis tubing
strong sugar solution (corn syrup or molasses)
50 cm glass tubing
twist tie
rubber stopper
beaker of distilled water
stand and clamp
small piece of masking tape

Procedure

1. Soak the dialysis tubing in water for a few minutes. Rub it between your thumb and forefinger to open it.
2. Tie a tight knot in one end of the tubing, about 1 cm from the end.
3. Fill the tubing with the strong sugar solution.
4. Fit the open end over the rubber stopper into which the glass tubing has been inserted (Figure 13.8). Use the twist tie to attach the dialysis tubing to the glass tubing.
5. Rinse the outside of the dialysis tubing with water.
6. Lower the dialysis tube into the beaker of water and clamp the tube vertically.
7. Mark the level of the sugar solution on the glass tube (first level) with a small piece of masking tape.
8. Mark the level again after 10 min.

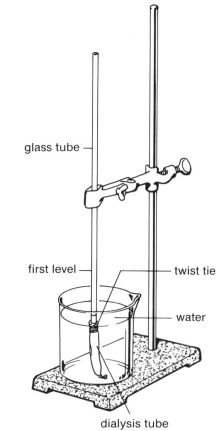

Figure 13.8 *Set-up for Activity 13B*

Observations

1. Describe what happens to the sugar solution in the dialysis tubing during the 10 min.
2. Was there any evidence that water diffused through the tubing? Explain.
3. Was there any evidence that sugar diffused through the tubing? Explain?

Questions

1. Use the particle model to explain what you observed in this activity.
2. How does this model relate to cell membrane action?

Osmosis—A Special Kind of Diffusion

Osmosis refers to the diffusion of water across a differentially permeable membrane. Figure 13.9 shows a model for osmosis.

Osmosis depends on the passage of water molecules through a membrane. There is a higher concentration of water molecules on the left side of Figure 13.9 than on the right side. The water molecules on the left side tend to move (diffuse) to an area where there are fewer water molecules—the right side. Thus, the water molecules cross the membrane into the sugar solution. The sugar molecules cannot pass through the membrane because they are too large. Therefore, you can see that water passes from an area of high concentration to an area of low concentration through the differentially permeable membrane. This is osmosis.

Self-check

1. (a) What is diffusion?
   (b) Give two examples of diffusion in cells.
2. Sugar is placed in a cup of coffee. Why does the coffee taste sweet to the last drop?
3. What are differentially permeable membranes?
4. (a) Define osmosis.
   (b) Give one example of where osmosis occurs in cells.

Challenge

You have just finished a delicious meal of stir-fried beef and vegetables. The thin strips of beef had been marinating the night before in a spicy, flavourful sauce. Use your knowledge of diffusion to explain why the cooked beef was so tasty.

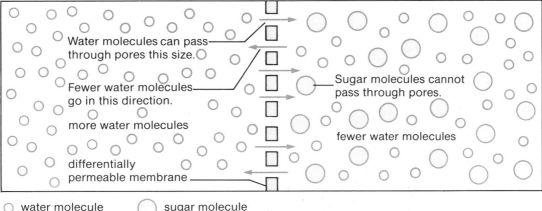

○ water molecule ◯ sugar molecule

Figure 13.9 *A model to explain osmosis. The arrows show direction of movement of water molecules.*

The Artificial Kidney Machine

Chemical reactions take place in your cells. These reactions produce energy which keeps the cells, and you, alive. The chemical reactions also produce waste substances which must be removed from your body. The kidneys remove wastes by cleaning your blood.

When blood that contains wastes flows through the kidneys, the waste substances are filtered out and concentrated. These wastes are eliminated, in the form of urine, from your body. The waste substances in urine include water, urea, and salts. People who suffer from kidney diseases cannot have these wastes removed from their blood. Dangerous levels of wastes can build up and cause death.

You already know that cell membranes are differentially permeable. This simple property of membranes has been applied to the field of medicine, resulting in the artificial kidney machine. The artificial kidney machine does the job of the kidney by cleaning the blood of wastes. The machine has two main parts. One part consists of cellophane tubing arranged in a compact coil. This tubing acts as a differentially permeable membrane. The other part is a tank which holds a cleaning solution and the coiled tubing.

How does the artificial kidney machine work? Blood containing waste substances flows from an artery in the person's arm into the tubing. Waste substances diffuse through the tube membrane into the cleaning solution and are removed. After the blood has passed through the many coils of tubing, it reenters a vein in the person's arm. To clean the blood thoroughly usually takes three to six hours. This process saves thousands of lives each year.

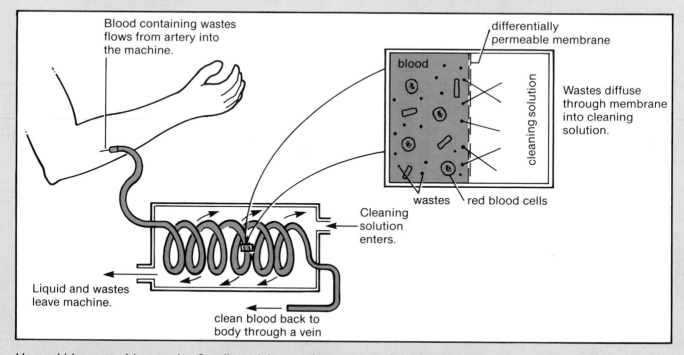

How a kidney machine works. Small particles, such as urea and water, can move across the membrane between the blood and the cleaning solution. Larger particles and blood cells stay in the blood.

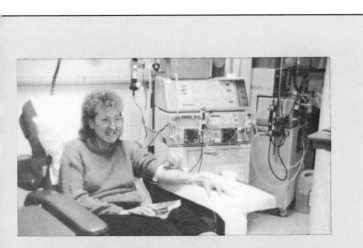

The person shown here uses the artificial kidney machine in a hospital. She has to remain hooked up to the machine for three to six hours. She usually visits the hospital three times a week for this treatment.

Instead of going to a hospital three times every week, this person can clean her blood at home using a different method. The membrane which covers the intestines is used to act like the cellophane tubing in a kidney machine. The cleaning solution is added and removed through a small tube in the abdomen. This process is repeated four times a day. Using this method, a person can even travel away from home.

Cells and Their Environment

About 70% of a typical cell is water. Other substances such as foods are dissolved in this water. How are cells affected by a change in the concentration of the solution that surrounds them? Examine Figure 13.10 and discuss the following questions in class:

1. The cell is placed in a solution whose concentration equals that of the cytoplasm. What might happen to the cell's size?
2. The cell is placed in a solution containing more water than the cell. What might happen to the cell's size?
3. The cell is placed in a solution containing water less concentrated than the cell. What might happen to the cell's size?

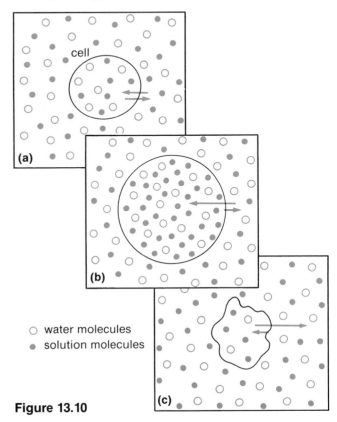

○ water molecules
● solution molecules

Figure 13.10

199

A change in the concentration of solution has a dramatic effect on plant and animal cells. Organisms or their individual cells are often surrounded by fluids that are less or more concentrated than their own.

The cell wall in plant cells is permeable to water and dissolved substances. That is, it allows these substances to pass through. The cell membrane is differentially permeable. When a plant cell is placed in an environment of pure water, water will pass into the cell. The cell will swell but will not burst because of the rigid cell wall. If the cells of a plant do not contain enough water, the plant wilts (Figure 13.11).

Figure 13.11

(a) *Plants showing evidence of wilting*

(b) *The same plants a few minutes after being watered*

Problem

What will be the effect of brine on onion cells?

Materials

onion
microscope slide
coverslip
paper towel or tissue
dropper
microscope
salt solution (brine)

Procedure

1. Prepare a wet mount of onion membrane as you did in Activity 12A.
2. Draw at least three cells, making particular note of the position of cytoplasm and cell wall.
3. Place one or two drops of brine at the right edge of the coverslip using the dropper. Hold a piece of paper towel at the left edge of the coverslip to draw out the water (Figure 13.12). Wait 5 min.
4. Examine the slide again under the microscope.
5. Draw the three cells as they appear now.

Figure 13.12 *The paper towel will draw out the water and allow the brine into the slide.*

Observations

1. How had the onion cells changed?

Questions

1. What caused the effect on onion cells when they were placed in brine solution? Explain your answer using diagrams.
2. Give one example of where this might occur. (Hint: Think of grass near the road in winter.)
3. Animal cells behave in a manner similar to plant cells. But animal cells will burst when placed in pure water. Why do you think this is true?

Ideas and Applications

A person who is stranded at sea for long periods of time without fresh water faces the danger of dying from dehydration. Drinking sea water can be fatal because of the high concentration of salt that it contains. This high concentration of salt causes water to leave body cells, so they become more dried out than they were before. However, one liquid that can help people who are stranded at sea survive is . . . cod urine! Cod urine has less salt than human body fluids, and drinking it can prevent dehydration.

Problem

What are some environmental factors that will affect other living things?

Materials

amoeba or paramecium	dropper
4 microscope slides	coverslips
brine (salt solution)	paper towel or tissue
alcohol	compound microscope
source of strong light	

Procedure

1. Prepare four slides of amoeba or paramecium.
2. Examine each slide under the microscope.
3. To one slide, add one or two drops of brine at the right edge of the coverslip.
4. Hold a piece of paper towel at the left edge of the coverslip to draw out the water. Wait 5 min, and then examine the slide under the microscope again.
5. To the second slide, add one or two drops of alcohol and repeat the procedure in step 4.
6. Allow the third slide to sit under strong light for 15 min. Examine the slide under the microscope.
7. Use the fourth slide as the *control* for comparison with the other slides.

Observations

1. What changes did you observe in each of the slides?
2. What caused the effect on the organisms you studied?

Questions

1. What effect would the addition of foreign substances to the environment have on an organism?
2. What effect would the removal of foreign substances from the environment have on an organism?

201

Cells, Environmental Changes, and Cancer

Environmental factors such as pollutants and harmful radiation affect cells in different ways. For example, chemical irritants, such as PCBs, dyes, and the tar found in cigarette smoke, as well as radiation irritants, such as nuclear radiation and excesses of X rays, or even too much sunlight have been linked to the production of cancers in organisms.

Cancers are formed when cells in the body start to divide in an uncontrolled way (Figure 13.13). Some cancers are not harmful growths. These do not usually spread throughout the body or cause death. If they are in the brain (brain tumours) they could cause death unless detected early. This is because the brain controls many vital body functions. Other cancers can cause harmful (malignant) growths. These can grow and spread very quickly throughout the body. Treatment of cancer today is confined mainly to control of the malignancy. Certain lifestyles may either increase or reduce the risk of developing cancer. Scientists have not yet discovered a cure.

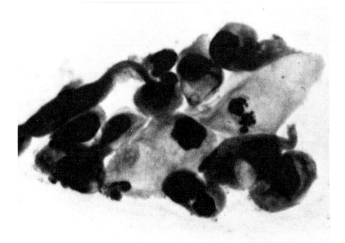

Figure 13.13 *These smaller dark cells are cancerous human cells. Cancer cells have long dense nuclei.*

Chapter Objectives

NOW THAT YOU HAVE COMPLETED THIS CHAPTER, CAN YOU DO THE FOLLOWING?	FOR REVIEW, TURN TO SECTION
1. Explain why cells need to divide.	13.1
2. Explain the process of cell division.	13.1
3. Identify the stages of plant cell division.	13.1
4. Define the terms tissue, organ, and organ system, and give examples of each.	13.2
5. Explain how materials enter and leave cells.	13.3
6. Define osmosis and explain why this process is important for living organisms.	13.3
7. Explain how cells are affected by changes in concentration of the liquids around them.	13.4
8. Explain how cells react to environmental change.	13.4

Words to Know

cancer
mitosis
tissue
organ
organ system
differentially permeable membrane
diffusion
osmosis

Tying It Together

1. When do normal cells divide?
2. Explain the process of cell division. (Using diagrams may be helpful.)

3. How does animal cell division differ from plant cell division?
4. (a) What is a tissue?
   (b) Give two examples of tissues.
5. (a) What is an organ?
   (b) Give two examples of organs.
6. (a) What is an organ system?
   (b) Give two examples of organ systems.
7. (a) How does osmosis differ from diffusion?
   (b) How are the two processes similar?
8. Describe the process of osmosis in terms of the particle theory.

Applying Your Knowledge

1. During mitosis, cell chromosomes duplicate exactly. Describe what might occur if the chromosomes did not duplicate properly.
2. Lettuce often goes limp in the refrigerator. What type of solution could be used to restore the crispness to the lettuce? Explain your answer.
3. The usual concentration of salts in blood is equivalent to 0.9% salt solution. This is called normal saline. After an operation, a patient is often injected with this solution.
   (a) What will happen to the patient's blood cells when normal saline is added?
   (b) What would happen to the patient's blood cells if the saline solution contained 1.5% salt?
   (c) What would happen to the patient's blood cells if the saline solution contained 0.009% salt instead of the normal saline?
4. What problems would an organism have moving from
   (a) sea water to fresh water?
   (b) fresh water to sea water?

Projects for Investigation

1. Select one environmental factor, such as acid rain, radiation, or a poisonous (toxic) chemical, that can affect cells. Find out how cells are affected. How can cells survive exposure to this environmental factor?
2. Human cells are affected by drugs. Select one drug, such as alcohol, nicotine, or aspirin, and research its effect on human cells.
3. You have learned that many cells are specialized for a particular function. Imagine how the structures of certain cells might have become adapted to carry out specific jobs. Choose two of the situations below. Then design a cell that will meet the needs of each situation. Write down the features that you think the cell will require. Then do some research to find out how some cells are actually adapted for the purposes that you have selected. Prepare a report for the class, showing your designs and the actual structures that you have found.
   (a) A cell in an organism living in the desert stores water for survival.
   (b) A cell carries messages from your brain to your hand.
   (c) A cell carries water upward in a plant stem.
   (d) A cell helps an animal move. This cell contracts (gets shorter) and needs a good supply of energy to do this work.
   (e) A single-celled organism lives in water and eats other single-celled organisms.

Unit Four: The Cell and Its Systems

MATCH

In your notebook, write the letters (a) to (i). Beside each letter, write the number of the word in the right column that corresponds to each description in the left column.

(a)	control centre of a living cell	1. cell membrane
(b)	fluid within which all cell organelles are found	2. cell wall
(c)	outer living surface of a cell	3. chloroplast
(d)	organelle containing chlorophyll	4. cilia
(e)	organelle made of RNA	5. cytoplasm
(f)	organelle responsible for releasing energy from food	6. flagellum
(g)	non-living layer that surrounds most plant cells	7. mitochondrion
(h)	short, hair-like projections from the surface of some cells	8. nucleus
(i)	whip-like projection from the surface of some cells	9. ribosome
		10. chromatin

MULTIPLE CHOICE

In your notebook, write the numbers 1 to 10. Beside each number, write the letter of the best choice.

1. While you are using the compound light microscope, focussing away from the slide will
   (a) break the cover slip
   (b) bring the object and objective as near as possible
   (c) increase the working distance
   (d) regulate the light.

2. When you are viewing an object under the low power objective, the object seems to be to the left. To bring the object back to the centre of the field of view, you must move the slide
   (a) to the left
   (b) to the right
   (c) away from you
   (d) toward you.

3. A typed letter was placed in the field of view in this position: R. How would the magnified image appear?

   (a) Я
   (b) R
   (c) Я
   (d) Я

4. The scanning electron microscope produces images that appear to be
   (a) three-dimensional
   (b) two-dimensional
   (c) spherical
   (d) limited.

5. Which of the statements about cells is *not* true?
   (a) All organisms consist of many cells.
   (b) Cells vary in size and shape.
   (c) The cell is the basic unit of plants and animals.
   (d) Cell structure and function depend on location.

6. Plant cells can be distinguished from animal cells because only plant cells have
   (a) cell membranes
   (b) a nucleus
   (c) vacuoles
   (d) cell walls.

7. When you are viewing white onion cells through the microscope,
   (a) the nucleus and cell wall are clearly visible
   (b) the cells must be stained with iodine for the nucleus to be seen
   (c) the cells are separated
   (d) the chloroplasts make the cells visible.

8. A dialysis bag is filled with water and placed in a beaker containing sugar solution. The bag will
   (a) burst
   (b) expand but not burst
   (c) shrink
   (d) remain the same.
9. Cells need to divide when
   (a) the surface area is too large
   (b) the chromosomes split
   (c) they get too large
   (d) they spread out.
10. The cell theory states that
   (a) cells are the basic structural unit of life
   (b) all life functions are based on cell functions
   (c) all cells are formed from other cells
   (d) all of the above are correct.

TRUE/FALSE

Write the numbers 1 to 9 in your notebook. Beside each number, write T if the statement is true and F if the statement is false. For each false statement, rewrite the statement so it becomes true.

1. A compound microscope uses a series of lenses and a light source to view an object.
2. The total magnification of a microscope is engraved on the base of the microscope.
3. When focussing a microscope, begin your examination with the high power objective lens.
4. Each cell has three main parts responsible for its basic functioning: the cell membrane, the cytoplasm, and the nucleus.
5. Chromosomes are made up of acids called ascorbic acids.
6. The greater the need for energy in a cell, the greater the number of vacuoles found in the cell.
7. The cilia of a *Euglena* help it to move through water.
8. The kidney is an organ of the excretory system.
9. The cell theory can be used to explain how substances diffuse through liquids and gases.

FOR DISCUSSION

Read the paragraphs and answer the questions that follow.

The most important function of blood is to carry oxygen from the air we breathe to the cells in the body. If cells do not receive oxygen for about seven minutes, the body will die. But a few years ago, an animal survived for several hours after all its blood was removed. Its blood had been replaced by a milky liquid made in the laboratory. This was the first time artificial blood had been developed and used.

The main ingredients of this artificial blood are fluorocarbons. Fluorocarbons are chemicals that were first used as parts of the atomic bomb project during World War II. Fluorocarbons are good carriers of oxygen. They are not poisonous, do not damage the lungs, and will carry carbon dioxide away from body cells. But this artificial blood will not carry out all the functions of the blood. It won't supply body cells with food and hormones. It won't carry all waste products, and it won't fight disease.

Artificial blood is still not ready to be used by doctors. But researchers are close to a breakthrough.

1. Blood is carried in the circulatory system of the body. What organs make up this system?
2. Oxygen dissolves in blood. Describe how this process might take place.
3. List as many functions of the blood as you can. There should be at least five functions.
4. How many of the functions you listed in question 3 can be carried out by artificial blood?
5. Name three properties that artificial blood should have in order to be effective.

The Importance of Green Plants

Green plants cover the earth. Some of them tower far above human structures. Some are little more than a few millimetres high. They grew on land before any birds, mammals, or humans existed. After we stop using any area, plants cover the traces of our presence. Why are the green plants so successful?

Without plants, our planet would be lifeless. These green organisms produce the foods that every living thing needs to survive. Plants also supply oxygen (which we need) to the atmosphere and use up carbon dioxide (which we don't need). Have you thanked a green plant today?

In this unit, you will find out why plants are so successful, how they are able to make foods, how they live, and how they may be used.

The Role of Green Plants

Key Ideas

- Green plants produce food and oxygen for all the living organisms on earth.
- Plant products are used in many aspects of our lives.
- The sale of wheat and forest products is important to Canada.
- Plants grown for enjoyment increase the beauty of our surroundings.

Green plants can do something that humans and other animals cannot do. They can trap the energy of sunlight. When plants do this, they store the energy in chemical compounds. Perhaps you have already learned that *photosynthesis* is this energy-storing process. In this chapter, you will look at the many important materials that result from photosynthesis—from lumber for making ships to medicines and clothing.

Food from Plants

Consider Figure 14.1. Here are the outlines of seven important food plants from around the world. These seven combine to produce almost two-thirds of all the foods humans eat. Do you notice any similarities among the seven? They are closely related. They all are members of the grass family. The grasses produce grains that are rich in starch and protein, which are important nutrients in our diet. Besides human food, the grasses supply food for cattle. When we eat beef, some of the energy the cattle gained from the grasses passes along to us.

What other plants or plant families provide a substantial part of the human diet? Consider Figure 14.2. Another important group of food-producing plants is the legume family. Legumes include peas, the many kinds of beans, peanuts, clover, and alfalfa. Cattle eat the clover and alfalfa, and we get their nutrients later. Legumes are a good source of protein in our diets. Other important food plants, potatoes and tomatoes, are members of the nightshade family. Potatoes are a good source of starch for energy. What other groups of food plants can you identify from the diagram?

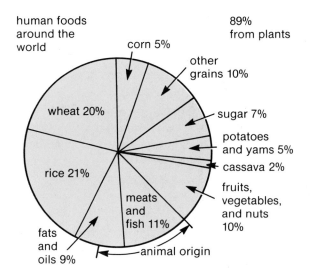

Figure 14.2 *The sources of foods that humans eat*

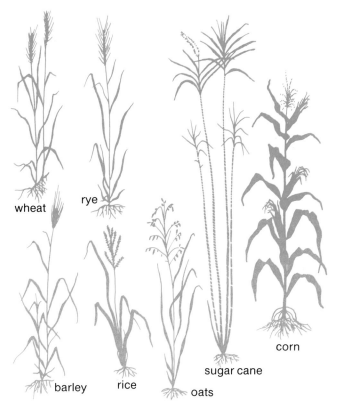

Figure 14.1 *Food plants from around the world. What important nutrient (food substance) do these plants supply to the human diet?*

Activity 14A

PART 1

Problem

How can we tell if starch is present in a substance?

Materials

a number of plant products, such as corn starch,
  wheat flour, potato flour, rice flour, arrowroot flour,
  oat flour, soya flour, rye flour, sugar, olive oil
slices of fresh vegetables, such as potato, cabbage,
  onion, carrot, turnip, squash, peas
slices of fresh fruits, such as apple, banana, orange
iodine solution in potassium iodide solution (Lugol's
  iodine solution)
test tubes
dropper (or dropping bottles)

> **CAUTION** Iodine is a corrosive liquid
> that will stain the skin,
> clothing, desk tops, and floors. Take care to
> avoid spilling. Quickly wipe up any accidental
> spills, and rinse with water.

Procedure

1. Prepare a table in your notebook for your
   observations, similar to Table 14.1. Record all of
   your observations in the table.
2. Use the corn starch first, as a control, because it
   is a substance known to be starch.
3. Place a small amount of each substance to be
   tested into a test tube. Add about 10 mL of water,
   and shake to mix thoroughly.
4. Add one drop of the iodine solution to the test
   tube and record the result.

Observations

Table 14.1 *Results of Testing with Iodine*

SUBSTANCE TESTED	OBSERVATIONS
Corn starch	
Etc.	

SAMPLE ONLY

Questions

1. What happens when iodine and a starch come
   together?
2. What substances did not react with iodine?
3. What is the test for starch?
4. List the substances you tested that contain
   starch.
5. List the substances you tested that do not
   contain starch.
6. Make a conclusion about the presence of starch
   in foods.

PART 2

Problem

Why do plants store food in seeds?

Materials

bean seeds, soaked overnight
bean seedlings that have developed their primary
  leaves
Lugol's iodine solution
scalpel or razor blade
petri dish or watch glass

Procedure

1. Place a soaked bean seed into a petri dish or watch glass.

> **CAUTION** The scalpel is very sharp. Use it very carefully.

2. Cut through the halves (cotyledons) of the bean seed to expose a fresh surface. See Figure 14.3(a).

> **CAUTION** Remember the warning about the iodine solution in Part 1.

3. Add a drop of the iodine solution to the cut surface of the bean.
4. Locate the cotyledons of the young bean plant that grew from a seed. See Figure 14.3(b). Describe the appearance of a cotyledon.
5. Place one cotyledon into the dish, and cut across it with the scalpel.
6. Add a drop of iodine solution to the cotyledon of the bean plant.

Figure 14.3

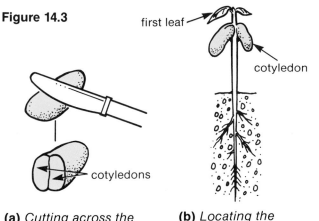

first leaf
cotyledon
cotyledons

(a) *Cutting across the cotyledons of a soaked bean seed*

(b) *Locating the cotyledons in a young bean seedling*

Observations

1. What did you observe when you added iodine to the cut surface of the bean seed?
2. Compare the appearance of the cotyledon in the bean seed with that of the young plant.
3. What did you observe when you added iodine to the cotyledon of the young plant?

Questions

1. What food nutrient must have been present in the bean seed?
2. Was this nutrient still present in the cotyledons after the seed had grown into a young plant?
3. What had happened to the nutrient that was stored in the seed?
4. What had happened to the chemical energy of the starch?
5. Why do you think that the seed plants developed a supply of stored food in their seeds?

Self-check

1. Why are green plants so important for life on earth?
2. Name the process that you described in your answer to question 1.
3. What plant family produces two-thirds of all human food?
4. Name another family of plants that provides humans and livestock with a large number of protein-rich foods.
5. What is the value to plants of storing starch, protein, and oils in seeds?
6. When seeds sprout, what becomes of the stored energy?
7. Describe the test for the presence of starch.
8. Name five foods that contain starch, and three foods that do not contain starch.

Useful Plant Products

Plants and Medicine

Long ago, native peoples of North America discovered the medicinal use of the willow. They chewed willow twigs to reduce aches and pains. When modern scientists identified the pain-killing ingredient in willow twigs, they named it "salicin," after the scientific name of the willow, *Salix*. Today we use acetyl*salicy*lic acid (ASA), or "aspirin," to relieve pain.

For thousands of years, people around the world have used plants to treat a variety of illnesses. In fact, written records of medicinal uses of plants date from at least 400 B.C. These records occur in Egyptian, Hebrew, and Chinese documents. Table 14.2 summarizes information about a few of the plants used for medicinal purposes around the world.

Insecticides

Many insects eat plants, sometimes killing the plants. Can plants save themselves by killing insects? In fact, several kinds of plants have developed chemicals, called **insecticides**, that kill insects. The best known insecticide derived from a plant is **pyrethrum**. People have been using it to kill insects for 200 years. It is made from the flowers of a plant that looks like a common daisy (Figure 14.4).

Another common insecticide obtained from a plant is called **rotenone**. Rotenone is made from *Derris*, a plant that grows in Malaya. (Incidentally, *Derris* is also the source of poisons that hunters in the jungles used on their arrows.)

Why do we still use poisons produced by plants when chemists can make insecticides? Since pyrethrum and rotenone are produced by living things, they are easily broken down by other living things. They will not last forever.

Table 14.2 *Some Drugs Obtained from Plants*

DRUG	PLANT SOURCE	ORIGIN OF THE PLANT	MEDICINAL USE
Aconite	*Aconitium napellus** (monkshood) roots	mountains of Europe and Asia	used externally for rheumatism and internally for fever and pain
Ginseng	*Panax schinseng* (ginseng) roots	Eastern Asia	used as a stimulant, "a wonder cure"
Ipecac	*Cephaelis ipeccuanha* roots	Latin America	used to treat amoebic dystentery and diarrhea
Quinine	*Cinchona calisaya* bark	South American Andes	used to prevent and control malaria
Cocaine	*Erythroxylon coca* leaves	Peru and Bolivia	used as a local anaesthetic
Digitalis	*Digitalis purpurea* (foxglove) leaves	Europe and North America	used to treat heart disorders
Atropine	*Datura stramonium* (Jimsonweed) leaves	North America	relief of colds and hay fever, dilates pupils
Scopolamine	*Datura stramonium* (Jimsonweed) leaves	North America	anaesthetic, used for motion sickness and relief of Parkinson's disease

*Scientific names of organisms are two words in Latin form. The first name is always capitalized. The second is never capitalized.

Figure 14.4 *A field of pyrethrum flowers growing in East Africa. These daisy-like flowers make the chemical, pyrethrum, that is used as an insecticide.*

Figure 14.5 *Natural insecticides that contain rotenone and pyrethrum*

They will not be able to poison the environment forever. Scientists call this property **biodegradable**. Bacteria and fungi, organisms that cause decay, break down ("degrade") such substances into simple materials that are harmless. On the other hand, synthetic chemicals are not always easily broken down and may last for a long time in the environment. Bacteria and fungi cannot break down all synthetic chemicals. Synthetic chemicals may add to the pollution of our soils and waterways. To avoid this pollution, many gardeners and agriculturists prefer to use natural insecticides (Figure 14.5).

Wood

Why is wood so desirable for building? It has been used for centuries because it can be easily cut and shaped. We can quickly construct homes of wood. Besides, wood is a good insulator because it prevents the rapid transfer of heat. Wooden structures will last a long time if they are kept dry.

But wood does not last forever—it is biodegradable. Water allows the organisms of decay, bacteria, and fungi to attack the wood, making it soft. Certain insects like carpenter ants, termites, and many beetles make tunnels through wood and weaken it.

For hundreds of thousands of years, people have burned wood to get warm and to cook food. Wood is a source of energy. Wood is a **renewable resource**. We can always grow more trees to replace those we cut down. Today, however, humans are destroying forests faster than they are replanting them. In many parts of the world, wood is the only source of heat for people. As the population outgrows the resource, people go farther and farther to cut trees. Once the trees are gone, the soil is soon eroded. In many places, humans have created deserts where forests once stood.

WOOD PRODUCTS

Besides lumber and fuel, there are many other uses of wood. Figure 14.6 shows some of the many products that can be obtained from wood. Today, petroleum is used instead to make many of these products. Unfortunately, petroleum is a **non-renewable resource**. That is, the amount of petroleum in the world is becoming less. No new petroleum is being made to replace that which is being used up. Perhaps in your lifetime we will have to return to making these products from wood.

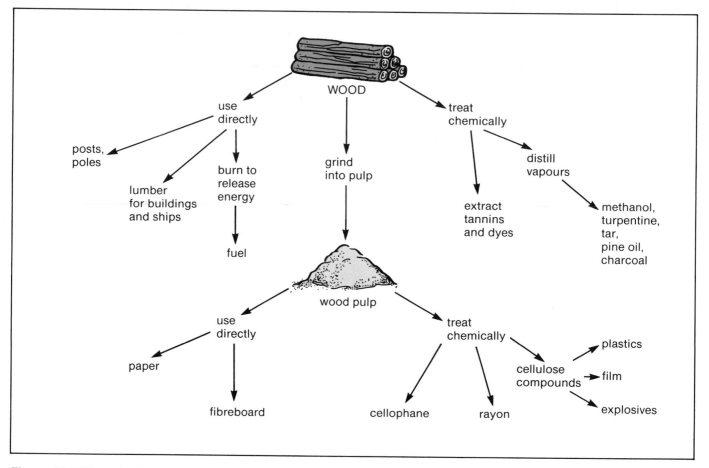

Figure 14.6 *Wood is the source of many useful products.*

You learned about the process of distillation in Chapter 7. In this demonstration, you will see how useful products are obtained from wood, using a process called *destructive distillation*. It is called "destructive" distillation because the wood is destroyed.

Problem

How can distillation separate useful products from wood?

Materials

wooden splints
large Pyrex test tube
safety goggles
gas-collecting bottle
Bunsen burner
stand
adjustable clamp
one-hole stopper to fit test tube
two-hole stopper to fit gas bottle
glass tubing as shown in Figure 14.7
litmus or pH test paper

CAUTION Wear safety goggles during this activity.

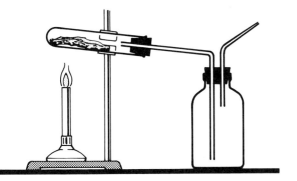

Figure 14.7 *Apparatus for the destructive distillation of wood*

Procedure

1. Your teacher will break the wooden splints in half and place them into the test tube.
2. The apparatus will be set up as shown in Figure 14.7. The test tube should tilt slightly downward. The gas-collecting bottle should be connected so that the delivery tube from the test tube extends almost to the bottom.
3. Your teacher will heat the wooden splints strongly.
4. Note any residues that appear in the test tube and gas bottle as your teacher heats the splints.
5. Your teacher will use a lighted wooden splint to try to ignite any gas that may escape through the glass jet from the gas bottle. Then the Bunsen burner will be turned off.
6. Any liquid in the gas bottle will be tested with litmus or pH paper.
7. When the contents of the test tube have been emptied, your teacher may try to ignite them with a Bunsen burner flame.

Observations

1. Describe the appearance of the residues in the test tube and the gas bottle.
2. Describe what you saw when a lighted splint was brought to the glass jet.
3. Describe the effect of the distillate on the litmus or pH paper.
4. What happened when the remains of the wood were held in the flame of the burner?

Questions

1. What gas might have been ignited at the jet?
2. What were some of the liquid products in the distillate?
3. What solid was left after distillation?

PAPER MAKING

One of the products of wood in Figure 14.6 is pulp, which is used to make paper. Black spruce trees (Figure 14.8) provide much of the Canadian wood pulp. Other plant fibres have been used in making paper, the oldest being papyrus. The long stems of papyrus reeds were used to make a kind of paper in Egypt over 4000 years ago. To make smooth, high-quality papers today, cotton and linen fibres are mixed with the wood fibres.

The wood fibres of used paper can be used again to make recycled paper. This helps to save cutting down so many trees. Recycled paper is often used in exercise books and for "second copy" typing paper. There are still some problems with the use of recycled paper. Because recycled fibres are shorter, the quality and strength of recycled paper is not as good as that of the original paper. It is also difficult to remove inks from paper during recycling. For this reason, many recycled paper products are grey or coloured.

Other Products from Plant Materials

What do these products have in common: clothing, rope, varnish, squash balls, some paints? They all are made—in whole or in part—from plant materials. Table 14.3 lists the sources, properties, and main uses of plant materials you use every day. Figure 14.9 shows several of the sources of these materials.

Figure 14.8 *Black spruce trees provide the fibres from which most paper is made today.*

Table 14.3 *Sources, Properties, and Main Uses of Plant Materials*

PLANT MATERIALS	SOURCE	PROPERTIES	MAIN USES
Cotton	Fibres from flowers of cotton plant	Keeps its shape well (but wrinkles easily); lets moisture escape readily	Clothing, thread
Linen	Fibres from stem of flax plant	Strong (but wrinkles easily), long-lasting	Tablecloths, thread, fish nets, handkerchiefs, fire hoses
Sisal	Fibres from leaves of sisal plant	Coarse, strong	Baskets, mats, rope, twine
Jute	Fibres from leaves of jute plant	Coarse, strong	Mats, burlap sacks, carpet backing, rope, twine
Rubber	Milky juice (latex) found in many different plants (including rubber plant)	Elasticity, water-resistance	Tires, balls for sports, other inflatable objects
Gums	Sap from trees such as gum tragacanth, balsam, and acacia	Forms a colloid that, when mixed with water, thickens and becomes sticky when dry	Ice cream smoother, adhesives, paper stiffener, paints
Resins	Sap from trees such as pine and spruce	Water-resistance; forms chains of molecules that can be shaped by heating (plastics)	Varnishes, sealing wax, paper stiffener, protective coatings
Dyes	Stems, seeds, shoots, leaves, berries, roots, bark of many kinds of plants and trees	Adds colour to products; binds with fibres	Colour for paints and clothing and other fabrics

Figure 14.9

(a) *The cotton flower produced the fibres for making cotton thread and cloth.*

(b) *A sisal plant growing in East Africa*

(c) *The rubber plant, like commercial rubber trees, produces natural rubber.*

(d) *Commercial gum trees*

Ideas and Applications

Sometimes, dyes only develop their full colour when they are mixed with air to allow oxygen to react with the materials. An example of this is the deep blue dye produced from the indigo plant. (This is the dye used to make "blue jeans" blue.) The boiled extract of the leaves and stems is yellow at first. After it reacts with oxygen in air, it turns indigo blue.

Self-check

1. What plants have a natural pain killer in their twigs?
2. Name three drugs or medicines that plants produce.
3. (a) Name two insecticides that plants produce.
   (b) Why would it be an advantage to a plant to make an insecticide?
4. Why do some people prefer to use natural insecticides rather than the synthetic kind?
5. What properties of wood make it desirable for building?
6. List four uses for wood pulp.
7. List four products obtained from wood (not including wood pulp).
8. Name three plants that supply us with fibres that can be used for making cloth, rope, or twine.
9. What are gums and resins used for?

Plants and Canada's Economy

Canada's Wheat

Canada is one of the world's leading producers of wheat (Figure 14.10). Until 50 years ago, wheat ranked first in dollar value of all Canada's exports. In 1982, Canada's wheat exports brought in over four billion dollars. Wheat made up 46% of all the agricultural products exported in 1983.

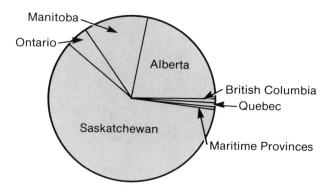

Figure 14.10 *Canada's wheat production by province. Which provinces produce most of Canada's wheat?*

There are two major kinds of wheats: hard wheats and soft wheats. The hard wheats are good for making pasta products like macaroni, noodles, and spaghetti. Soft wheats become flour for use in baking bread and pastry.

Canadian scientists have been active in breeding new wheat varieties that will resist plant diseases and parasites, and varieties that will mature earlier, before the frost. In 1904, Dr. Charles Saunders, a scientist with the Canadian Department of Agriculture, crossed Red Fife wheat with Red Calcutta wheat from India. The result, **Marquis wheat**, became the leading crop for many years. But wheat diseases, like human diseases, keep changing. Scientists must keep developing new varieties of wheat to resist the latest varieties of disease and insect pests.

Canada's Forests

Canada has always been a major exporter of forest products. In 1985, exports of newsprint, paper goods, pulp, and lumber brought in $13 billion and made up 16% of all Canada's exports. Forests occupy 44% of the total area of Canada (Figure 14.11). British Columbia is the leading forest province. In 1982, B.C. produced 65% of all the lumber and most of the softwood plywood. Ontario and Quebec are the next largest producers, exporting most of the ground wood pulp, newsprint, and hardwood plywood.

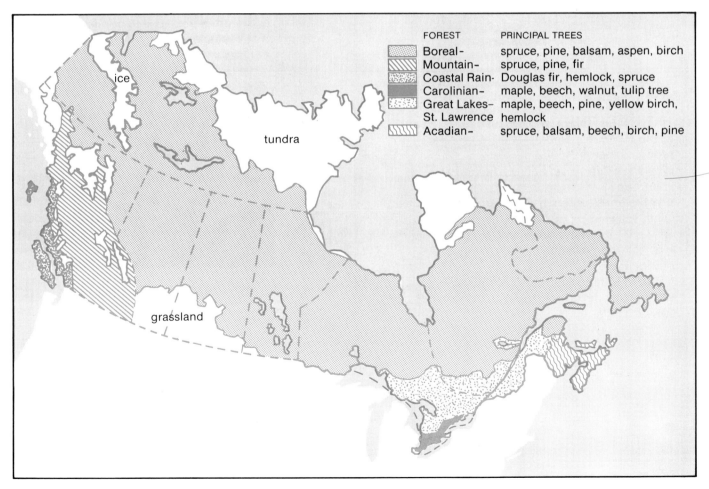

Figure 14.11 *The forest regions of Canada*

Softwoods include pine and fir, which are usually used for constructing buildings. Hardwoods, such as oak, birch, and maple, are used to make furniture.

Canada's forests provided jobs for 64 000 workers in 1985. Many workers are still needed to fight forest fires and to plant new trees. After the trees are cut, the forests can be renewed. In Canada, however, they have not been renewed. To make our forests a renewable natural resource, Canadians must ensure that the planting of new trees makes up for the harvested trees. Researchers continue to seek effective ways of controlling insect pests, such as the spruce budworm, and of controlling fungous disease, such as the pine rust.

Self-check

1. Why is wheat important in Canada's economy?
2. Name two major kinds of wheats, and name two products that are made from each.
3. Why were Fife wheat and Marquis wheat important?
4. When plant breeders try to improve wheat varieties, what properties do they look for?
5. Why are forests important in Canada's economy?
6. Name the three leading provinces in forest production.
7. Name four important forest products.
8. What problems do forestry researchers try to solve?

Science at Work

Dr. Neil Towers is a world-renowned expert on plants at the University of British Columbia. He has found that some of the natural pesticides plants make to protect themselves are activated by light. In experiments, Dr. Towers applied marigold root juice to containers filled with water and mosquito larvae. Then the containers were exposed to light. When the larvae surfaced, the poison juice they had taken into their bodies was activated by the sunlight, and the larvae were killed. It may be that one day, light-activated, natural pesticides will be available for commercial use to control the numbers of pest species.

Dr. Neil Towers

Protecting Plants in the Home

Plants add to the beauty of our homes and communities (Figure 14.12). It is hard to think of a world without green plants. However, like any other organism, plants may become diseased. Just as with humans and animals, diseases of plants are caused by bacteria, fungi, viruses, and parasitic worms. And as with humans and animals, the best way to prevent diseases of plants is to prevent the disease organisms from reaching their hosts.

You can prevent the soil from bringing disease to plants by sterilizing it in an oven at 80°C for 20 minutes. Another way is to pour enough boiling water through the soil until its temperature rises to 75°C or more. In each of these methods, you would use a thermometer, and then let the soil cool to room temperature and dry out before planting. Or, you can buy sterilized potting soil in garden stores, hardware stores, or supermarkets.

Figure 14.12 *Plants bring beauty and life into our surroundings.*

You can prevent plants from bringing disease by keeping each new plant away from your other plants until you are sure it is healthy. Use only healthy plants for making cuttings.

The way you water plants is important for preventing disease. If plants are left standing in water, the roots cannot get oxygen. Then fungus or bacteria can attack the roots and make them rot. If the soil becomes too dry, or if the air is too hot or dry, the plant will weaken and disease may strike. Water that is splashed on leaves may favour the growth of grey mould or white powdery mildew (Figure 14.13). If you see any leaves with grey mould or white mildew, you should cut them off and get rid of them. A light sprinkling of powdered sulphur will kill these fungi. Be sure to get rid of any badly diseased plants.

There are some insects and other small animals that will harm your house plants by eating holes in the leaves or sucking the plant juices. Do you remember the natural insecticides, pyrethrum and rotenone, discussed in Section 14.3? These can be dusted on plants to kill white flies and red spider mites. Mealybugs, aphids, and scale insects are better treated with a spray of soapy water or liquid soap containing 50% methanol or ethanol. For soil pests, such as

Figure 14.13 *The fungus, powdery mildew, has attacked the leaf on the left.*

Table 14.4 *Protecting Plants from Pests*

PEST		DESCRIPTION	REMEDY
Aphids (3-10 mm)		Tiny plant "lice," green, pink, brown, black, or white, that drop a sticky solution	Bathe or spray plant with warm soapy water, or dust with pyrethrum or rotenone.
Fungus gnats (5-10 mm)		Small black flies, the size of fruit flies, with white larvae	Drench soil with diazinon or Malathion.
Mealybugs (2-3 mm)	waxy cover	Cottony flecks of white wax on leaves and stems	Bathe or spray with soap and alcohol.
Mites (0.3-0.8 mm)		Tiny specks that make cobwebs around and under leaves	Spray with Kelthane every week for four weeks
White flies (2-3 mm)		Insects that fly off leaves like puffs of white dust.	Same as for aphids.

fungus gnats, drench the soil in diazinon solution. Table 14.4 summarizes ways of protecting house plants from insects.

Farmers and other commercial growers are also concerned with plant diseases and pests. You will find out more about their methods of pest control in Chapter 18.

Self-check

1. What kinds of organisms cause plant diseases?
2. What are five ways of preventing plant diseases?
3. Name four kinds of insects that attack house plants.
4. Describe three ways of killing insects on house plants.

Chapter Objectives

NOW THAT YOU HAVE COMPLETED THIS CHAPTER, CAN YOU DO THE FOLLOWING?	FOR REVIEW, TURN TO SECTION
1. State why plants are important for all forms of life.	14.1
2. List seven ways that plants are used in human lives.	14.2.
3. Describe how to test for the presence of starch.	14.2
4. Name three drugs produced from plants, and tell what disease or condition each is used to treat.	14.3
5. Define *insecticide*, name two natural insecticides, and tell why natural insecticides are important.	14.3
6. List four uses of wood and four products that can be obtained from wood.	14.3
7. Explain the meaning of *renewable resource* and *non-renewable resource*, and give an example of each.	14.3
8. Name four plants from which fibres are obtained for making into fabrics, rope, or twine.	14.3
9. Name two plants from which latex is obtained, and tell which of them can be used commercially for the production of rubber.	14.3
10. Distinguish between a gum and a resin, and state two uses of each.	14.3
11. Explain the importance of wheat in Canada's economy.	14.4
12. Explain the importance of forests in Canada's economy.	14.4
13. Name three kinds of organisms that cause diseases of house plants, and describe four ways that you can reduce the chance of house plants getting diseases.	14.5
14. Name three kinds of insect pests that attack house plants, and describe three ways of killing the insects.	14.5

Words to Know

insecticide
pyrethrum
rotenone
biodegradable

renewable resource
non-renewable resource
Marquis wheat

Tying It Together

1. List five ways that humans use plants.
2. Name five plants of the grass family.
3. What would you have to do to test a substance to see if it contains starch?
4. List five plant products that contain starch and three plant products that do not contain starch.
5. What happens to the starch in a seed when the seed grows into a new plant?
6. What pain-killing drug did North American Indians obtain when they chewed willow twigs?
7. What unusual properties are shared by the plants *Pyrethrum* and *Derris*?
8. Why do some people prefer to protect their plants with biodegradable insecticides rather than with synthetic chemicals?

9. (a) Why is wood such a convenient fuel?
   (b) Could wood ever become a renewable source of fuel? Explain.
10. (a) What four properties of wood have made it desirable for building?
    (b) What are two disadvantages of using wood for building?
11. Besides fuel and lumber, what are three other uses for wood?
12. List three uses of plant fibres. For each, give an example of a plant that produces the fibres.
13. (a) What plant is the source of natural rubber?
    (b) What substance in this plant provides the material from which rubber is made?
    (c) What other plant contains the material you named in (b)?
14. What did Charles Saunders contribute to the growing of wheat in Canada?
15. (a) Name three kinds of organisms that cause disease in house plants.
    (b) How can you prevent or control diseases in house plants?
16. (a) Name three kinds of insects that attack plants.
    (b) How can you control insects that attack plants?

Applying Your Knowledge

1. Why might some plants, such as *Pyrethrum*, and *Derris*, produce natural insecticides?
2. Wood is classified as hard or soft depending on its hardness. Hardwoods are harder to cut and shape and are more resistant to wear. Hardwoods are produced by such trees as oak, birch, and maple. These are broad-leafed deciduous trees (trees that drop their leaves each fall). Softwoods, such as pine, fir, and spruce are produced by conifers (cone-bearing trees with needle-like leaves that remain over two or three years). List four uses each for hard and soft wood.
3. Wheat rust is a fungous disease that attacks the stems, leaves, and kernels of wheat. In a typical year, this disease reduces the value of the wheat crop by about 10%. How would scientists go about producing varieties of wheat that would resist the disease?

Projects for Investigation

1. Spruce trees grow in the Boreal forest, a broad zone that extends right across Canada (Figure 14.11). They are an important source of the pulp that is used in making paper. The spruce budworm (Figure 14.14) burrows into the buds at the tips of the stems, killing many trees. Consult some reference books to find out how foresters attempt to control the spruce budworm. What is meant by "biological control" of an insect pest, in contrast to chemical control? How could the biological control of the spruce budworm be accomplished over such a vast forest?
2. Recently, some anthropologists (scientists who study human origins, development, and customs) have suggested that barley was the plant that got humans to settle down and to stop being wandering hunter-gatherers. Barley seeds might have fermented naturally to produce beer. Find out how beer is made today. Suggest how primitive humans might have made it, and why it might have changed their lifestyle.
3. People have been growing wheat for only about 8000 years. Before the beginning of agriculture, how did people obtain their food? How did the beginning of agriculture change human lifestyle and population? What later changes affected lifestyle and population greatly?

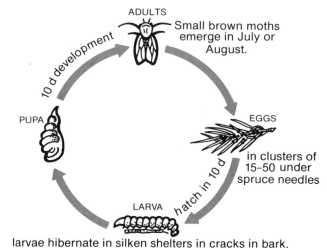

Figure 14.14 *The life cycle of the spruce budworm*

Plants at Work

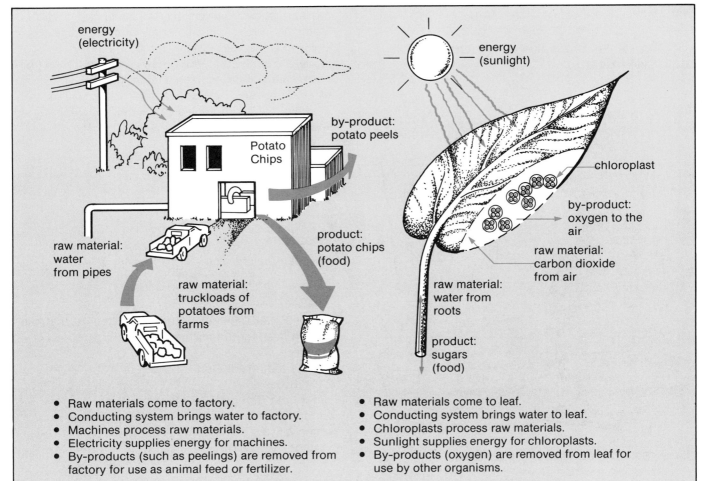

- energy (electricity)
- Potato Chips
- by-product: potato peels
- raw material: water from pipes
- raw material: truckloads of potatoes from farms
- product: potato chips (food)
- energy (sunlight)
- chloroplast
- by-product: oxygen to the air
- raw material: carbon dioxide from air
- raw material: water from roots
- product: sugars (food)

- Raw materials come to factory.
- Conducting system brings water to factory.
- Machines process raw materials.
- Electricity supplies energy for machines.
- By-products (such as peelings) are removed from factory for use as animal feed or fertilizer.

- Raw materials come to leaf.
- Conducting system brings water to leaf.
- Chloroplasts process raw materials.
- Sunlight supplies energy for chloroplasts.
- By-products (oxygen) are removed from leaf for use by other organisms.

Key Ideas

- The green parts of plants produce starch by photosynthesis.
- Leaves are especially adapted to carry out photosynthesis.
- Stems and roots also have functions that are essential to the life of plants.

Plants produce the foods that every living thing needs. How do plants make food? Another way to ask this question is to ask what a leaf and a potato chip factory have in common. It may seem odd to compare a leaf with a potato chip factory. But this comparison will help you begin to understand how plants make food. In this chapter, you will look at the details of how leaves do their work and how stems and roots help them.

Leaves and Photosynthesis

Comparing a leaf to a potato chip factory gives you all the pieces to the puzzle of how leaves do their work. See Figure 15.1. By the end of this section, you will know how to put the pieces of the puzzle together.

How Does Light Make Photosynthesis Go On?

What Products Do Plants Make During Photosynthesis?

What Materials Do Plants Need For Photosynthesis?

What Does Chlorophyll Do In Photosynthesis?

What Is The Source Of The Oxygen Given Off In Photosynthesis?

Figure 15.1 *How do the puzzle pieces fit together?*

 **Activity 15A**

Think back to Activity 14B in Chapter 14. In that activity, you learned how to identify starch. (Iodine solution, added to starch, produced a blue-black colour.) Now you will use this test to find out what leaves need to make starch. Before you begin, you must know two things. First, the living cell membranes of a leaf will not let iodine solution pass easily into the cells. Therefore, you will have to kill the leaf first. Second, the green colouring of the leaf —the **chlorophyll**—makes it difficult to see the blackish colour of the starch–iodine compound. Thus, you will try to dissolve the chlorophyll in a solvent and remove it from the leaf.

PART 1

Problem

How can the starch test be applied to a leaf?

Materials

leaf of a plant, such as geranium or *Coleus*, that has been in sunlight for 2 h or artificial light for 24 h
ethanol
Lugol's iodine solution
2 test tubes
water bath (hot plate and beaker)
test tube holder
watch glass
forceps

Procedure

1. In your notebook, make a table similar to Table 15.1 for your observations.
2. Hold the leaf stalk with the forceps. Dip the leaf into boiling water for a few seconds (Figure 15.2).

CAUTION The vapour is flammable. There should be no open flame nearby.

Figure 15.2 *Steps in testing a leaf for starch*

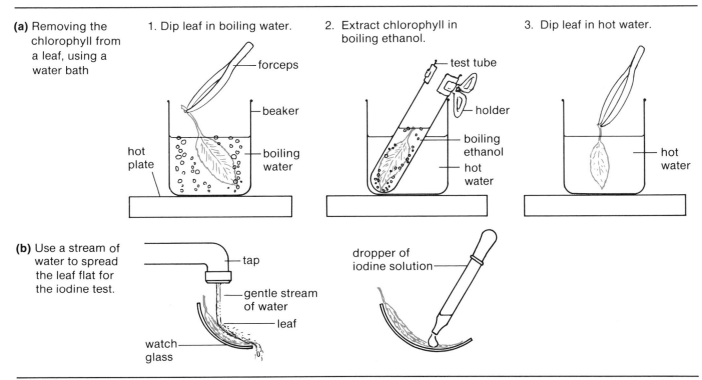

(a) Removing the chlorophyll from a leaf, using a water bath

1. Dip leaf in boiling water.

forceps

beaker

hot plate

boiling water

2. Extract chlorophyll in boiling ethanol.

test tube

holder

boiling ethanol

hot water

3. Dip leaf in hot water.

hot water

(b) Use a stream of water to spread the leaf flat for the iodine test.

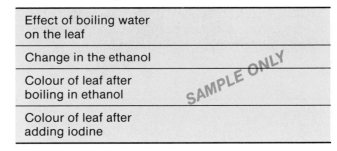

tap

gentle stream of water

leaf

watch glass

dropper of iodine solution

3. Roll up the leaf, and place it into the bottom of a test tube. Cover the leaf with ethanol. Place the test tube into the water bath, and leave it until the ethanol boils for a few minutes.
4. When the ethanol becomes coloured, use the test tube holder to remove the test tube from the hot water. Pour the ethanol into the second test tube, and observe its colour. Now pour the ethanol into the container provided by your teacher.
5. With the forceps, remove the leaf from the test tube and spread it out flat across the watch glass. Use a gentle stream of water from a tap to help spread the leaf. Record its appearance.
6. Recall the CAUTION about the use of iodine in Activity 14A. Add a few drops of Lugol's iodine solution to the leaf. Record any change in the appearance of the leaf.

Observations

Table 15.1 *Testing a Leaf for Starch*

Effect of boiling water on the leaf	
Change in the ethanol	
Colour of leaf after boiling in ethanol	
Colour of leaf after adding iodine	

SAMPLE ONLY

Questions

1. Why was the leaf dipped into boiling water?
2. Is chlorophyll soluble in water?
3. Why was the leaf boiled in ethanol?
4. Is chlorophyll soluble in ethanol?
5. What can you infer from the colour change after the iodine solution was added to the leaf?

PART 2

Problem

Is light needed for the leaf to make starch?

Materials

leaf of a *Coleus* or geranium that has been in the
  dark for 2–3 d
other materials as for Part 1

Procedure

1. In your notebook, make a table similar to Table
    15.2 for your observations.
2. Test the leaf for starch as you did in Part 1, Steps
    2 to 6.

Observations

Table 15.2 *Testing a Leaf from the Dark for Starch*

Colour of leaf after adding iodine

Questions

1. What does the starch test show about this leaf
    from the dark?
2. Compare the starch test of Part 2 with that of
    Part 1. What statement can you make about the
    need for light in the making of starch?
3. In this activity, what is the "control"?

Challenge

Part 1 and Part 2 of Activity 15A established the
need for light in making starch. Design a
procedure in which you could combine these
two parts into ONE experiment to do the same
thing. What is the control in your experiment?

Light is necessary for plants to make starch. Light is
a form of energy. Starch, the final product of
photosynthesis, is a chemical compound that
contains stored energy. In this activity, you will find
out which cells in a leaf carry out photosynthesis. In
effect, you will be looking for the "machine" that
traps the energy of light.

Problem

In what parts of a leaf does photosynthesis occur?

Materials

variegated leaf (one that is partly green and partly
  white) that has been exposed to sunlight for 2 h or
  to artificial light for 24 h (*Coleus, Impatiens*, or
  geranium plants are suitable.)
materials for testing a leaf for starch, as in Activity
  15A, Part 1

Procedure

1. In your notebook, prepare a table for your
    observations similar to Table 15.3.
2. Make a sketch of the variegated leaf in your
    observation table. Label the areas that are green
    and those that are not.
3. Test the leaf for starch, following steps 2 to 6 of
    Activity 15A, Part 1.
4. Make a sketch of the leaf showing the areas that
    gave a positive starch test and the areas that
    gave a negative test.

Observations

Table 15.3 *Testing a Variegated Leaf for Starch*

APPEARANCE OF LEAF BEFORE TEST	APPEARANCE OF LEAF AFTER STARCH TEST
SAMPLE ONLY	

Questions

1. What can you infer about the green areas of the leaf?
2. What can you infer about the parts of the leaf that were not green?
3. The chemical compound that makes the green colour in plants is called chlorophyll. What do you think is the relationship between chlorophyll and the process of photosynthesis?

The Role of Chlorophyll in Photosynthesis

The green chemical compound, chlorophyll, has a special ability that not many other substances have: it can trap the energy of light. That is why the green cells of plants are able to carry on photosynthesis. Besides plants, algae and cyanobacteria also make chlorophyll and carry on photosynthesis. (Algae are the pond scums and seaweeds consisting of one cell or a number of similar cells. Cyanobacteria are simple, one-celled organisms that lack organelles.) All forms of life depend on these producers for the energy they must have to keep alive.

The Chemistry of Starch

What materials are needed to make starch? Starch is a large molecule, made up of dozens of sugar-like units (Figure 15.3.). Sugar and starch can each be converted into the other. Sugar is the soluble form, used for moving energy-rich materials from place to place. Starch, the insoluble form, is used for storage in plants.

Figure 15.3

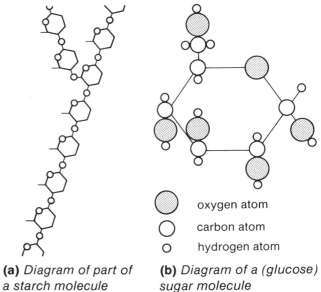

- oxygen atom
- carbon atom
- hydrogen atom

(a) *Diagram of part of a starch molecule*

(b) *Diagram of a (glucose) sugar molecule*

Starches and sugars contain only carbon, hydrogen, and oxygen. Where does the plant get these elements to make into the larger compound?

Two possible sources of the raw materials for photosynthesis are the air around the plant and the water the plant absorbs. In Activity 15C, you will use a plant that normally grows under water. In this way you can concentrate on the role of the gases that dissolve in the water. Carbon dioxide from the air dissolves in water to form carbonic acid. The acid/base indicator bromothymol blue will show when carbon dioxide is present (it turns yellow), or when it has been used up (it turns green or blue).

Activity 15C

Problem

Is carbon dioxide the source of the carbon that plants make into starch?

Materials

healthy water plants, such as *Elodea, Cabomba,* or *Myriophyllum*
4 test tubes
test tube rack
4 stoppers to fit test tubes
water that has been boiled to drive out dissolved gases and then cooled to room temperature
bromothymol blue indicator solution
2 straws

Procedure

1. In your notebook, make a table similar to Table 15.4. Use this table to record all your observations.
2. Fill each of the test tubes nearly full with water that has been boiled and cooled. Label the test tubes A, B, C, and D as in Figure 15.4.

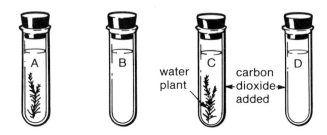

Figure 15.4 *Apparatus set up for Activity 15C*

3. Add five drops of the bromothymol blue indicator to each of the test tubes. It should appear pale green in each tube.
4. Blow gently through the straws to bubble carbon dioxide gas through the water of test tubes C and D. Keep blowing gently until the colour of the indicator just turns from green to yellow.
5. Select two healthy sprigs of water plant that appear to be the same, and place them into test tubes A and C.

6. Seal all four test tubes with their rubber stoppers.
7. Place all the test tubes where they will get fairly bright light, but not direct sunlight.
8. Observe the colours of the indicator in the four test tubes after a day in the bright light. Look closely at each of the tubes and the water plants for any other changes.

Observations

Table 15.4 *Observations of Changes in Activity 15C*

TEST TUBE	A	B	C	D
Colour at start				
Colour after 1 d in light				
Other observations				

Questions

1. What are the controls in this experiment?
2. What can you infer from any colour changes?
3. What can you infer from any tubes that did not change?
4. Why was the water boiled for this activity?

Challenge

Suppose that the four test tubes of Activity 15C had been placed in the dark instead of in the light. What observations would you expect to make? What might you expect to conclude?

In Activity 15C, two of the test tubes contained healthy green plants, and both were exposed to the light. The indicator showed that a chemical change took place in only one tube. This was test tube C, the one that had some carbon dioxide. Only in test tube C was carbon dioxide disappearing. From this you may infer that carbon dioxide is necessary for photosynthesis to occur. Did you make any other observations of changes in test tube C?

Perhaps there was an indication of another product of photosynthesis in the form of a gas that may have appeared as tiny bubbles. This gas is oxygen. It would take a lot of healthy water plants before you could collect enough oxygen to identify. Do you remember the glowing splint test for oxygen from Unit Three? Under ideal conditions, this test can be used to show that plants release oxygen when photosynthesis is going on (Figure 15.5). If the oxygen came from water molecules, what element of the water must the plant have used to make starch?

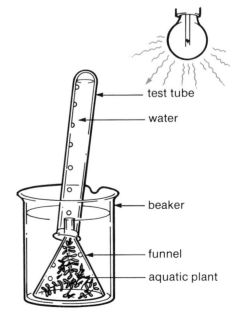

Figure 15.5 *Production of oxygen by a water plant during photosynthesis*

Summary: How Does Photosynthesis Work?

The activities of this chapter have shown that light, chlorophyll, and carbon dioxide are needed for photosynthesis to take place. The products of photosynthesis are starch and oxygen.

No mention has yet been made of water. Without sufficient water, the plant wilts and photosynthesis stops. Complex experiments have shown that water is indeed needed for the process. Water molecules are split when light strikes the chlorophyll molecule. Water is the source of the hydrogen that goes into starch molecules (Figure 15.6).

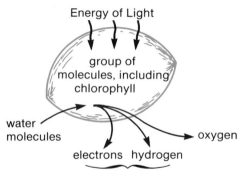

Figure 15.6 *Energy is trapped when light strikes a chlorophyll molecule.*

The process of photosynthesis can be stated in a word equation:

(Raw Materials   Energy Source   Products)

$$\text{carbon dioxide} + \text{water} \xrightarrow[\text{(chlorophyll)}]{\text{light}} \text{starch} + \text{oxygen}$$

The name, *photosynthesis*, is derived from three Greek words:

photos = light
syn = together
thesis = putting

Thus, we may define photosynthesis as the putting together of complex substances in the presence of light and chlorophyll.

Self-check

1. Copy Table 15.5 into your notebook, and then complete it.

Table 15.5 *Experimental Evidence of Photosynthesis: A Summary*

CONCEPT	WHICH ACTIVITY ESTABLISHED THE CONCEPT?	
	NUMBER	METHOD
1. Light is needed for photosynthesis.		
2. Chlorophyll is needed for photosynthesis.		
3. Carbon dioxide is needed for photosynthesis.		
4. Starch is produced during photosynthesis.		
5. Oxygen is produced during photosynthesis.		

2. Table 15.6 compares photosynthesis with a manufacturing process that might go on in a factory. Copy the table into your notebook and complete it.

Table 15.6

	FACTORY	GREEN CELLS OF A PLANT
Raw materials		
Source of energy		
Machines		
Products		
Working hours		

Looking Inside a Leaf

Photosynthesis occurs in all cells that contain chlorophyll. Chlorophyll occurs in the green cells of leaves, stems, and some parts of flowers and fruits. Green cells also include most of the cells of the simpler plants: mosses and ferns. And they include the cells of algae: the pond scums, seaweeds, and lichens. But leaves are the parts of the higher plants specialized for photosynthesis.

The Parts of a Leaf

Most leaves consist of a broad thin part called the leaf **blade**. Because the leaf is broad, it has a larger area for catching the light. And because the blade is thin, none of its cells are very far away from the air or the light. The blade is supported by a main vein, or **midrib**, and often with a branched network of finer veins. The veins bring the water close to every cell and carry away the sugars. The midrib usually becomes a leaf stalk, or **petiole**, that joins the stem of the plant (Figure 15.7). The petiole turns the blade to catch the most light.

Just where is the machinery that carries out all that chemistry? How do materials and products get into and out of the leaf? How does photosynthesis go on in particular leaf cells? In Activity 15D, you will investigate how the structure of a leaf is adapted to its functions. For this study, you will use the microscope to examine the ways the cells of a leaf are specialized. You will look first at sections cut through leaves.

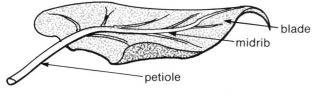

Figure 15.7 *The parts of a leaf*

Problem

How are the cells of a leaf adapted for their functions?

Materials

fresh leaf
prepared stained slide of leaf sections
razor blade, single edged
microscope
microscope slide
dropper

> **CAUTION** Use the razor blade very carefully to avoid cutting yourself.

Procedure

1. With the razor blade, cut as thin a section as possible across the leaf (Figure 15.8). Place the section on its edge on the microscope slide, so that you will be able to look into its interior. Add a drop of water to the section.

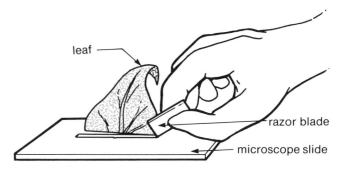

Figure 15.8 *Preparing a thin section of a leaf*

2. Examine the cross section of the leaf under the microscope, using the low- and medium-power objectives (4x and 10x). Sketch what you see. Note which cells contain chlorophyll and which cells do not.

3. Replace your section of the leaf with a prepared slide that has been cut thinly for easier viewing. Move the slide along as you view with the lowest-power objective. Notice the "bumps" where the leaf is thicker. These are the veins that form a branching network, carrying both water from the roots and the sugars produced in photosynthesis. Identify the top and bottom of the leaf. (The top will have less of a bulge at each vein. Leaves are bumpier on the bottom.) Place the slide so that the top of the leaf is away from you.

4. See how each vein is surrounded by a layer of thick fibres for support. For this reason, veins are often called **fibrovascular bundles**. The "vascular" part of the name refers to the **vessels**, which are thick-walled hollow tubes of **xylem** that carry water. Xylem is named for the Greek word for *wood*, since it makes up the wood of trees. Describe the location of the xylem vessels. Examine the remaining cells in the fibrovascular bundle. Describe their location with respect to the xylem vessels. These cells make up the **phloem**, the tissue that carries food materials in solution (sugars and amino acids) from the leaf to other parts of the plant. Describe the structure of the phloem.

5. Move the slide so that you can examine a typical part of the cross section of the leaf between the veins. Look first at the top row of cells: the **upper epidermis**. Look for a clear layer of waxy secretion, the **cuticle**, above the upper epidermis.

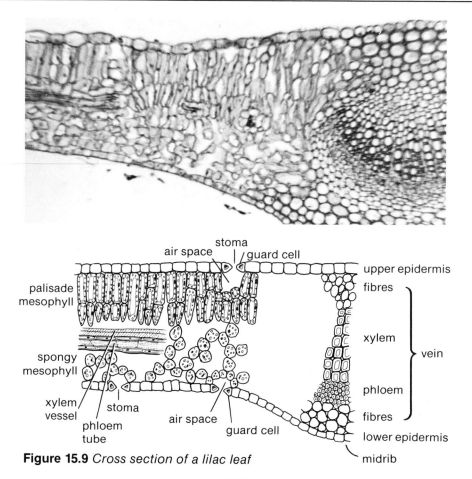

Figure 15.9 *Cross section of a lilac leaf*

234

Such a layer would be found on any shiny leaf but may be missing if your leaf is dull. Describe the way the cells of the upper epidermis fit together. Look for spaces between them. These openings are the **stomata**, through which carbon dioxide can enter and oxygen escape. See Figure 15.10. Each **stoma** (singular of stomata) is located within a pair of **guard cells**. If you don't find stomata, don't worry. Perhaps your leaf grew in the sunshine. Many plants have stomata only on the lower epidermis, where they will be shaded.

6. Look next at the bottom layer of cells, the **lower epidermis**, and search for stomata there. Describe how the guard cells differ in structure from other epidermal cells.

7. Between the upper and lower epidermis is the **mesophyll**, the middle leaf tissue. Compare the shapes of these cells with those of the epidermal layers. Compare what you see with Figure 15.9. Note that the mesophyll is arranged in two layers: the upper *palisade* layer of mesophyll (so called because the cells are arranged like a line of stakes), and the *spongy* layer of mesophyll (which is full of air spaces, like a sponge). If the leaf you are examining grew in strong sunlight, it may have more than one layer of palisade cells.

All of the mesophyll cells are thin walled so that the raw materials and products of photosynthesis can move through them readily by diffusion. Look for other ways mesophyll cells are specialized for photosynthesis. Look for **chloroplasts**, green bodies that contain the chlorophyll. Make note of which cells contain chloroplasts. Describe how the air spaces of the spongy layer would allow gases to reach or leave all cells easily.

Observations

Make a labelled drawing of the cross section of a leaf. Show the outline of the various tissues you have located, including a small vein, and carefully draw the detailed structure of three cells of each kind. Be sure to include the magnification of your drawing.

Questions

1. List the kinds of cells that (a) contain chlorophyll and (b) do not contain chlorophyll.
2. List the parts of the leaf that reduce water loss.
3. What part of the leaf acts as its skeleton, supporting it?

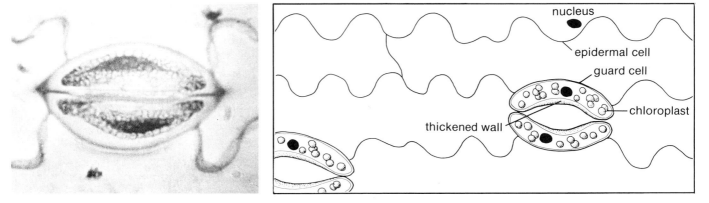

Figure 15.10 *Structure of a stoma of a lily leaf*

The Adaptations of Leaf Structures

The cells of a leaf are especially suited (adapted) to carry on photosynthesis. *Adaptation* is the special relationship between structure and function that makes living things better able to survive in a given environment. Use Table 15.7 as an aid to remembering the adaptations of the tissues of a leaf.

Table 15.7 *A Summary of the Adaptations of the Tissues of a Leaf*

PART	STRUCTURE	LOCATION	ADAPTATION
Cuticle	Waxy layer	On upper epidermis of some leaves	Transparent to let light in, waterproof to reduce water loss
Upper epidermis	Layer of flat interlocking cells without chloroplasts	Upper surface	Transparent to let light in, reduces water loss
Palisade mesophyll	Long, narrow cells with chloroplasts and thin walls	Upper layer of middle of leaf	Lets light penetrate to centre of leaf, carries on photosynthesis
Spongy mesophyll	Loose, round cells with chloroplasts and thin walls	Lower layer of middle of leaf	Has air spaces for carbon dioxide, oxygen, and water vapour to diffuse; carries on photosynthesis
Fibrovascular bundle	Sheath of fibres around woody vessels of xylem and narrow phloem tubes	In middle of mesophyll, forms a network	Skeleton to support; carries water, solutes, and sugars
Lower epidermis	Layer of flat interlocking cells without chloroplasts	Lower surface	Reduces water loss
Guard cells	Paired bean-shaped cells with chloroplasts and thick inner walls	In epidermis	Control opening and closing of stoma
Stomata	Openings between guard cells	In epidermis	Allow gases to diffuse into or out of leaf

Self-check

1. (a) What are fibrovascular bundles?
   (b) What types of cells make up the fibrovascular bundles?
   (c) What function do each of the kinds of cells of a fibrovascular bundle perform?
2. (a) What is the mesophyll and what is its function?
   (b) What are the two layers of the mesophyll, and how do they differ in structure and function?
3. (a) Describe the epidermis of a leaf.
   (b) What are the functions of the epidermis?
4. Describe the structures of a leaf that are involved in each of the following. In each case, state how the structure is adapted to performing the function. How does
   (a) the light get in to the chloroplasts in the spongy cells of the mesophyll?
   (b) water get to the spongy cells of the mesophyll?
   (c) carbon dioxide get into the palisade cells of the mesophyll?
   (d) the leaf reduce the loss of water from the mesophyll?
5. Describe the structure of a pair of guard cells, and explain how they carry out their functions.

How Stems Support Photosynthesis

Think of some of the tallest trees you have ever seen (Figure 15.11). How would water get to the tops of such trees? And how would the trees be strong enough to hold the leaves up to the light, in spite of the force of the winds? In this section, you will look at two main functions of stems: support for the leaves and transport of water to the leaves. In these two functions, the stems assist the leaves to carry on their job of photosynthesis.

Figure 15.11 *Some of the tallest trees in the world are over 100 m high. How would water rise to the leaves for photosynthesis?*

The Stem and Plant Transport

Look at the picture of the flower shown in Figure 15.12. Why are only parts coloured, and not the whole flower? What does this tell you about the arrangement of the conducting vessels in the stem? Food colouring was placed in some of the water rising up the stem. This was done by slitting the stem and placing parts of it into different solutions.

Figure 15.12 *This white flower has been partly coloured with food colouring.*

In the stems of most plants, the fibrovascular bundles are located in a circle, near the outside (Figure 15.13). In each bundle, the fibres are packed around the outside, making the bundle strong. Xylem vessels make up the larger part of the bundle. They carry water and dissolved materials upwards, and their thick woody walls strengthen the stem. The phloem tubes are located closer to the outside of the stem. Phloem carries the sugar solutions produced by photosynthesis.

Near the outside of a green stem, just inside the epidermis, the cells contain chloroplasts and carry on photosynthesis. These cells are packed together more tightly than the cells of the leaf mesophyll. The green stem cells have thickened corners to make the stem stronger. Because these green cells need to take in carbon dioxide for photosynthesis, there are some stomata in the epidermis of the stem, with air spaces inside.

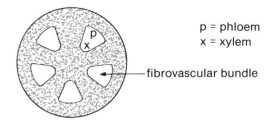

p = phloem
x = xylem

fibrovascular bundle

Figure 15.13 *The arrangement of the conducting tissues in a young stem*

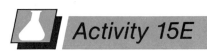
You have discovered something about the "plumbing" of plants: the xylem vessels are the pipes through which water flows. But where are the "pumps"? In this activity, you will begin to explore the forces that pump water up to the tips of plants.

Problem

Is a force created by leaves to help water rise through plants?

Materials

3 leafy stalks of celery
food colouring
3 beakers (400 mL)
clear plastic bag
elastic band

Procedure

1. Cover one leafy stalk of celery with a clear plastic bag, secured with the elastic band. Place the bottom of the stalk into a beaker of water containing some food colouring.

2. Place the second leafy stalk of celery into the second beaker of water and food colouring.
3. Remove most of the leaves from the third stalk of celery, and place it into a beaker of water and food colouring.
4. Place all three stalks in a sunny place or under a bright light, as shown in Figure 15.14.
5. The following day, cut a number of cross sections to see how far the food colouring rose up the stalks.

Observations

1. How far did the colouring rise in each of the stalks?
2. Describe the appearance of the inside of the plastic bag after a day.

Questions

1. What effect do leaves have on the movement of liquid through the conducting system of a plant?
2. What evidence from inside the plastic bag suggests how a force was created?
3. What does each of the three stalks contribute to the experiment?

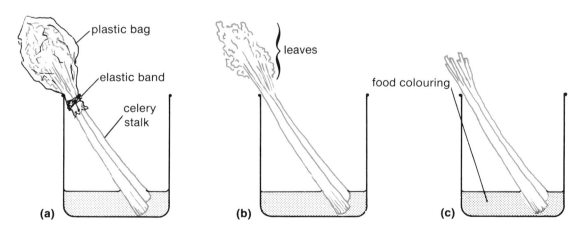

Figure 15.14 *Celery stalks were exposed to three different conditions. Which is the control?*

Some Forces That Move Water through a Plant

Leaves are involved in the movement of water through a plant. Through the stomata, leaves lose water vapour to the surrounding air. As this water vapour escapes, the air spaces within the leaf become drier, permitting more water to evaporate from the mesophyll cells.

Transpiration is the loss of water in the form of vapour from the exposed parts of the plant. The mesophyll cells that have lost water now have less water than their neighbouring cells. By osmosis, water will move from cell to cell all the way from the nearest xylem vessel in one of the fibrovascular bundles. Thus, **transpiration pull** is one of the forces involved in the movement of water in a plant. In the next two sections, you will learn about other forces that help to move water through plants.

Self-check

1. Describe two important ways that stems help leaves to do their job of photosynthesis.
2. In Figure 15.12, why did the food colouring only colour part of the flower, and not all of it?
3. Sketch a cross section of a green stem, and show how the fibrovascular bundles are arranged.
4. Name the three types of cells that make up the fibrovascular bundle of a green stem, and state the function of each.
5. How is a green stem able to carry on photosynthesis?
6. What is transpiration? How does transpiration create a force that helps to raise water in a plant?

The Role of the Roots

How can roots get enough water up from the soil to the leaves to keep the cells alive and provide for photosynthesis? A forest of maple trees would give off enough water by evaporation during the growing season to cover the ground to a depth of 70 cm (if it all lay on the ground without soaking in or evaporating). Just one corn plant in four months of one summer gave off 250 L of water. How do the roots manage to collect all that water?

Roots and their Functions

More of a plant is out of sight under the ground than is visible in the air. Scientists have carefully removed the earth from plants to measure the extent of their root systems. One experimenter found that a sugar beet, by the end of its first summer's growth, had roots that penetrated 6 m³ of soil, reaching a depth of nearly 2 m and spreading to a diameter of nearly 2 m. A rye plant, after four months' growth in a wooden container, was found to have nearly 14 000 000 roots with a total length of 620 km. In contrast, it had only 80 shoots with 480 leaves.

Why should there be so many roots? One of the main functions of roots is the *absorption* of water and minerals for the plant. The extensive root system has a very large surface area in contact with the soil for absorption. The rye plant, mentioned in the paragraph above, had a root surface area totalling 600 m². This is 130 times as great as the surface area of the shoot and leaves.

Besides absorption, the roots provide *anchorage*, holding the plant in place and binding the soil together. Roots also provide *storage* for food materials, especially starch, that will supply the energy for new growth in the spring, following the winter period of dormancy.

Root Hairs

Just back of the growing tip of a young root are a very large number of root hairs. In the case of the rye plant mentioned in the opening paragraph of this section, scientists estimated that there were 14 000 000 000 root hairs on its roots and that their total surface area was 400 m². Each root hair is an outgrowth of an epidermal cell. The root hairs greatly increase surface area of the root for absorption (Figure 15.15).

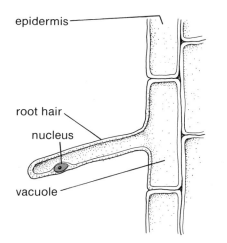

epidermis

root hair

nucleus

vacuole

Figure 15.15 *Structure of a root hair and its relationship to a cell of the root epidermis*

As roots grow in length, the delicate root hairs are damaged by being pushed past the soil particles. Thus the plant must always be developing new root hairs to replace those that are lost. There are no root hairs on the first few millimetres of a root tip, where the new cells are being produced. Root hairs form just back of the tip, where the young epidermal cells are most actively absorbing water and minerals.

Diffusion and Osmosis in Plants

In Chapter 3, you learned about the movement of substances by diffusion. Diffusion moves dissolved materials through the soil solutions, bringing the substances that plants need as nutrients to the root hairs and the walls of epidermal cells.

Diffusion also accounts for the movements of substances within cells. Mineral ions and the products of photosynthesis will diffuse through cell vacuoles to all parts of a cell.

All cells need oxygen to release the energy required for growth and other life functions. The oxygen must diffuse into roots from air spaces in the soil. In the exposed parts of plants (leaves, stems, flowers, and fruits), oxygen diffuses in through the pores (stomata). Entering the air spaces between cells, the gas will diffuse throughout each part. At the moist cell walls, oxygen will dissolve in water and diffuse in solution across cell membranes into every cell. Carbon dioxide diffuses out of the plant by these same routes, but in the opposite direction.

Osmosis, you will remember from Chapter 13, is a special case of diffusion. It is the diffusion of water across a selectively permeable cell membrane. Because it is still a form of diffusion, the water moves randomly from where there are more water molecules to where there are fewer water molecules. Osmosis explains how water enters the roots and how it moves from cell to cell across a root to the upward conducting cells. See Figure 15.16 on page 242. As water enters one cell, it dilutes the cell's contents, creating a difference in osmotic pressure between this cell and its neighbour. This difference in osmotic pressure keeps the water flowing from one cell to the next. Osmosis also causes the movement of water from cell to cell within the tissues of leaves, stems, flowers, and fruits.

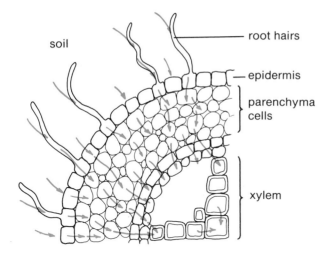

soil

root hairs

epidermis

parenchyma cells

xylem

Figure 15.16 *The movement of water across the cell of a root*

Activity 15F

Problem

How does the conducting system connect the stem, roots, and leaves of a plant?

Materials

flower pot or flat
tray or saucer to hold the pot or flat
seeds of radish or grass
food colouring or water-soluble dye
potting soil

Procedure

1. Prepare the flower pot with potting soil. Press the soil down well.
2. Scatter a few seeds on the top of the soil, and cover them with a thin layer of soil. Press the soil down.
3. Water the pot or flat from below by pouring water into the tray or saucer.

CAUTION Wash your hands after handling the soil and the dye.

4. Place the pot in a warm, dark place for two or three days. Check to see that the soil remains moist.
5. When the seedlings have sprouted, apply water containing the food colouring or dye. Move the pot or flat into the light.
6. Observe the plants for the next day or two.

Observations

1. What do you see that suggests that the coloured dye is moving from the soil through the plants?

Questions

1. What pattern does the dye take in the parts of the plant you can see? Explain why this pattern occurs.

The Functioning Plant: A Summary

The conducting system of a plant extends from the roots through the stems to the leaves. It also extends to the flowers or fruit. (Figure 15.12 showed a flower dyed by way of the fibrovascular bundles in its stem.) Thus, the conducting system of a plant (Figure 15.17) is a complete network of xylem vessels and phloem tubes that comes close to every living cell. This system brings the water and minerals the cells need. It also carries the soluble products of photosynthesis (the sugars) to where they are needed for energy and growth.

The Forces That Move Water through a Plant

Figure 15.18 shows all the forces at work that raise water in plants. But what exactly do they do?

If you did Activity 15E, you know that transpiration pull is a very strong force produced by the leaves. It involves lifting water up the plant. You have also seen earlier (in section 15.4) that water enters the root hairs from the soil by osmosis. Water then moves from cell to cell across the root until it enters upward-conducting vessels (the xylem).

Osmosis also produces a force. The combined force of all the root cells is often called root

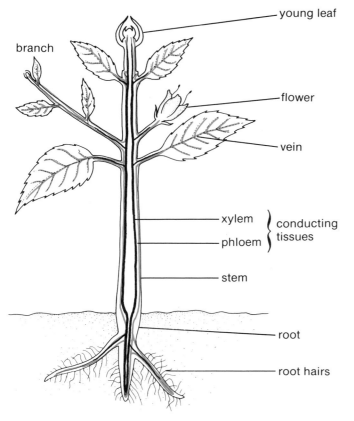

Figure 15.17 *The conducting system of a plant*

branch

young leaf

flower

vein

xylem
phloem } conducting tissues

stem

root

root hairs

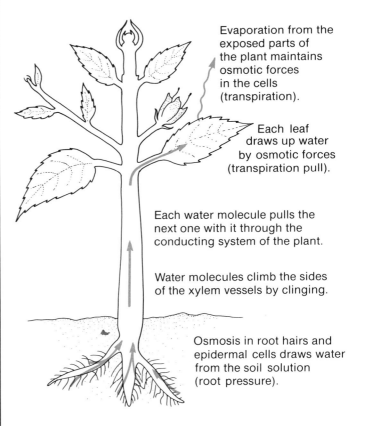

Figure 15.18 *A summary of the forces that raise water in plants*

Evaporation from the exposed parts of the plant maintains osmotic forces in the cells (transpiration).

Each leaf draws up water by osmotic forces (transpiration pull).

Each water molecule pulls the next one with it through the conducting system of the plant.

Water molecules climb the sides of the xylem vessels by clinging.

Osmosis in root hairs and epidermal cells draws water from the soil solution (root pressure).

pressure. When a tree is cut down in the spring, water flows out of the trunk because of root pressure. But is root pressure strong enough to raise water to the top of a tree? No. There must be other forces involved.

Water molecules have a strong attraction for one another. We see examples of this attractive force when drops of water run together on the surface of a car or window. In the xylem vessels, this force helps to pull the water molecules upward in an unbroken chain.

Water molecules also have a strong attraction for some other materials, such as the walls of xylem vessels. This other attractive force explains the way water pulls itself into tiny spaces. Xylem vessels are very fine water tubes. They are a continuous system of pipes from the root through the stem and its branches and through the leaves, flowers, and fruit. Even without roots or leaves, water will rise in xylem vessels.

Self-check

1. Where does most of the absorption of water and minerals take place in plants?
2. Describe the structure, location, and function of a root hair.
3. Why is it advantageous to a plant to have many root hairs?
4. Define (a) diffusion and (b) osmosis, as they relate to plants.
5. Explain why diffusion and osmosis occur in plants.
6. Use your understanding of diffusion and osmosis to explain why vegetables in the refrigerator go limp. What is the effective way of storing vegetables to prevent this?
7. Dye, applied to the soil, spreads through all parts of a plant. What does this show you about the conducting system of a plant?
8. In the cells of a root, what force is produced by osmosis?
9. Explain how water is raised in a plant. Drawing a diagram may help.
10. Why do all green plant cells need a steady supply of water?

Chapter Objectives

NOW THAT YOU HAVE COMPLETED THIS CHAPTER, CAN YOU DO THE FOLLOWING?	FOR REVIEW, TURN TO SECTION
1. Describe the steps in testing a leaf for starch, and tell why each step is needed.	15.1
2. Explain the role of light, chlorophyll, carbon dioxide, and oxygen in photosynthesis.	15.1
3. Write the word equation for photosynthesis.	15.1
4. List the main parts of a leaf and describe the function of each.	15.2
5. Name the different kinds of cells that make up a leaf, and tell how each is adapted for its functions.	15.2
6. State two ways that stems help the leaves to carry out photosynthesis.	15.3
7. Explain the role of leaves in moving water through plants.	15.3
8. Describe the arrangement of the fibrovascular bundles in a green plant stem.	15.3
9. Explain how water is absorbed by the roots of a plant.	15.4
10. Describe how water is raised from the soil to a leaf.	15.5

Words to Know

chlorophyll
photosynthesis
blade
midrib
petiole
fibrovascular bundle
vessel
xylem
phloem

upper epidermis
cuticle
stomata (singular: stoma)
guard cell
lower epidermis
mesophyll
chloroplast
transpiration
transpiration pull

Tying It Together

1. Copy the following table into your notebook, and complete it.

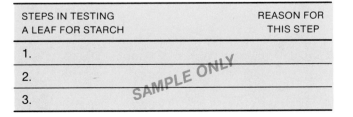

STEPS IN TESTING A LEAF FOR STARCH	REASON FOR THIS STEP
1.	
2.	
3.	

2. You have just tested a leaf for the presence of starch. What would you see if
(a) the leaf contained starch?
(b) the leaf did not contain starch?
3. You have applied the starch test to a variegated leaf that has been in the light for several hours. Figure 15.19 shows your results.
(a) What can you infer from the black parts of the leaf?
(b) What can you infer from the parts that are not black?
(c) What can you infer when you compare (a) and (b)?

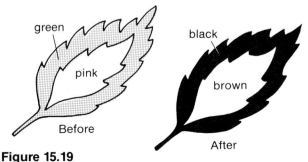

Figure 15.19

4. Define photosynthesis.
5. (a) What are the raw materials for photosynthesis?
   (b) What are the products of photosynthesis?
   (c) Write the word equation to represent photosynthesis.
   (d) Is photosynthesis a physical or a chemical change? Why?
6. Copy the following table into your notebook and complete it.

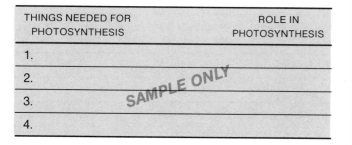

THINGS NEEDED FOR PHOTOSYNTHESIS	ROLE IN PHOTOSYNTHESIS
1.	
2.	
3.	
4.	

7. Describe how each of the following structures is adapted for making the leaf an especially efficient organ of photosynthesis:
   (a) the leaf blade
   (b) the midrib
   (c) the petiole
   (d) the network of veins in the leaf
8. Make a diagram to indicate the layers and structures you saw when you examined the cross section of a leaf with the aid of a compound microscope. Draw in two cells of each type. Label all the structures and types of cells.
9. How is the spongy mesophyll adapted for photosynthesis?
10. How are the stomata involved in photosynthesis?
11. Name two ways that stems help the leaves in photosynthesis.
12. Describe the adaptations of the cells in the outer part of a green stem that allow them to photosynthesize.
13. How do the leaves create a force to "pump" water upwards through the plant?

14. How does the root of a plant develop a huge surface for absorbing water?
15. In what parts of a plant are xylem vessels found? In what way do the xylem vessels make up a conducting system?

Applying Your Knowledge

1. Some of the cells of a variegated leaf lack chlorophyll. How would these cells get enough energy to stay alive?
2. Starch is used up in the cells of a plant that has been left in the dark. What chemical process is involved in using starch to get the energy for life? What are the products of this process?
3. Write a word equation for the process you named in question 2. Compare this with the word equation for photosynthesis. In what ways are the two processes similar? In what ways do they differ?
4. In many fruit and vegetable stores, the clerks spray the lettuce and other vegetables with water. What effect would this have? Why would it have this effect?

Projects for Investigation

1. In reference books, look up photographs of parts of plants that have been made with the aid of scanning electron microscopes. What do the details of such photographs add to the knowledge you gained using the compound light microscope?
2. Consult advanced textbooks of plant structure and function to find out how guard cells are able to open and close stomata. When water is scarce, how are the guard cells affected? When photosynthesis is going on, how do the guard cells act to let more carbon dioxide into the leaf?

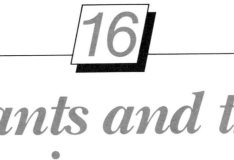

Plants and the Environment

Key Ideas

- The physical environment affects what plants grow in an area and how they are adapted for that area.
- Suitable acidity of the soil and suitable minerals are needed for plants to grow.
- Deforestation and acid rain are problems caused by human activities.
- There are many occupations for people who enjoy working with plants.

Why do different plants grow in different places? What determines whether a particular kind of plant will grow well or die? In this chapter, you will look at the conditions plants need in order to grow. You will also see how different environments can supply those needs.

The Physical Environment

The environment has both living and non-living parts. Together, the living and non-living parts form an **ecosystem**. (Figure 16.1).

The ecosystem of a lawn, for example, contains living organisms, such as grasses, weeds, earthworms, and spiders. The non-living parts of this ecosystem are the soil, the water, the air, and the heat. All the parts of the ecosystem form a delicate balance, interacting with one another.

Figure 16.1 *These students are studying the ecosystem of a school lawn.*

Soil

In any ecosystem, the soil is an important non-living component. The nature of the soil changes where people walk on it (Figure 16.2). By packing the earth, people crush the air spaces between the soil particles. These spaces are needed for oxygen to diffuse to the roots of plants. Also, rain water cannot soak into packed earth easily.

Figure 16.2 *Why does grass not grow where there is a path?*

Instead, the water runs off the surface. These are two reasons why plants do not grow where there is a path.

The kind of soil is another factor affecting the growth of plants. Soil is composed of various types of particles (Figure 16.3). *Clay* particles are very small, with little space separating them. When soil is mostly clay, it takes a long time for water to soak in. Drainage is also very slow. *Sand* particles are large with relatively large spaces between them. These particles allow water to drain through very quickly, often taking away nutrients in the process. *Silt* particles are larger than clay, but smaller than sand.

One other component of soil is missing—*humus*. You probably remember from earlier studies how humus from decaying plant matter helps to hold moisture and improves soils for plant growth.

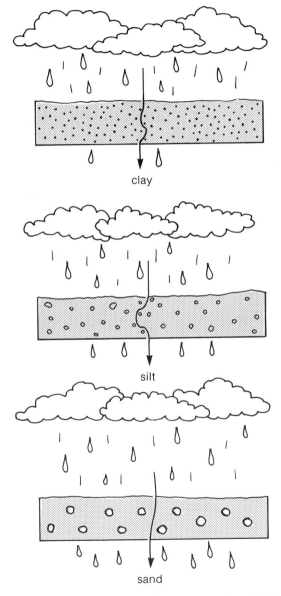

Figure 16.3 *Compare the three types of soil. What are the good characteristics of each? What are the poor characteristics of each?*

Climate

An important part of the physical environment that determines which plants will be able to grow in any ecosystem is the climate. *Climate* refers to the kinds of weather a place has over a period of years. Climate includes

- precipitation—How much rain, snow, hail, and sleet falls?
- humidity—How moist is the air?
- temperature—How hot or cold does it get?
- sunlight—How many hours does the sun shine in a year? How bright is it? Are the sun's rays direct or slanted?
- wind—How strong is the wind? Does it blow mainly from the same direction?

Each of these factors may have some effect on every organism. Together they determine which organisms can survive in a particular ecosystem. For example, precipitation is very important in determining which plants will grow in a particular place. If there is too little precipitation, there will be a desert with plants such as cactus and sage brush that are adapted to conserve the little moisture they get.

In the Canadian Arctic, the tundra has a cold, dry climate that limits plant growth to about two months per year.

A rain forest requires the most precipitation. Huge trees develop, and many different kinds of plants and animals flourish. The west coast of British Columbia is an example of a temperate (not too hot or too cold) rain forest. Grasslands or prairies, such as those of central North America, do not receive enough rain to support the growth of forests but do receive enough to grow grass.

Adaptations in Plant Structure for Different Environments

The leaf shown in Figure 16.4(a) may be thought of as a "typical leaf." Leaf structure, however, varies from one kind of plant to another. Even on the same kind of plant or on the same plant, leaf structure depends on where the plant grows or where the leaf is located. For example, on the same plant, leaves that grow in the shade are thinner and broader than leaves that grow in full sun. See Figure 16.4(b). If a geranium is moved from indoors out into the bright sun, the broad,

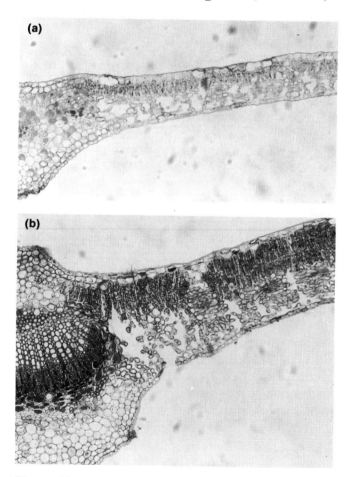

Figure 16.4 *Cross sections of two leaves from the same pear tree: (a) a leaf from the shade and (b) a leaf from the sun. How do they differ in structure?*

thin leaves soon dry out and die. The new leaves that form are smaller and thicker. If a geranium is moved indoors, the thick leaves soon die, and thinner leaves grow.

Plants grow best in the environments to which they have become adapted. **Adaptations** are special features of an organism that make the organism particularly well suited for living in certain conditions. For example, a large group of plants can survive in deserts or other dry places where water is very scarce (Figure 16.5). Cactus plants usually have their leaves so greatly reduced that they appear as needles or thorns. This cuts down on the surface area exposed and thus on the loss of water. Their stems have become green and fleshy to take over the function of photosynthesis and to store water.

You might not think of pine and spruce as trees that live in dry places. Yet they often are found among rocks on steep hillsides or growing in sand. Water drains quickly away from such places, and there is little humus to soak it up. Pines and spruces have compact, needle-like leaves to conserve water. They also produce a sticky waterproof resin that seals any broken leaves or stems, preventing loss of water.

Grasses that live in sand dunes often have leaves that curl up when they are very dry. This reduces the amount of surface area likely to experience water loss through evaporation (Figure 16.6). The grass leaf has a double epidermis to reduce water loss. The stomata are few in number and occur only in deep cracks in the lower epidermis where evaporation is greatly slowed.

Figure 16.5 *Some cactus plants. How are they adapted for conserving water?*

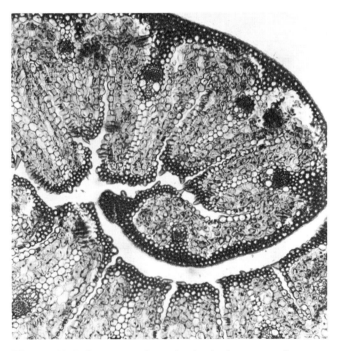

Figure 16.6 *Cross section of a leaf of a grass that lives in sand. It is curled to reduce evaporation. What other adaptations do you see?*

When you look for similarities in the structure of the leaves of plants adapted for life in dry places, you find such features as a thick, waxy cuticle, several layers of epidermal cells, few stomata, stomata deeply sunken in pits, matted hairs to keep air currents away from stomata, and compact mesophyll with few air spaces to reduce evaporation from the cell walls.

How do you suppose the leaves of plants adapted for life in wet places are constructed? These include floating plants such as duckweed and water lilies (Figure 16.7). Where do you expect you will find their stomata? How do you suppose their roots and stems will get air from the leaves? Plants that live underwater, such as *Elodea* (the Canada waterweed), have leaves that lack a cuticle and are only one or two cells thick. They also have very few veins. Each of their cells can get what it needs directly from the water that surrounds it.

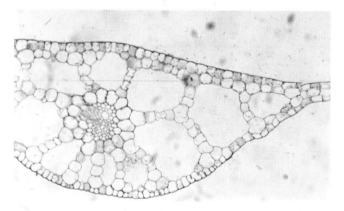

Figure 16.7 *Cross section of the leaf of a floating plant*

Challenge

How does leaf structure differ in plants adapted to life in different environments? With the aid of the microscope, examine prepared slides of cross sections of leaves from several different kinds of plants adapted to different habitats. Make sketches to represent what you see of their structure, showing only a few of each type of cells. Compare your sketches with the pictures in Figures 16.6 and 16.7. Make statements about how each species is adapted to survive in its particular environment.

Self-check

1. (a) What is an ecosystem?
    (b) State two examples of ecosystems.
2. What are the two main parts of ecosystems?
3. State two ways the soil might affect the growth of grass in a lawn.
4. (a) What is climate?
    (b) List five factors that make up climate.
    (c) Tell how one of the factors you listed in (b) might affect plant growth.
5. (a) What are adaptations?
    (b) How are leaves adapted for living in bright sunlight?
    (c) How are the leaves of some plants adapted for living in dry places?
    (d) What are two adaptations of the leaves of plants that float on the surface of a pond?

What Plants Need in Order to Grow and Reproduce

The plants and animals that live in any particular place are able to do so because the environment is "right" for them. What climatic factors does each organism need, or what climatic factors can it tolerate? What minerals do plants need to grow well? How does the soil or the amount of water make healthy growth possible? Knowing about these factors can help you understand different ecosystems. You will be able to explain why huge trees grow on the west coast of Vancouver Island, why wheat grows well on the prairies, and why blueberries flourish in the swamps and forest clearings of Nova Scotia and Newfoundland. You may even learn to grow beautiful plants or a successful garden! Most important, you will see how the interrelationships among all living things fit into the "balance of nature" you so often hear about.

Many plants can only grow well in a soil that is moderately acid (pH 5 to 7). Some examples of acid-loving plants are birch, blueberry, pine, spruce, and some oaks. In the garden, azalea, dogwood, gardenia, holly, hydrangea, lily of the valley, and rhododendron require acid soils. Peat moss or sulphur may be added to soils to make them more acidic. Many of the flowering house plants thrive when watered with rainwater, which is slightly acidic (Figure 16.8).

Another group of plants do better in slightly basic soils. Geraniums, petunias, sweet peas, lawn grass, and *Clematis* are examples. So are many of the garden vegetables (peas, beans, beets, and lettuce) and the fruit trees (peaches, pears, and plums). When you see mosses growing in lawns or gardens, you can be sure that the soil has become too acidic. You can correct this condition by adding ground limestone (pH 8).

Figure 16.8 *These house plants thrive under fluorescent lights. They are watered entirely with rainwater that is slightly acidic.*

Activity 16A

Problem

How can you determine the chemical nature of the soil?

Materials

soil-testing kit
samples of soils from different locations
distilled water
pH test paper
test tube

Procedure

1. In your notebook, make a table similar to Table 16.1 to record your observations.
2. Half fill a clean test tube with one soil sample.
3. Cover the soil with distilled water. Using your thumb to seal the test tube tightly, shake the soil and water together for about a minute.
4. Dip the end of the piece of pH test paper into the liquid in the test tube. Match the colour of the test paper to find the pH. Record the pH.

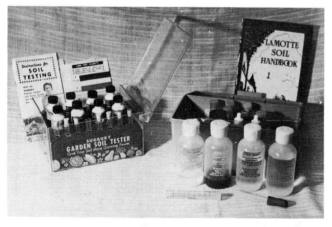

Figure 16.9 *Soil-testing kits*

5. Follow the instructions with the soil-testing kit (Figure 16.9). Measure the required amount of soil into each of the test bottles. Add the chemicals, and match the colours of the indicators. Record the amounts of nitrogen, phosphorus, and potassium present in the soil sample.
6. Wash your hands thoroughly after you finish handling the soil samples.

Observations

Table 16.1 *Results of Soil Tests*

Source of the soil:
pH of the soil
Amount of nitrogen
Amount of phosphorus
Amount of potassium

Questions

1. Check the pH against Table 16.2. Is the soil acidic, basic, or neutral?
2. According to the instructions in the soil-testing kit, should any minerals be added to this soil for healthy plant growth? If so, which ones should be added?

Using Soil Measurements

Soil testing is of great value to the gardener, farmer, and forester. The test results will suggest ways to grow better plants and increase crops. In Unit Six, you will consider in more detail the growing of food crops. Here you will begin to learn the meaning of some of the tests.

pH: The Degree of Acidity

The **pH scale** is a measure of how **acidic** or **basic** a substance is. The scale ranges from 0 to 14 (Table 16.2). The mid-point of the scale, pH 7, is called **neutral**. It is neither acidic nor basic. If the pH of a substance is less than 7, the substance is acidic. If the pH is greater than 7, the substance is basic.

Table 16.2 *The pH Scale*

RANGE	pH	EXAMPLE	SOME ORGANISMS THAT LIVE AT THIS pH
Acid	0	Sulphur	Sulphur bacteria
	1		A kind of fungus
	2	Lime juice / Stomach contents / Lemon juice	
	3	Vinegar	Some bacteria; "vinegar eels" (nematode worms)
	4	Peat bogs	Mosses; pitcher plant; black spruce
	5	Sour milk	Lactic acid bacteria (4.5)
	6		
Neutral	7	Pure water	Fish 7.4 — Human blood (6.7–8.6); Most field and forest plants; amoeba; yeast (4.5–8.5)
Basic	8	Sea water	
	9	Basic soil	Orchids
	10	Desert pool	Desert fish; a kind of mosquito
	11		A kind of mould
	12		
	13	Lye (concentrated)	
	14		

Problem

How is the germination of seeds affected by pH?

Materials

seeds of radish (or other plants)
paper towels
petri dishes with covers
glass-marking pen or crayon
solutions of different pH (such as 1, 3, 5, 7, 9, 11)
scissors
safety goggles

CAUTION Be careful when handling the pH solutions. Safety goggles must be worn.

Procedure

1. Make an observation table in your notebook similar to Table 16.3.
2. Using scissors, cut circles of paper towel to fit inside the petri dishes. Place three thicknesses of towel into each petri dish.
3. Moisten the paper towel in each dish with the solution of the pH that you are to use for that dish. Drain off the surplus liquid. Mark the dish with the pH number and your initials.
4. Count ten seeds into each petri dish and separate them from one another.
5. Place the covers on the dishes, and set them aside in a warm, dark place.
6. Each day for a week, observe your seedlings, and record how many have germinated in each petri dish. Add a little more of the correct solution to keep them from drying out. Be sure that water does not cover the seeds.

Observations

Table 16.3 *Observations of Germination at Different pH*

Kind of seeds	
pH of solutions used	
Number of seeds germinated on day number	
1	
2	
3	
4	
5	
6	

Questions

1. At which pH did most of your seeds germinate?
2. At which pH values did none of the seeds germinate?
3. What effect does pH have on the germination of these seeds?
4. Make a bar graph to represent the number of seeds germinated at each pH.

Soil Minerals

Soil minerals contain chemical elements essential for plant growth. The three most important elements that plants obtain from the soil are **nitrogen, phosphorus**, and **potassium**. Figure 16.10 shows the effect on plant growth when just one of the essential elements is lacking.

Nitrogen, in the form of nitrates and ammonia, promotes the growth of leaves and stems. Phosphorus, in the form of phosphates, stimulates the development of strong roots, flowers, and fruits. Potassium encourages active growth. It also increases the size and heightens the flavour of fruits.

The numbers on fertilizer bags (Figure 16.11) show the percentages of these elements in alphabetical order. For example, 20–3–4 contains 20% nitrogen, 3% phosphorus, and 4% potassium. This fertilizer would be applied to a lawn in the spring to stimulate rapid growth of leaves. In the fall, you might apply a 5–10–5 because the higher phosphorus content (10%) would

stimulate root growth. You might put 18–18–24 around your tomato plants to encourage active growth and flowering, and the production of fruits. A good fertilizer to keep house plants flowering is 15–30–15.

Figure 16.11 *What is the percentage of each of the important elements in each of the commercial fertilizers shown?*

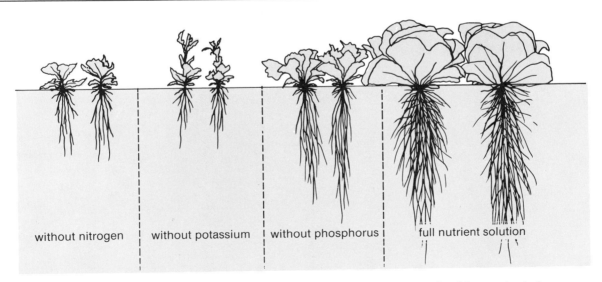

Figure 16.10 *The effects of a lack of essential minerals on the growth of cabbage plants in mineral solutions*

257

Self-check

1. What is pH, and how is it measured?
2. What is the neutral point on a pH scale, and what does neutral mean in terms of pH?
3. Classify the following pH readings as acidic, basic, or neutral: (a) pH = 3, (b) pH = 7, (c) pH = 9.
4. (a) Name two plants that thrive in acidic soils.
   (b) Name two plants that thrive in basic soils.
5. (a) What would you add to an acidic soil to make it more basic?
   (b) What would you add to basic soil to make it more acidic?
6. Name the three elements that are most important for plants to absorb from the soil.
7. Explain the meaning of the numbers 10-20-30 on a fertilizer bag.
8. How does the pH of their surroundings affect the sprouting of seeds?

People and the Environment

More than any other species, humans change the environment for their own purposes. In doing so, they often make life for plants (and animals) more difficult or impossible. On the other hand, humans protect and spread plants that have proven desirable. In this section, you will consider the human impact on the forests and on growing useful plants.

Forests and People

The original inhabitants of North America lived in harmony with their environment. They used only the plants and animals that they needed. Although some Native peoples grew food, they did not change the appearance of the land very much.

The Europeans came to North America with a different attitude. Their way of life was to exploit the natural resources for personal gain. They cut down forests for timber to send back to Europe for building ships (Figure 16.12). They cut down forests to open up land for farming and to get materials for building homes. They built mills on the rivers to saw lumber and to grind wheat. They changed the environment to suit their way of life. They expected the vast resources of the continent to renew themselves forever.

But the forests did not renew themselves. Why does a forest not always grow again when the trees are cut? The environment changes when the forest is removed. In a forest, in the shade of the trees, the earth remains cool, even on a hot day. Without the trees, the sun heats up the soil. The water in the soil evaporates. The dry soil is not as suitable for many tree seeds to sprout or for seedlings to grow.

Figure 16.12 *Loggers and the big trees of British Columbia in pioneer days*

In the forest, when a heavy rain falls, the leaves break the force of the raindrops. Water drips down gently from leaf to leaf and slowly soaks into the earth. The earth, rich in humus, acts as a sponge, letting the water out slowly and uniformly, and keeping the streams clear and running steadily. When the forest is removed, the rain strikes the soil with full force. More of the water runs off the surface than soaks in. The rivers run fast and muddy after a rain, overflowing their banks. As the water flows, the run-off carries soil particles and dissolved minerals away to the streams.

The forests also put a lot of water back into the air through transpiration. This means that forests help to ensure steady climate conditions that favour the growth of plants. **Deforestation** (removal of the forests) means that weather patterns change, becoming less suitable for the growth of plants.

Thus, there are many ways that deforestation affects the environment. Left alone, a deforested area will take perhaps 100 years to become a mature forest again. As you drive through the countryside, you will see many farms that are no longer in production. Many of them are on land from which the trees should never have been removed. The soil in these places is very thin, sandy, or full of rocks.

There is evidence from other parts of the world of human impact on the forests. For example, in the Sahara Desert there are huge tree trunks buried in the sand, showing that this desert was once a forest. The barren heath lands on the tops of the mountains in Britain were forested until people needed the trees as fuel to smelt iron from its ore. Today there is great concern about the destruction of the tropical forests in such countries as Brazil. Here many species of plants and animals are becoming extinct.

What can be done to help the forests grow again? People are finally beginning to realize that we must insist on planting new trees to replace every one that is removed. Governments and lumber companies are beginning to respond to this need. But there are many deforested areas that have not yet been replanted. Much remains to be done, and the task is growing larger every year.

Landscaping Our Parks

In almost any town or city, you can see and enjoy green spaces in many places—in front of the city hall, in parks, along the streets, and in river valleys. The people who are responsible for the green, open spaces may have many different kinds of jobs. All these people must have some knowledge of what green plants need to grow and thrive. Here are some of the activities that take place in and around parks.

Employees of the Parks and Recreation Department are responsible for building parks as well as maintaining them. Here, they are building a bridge over a creek. They must locate the concrete footings so the bridge will not be damaged during the spring flood.

People who work for the Parks and Recreation Department help to care for plants by weeding and cultivating the soil. Weeding reduces competition for nutrients and available light. Cultivation allows air and water to reach the roots of the plants.

In parks that are used by a lot of people, it makes sense to pave the pathways with asphalt. That way, people can ride their bikes, push baby carriages, and manoeuvre wheelchairs much more easily.

Some plants are ideal for shaping. These boxwood shrubs have many small branches and leaves. They have been trimmed into letter shapes. Even though they look artificial, they are still living plants.

Some planting beds in our parks need a lot of care. This formal bed must be clipped often to keep its neat appearance.

Parks provide recreation areas. Some playing fields have underground irrigation systems because the grass takes so much wear and tear. Fertilizer is used to maintain healthy, green grass.

Constructing sewers, paving, and grading of land outside a park can change the level of water underground. A subdivision was built outside this park. The builders did not plan properly. They did not allow for water to drain away. Unless the water is drained away, these park trees will die because air cannot reach their roots.

A playing field must be graded to ensure that water drains from it. Proper drainage is needed for good plant growth.

Rivers running through parks sometimes need protection from erosion. Large stones are used on the right bank of this river to absorb the force of the water. This is important because there is an old garbage dump under the soil and grass beyond the river bank. If the river is allowed to erode the land in this direction, old garbage will eventually be exposed.

Working with Plants

In this section, you have seen that trees need to be planted to replace those that are cut down. There are jobs in Parks and Recreation Departments, in tree nurseries, and in forestry planting the tree seeds, looking after the seedlings, and replanting the forests (Figure 16.13). Because the forests need to be managed, there are jobs fighting disease and fires that might destroy the trees. Workers are also needed in the lumbering industry to harvest the trees.

There are many greenhouses involved in producing plants for florists and for gardens. All of these need reliable workers who love plants. There are many retail outlets where people buy plants for their homes and gardens. There are landscaping contractors who hire workers to beautify the surroundings of homes, offices, shopping centres, and factories. Any of these may be looking for reliable employees. If you like working with plants, try some of these employers for part-time jobs that might lead to full-time employment when you finish your schooling.

Figure 16.13 *Students can work for Parks and Recreation Departments either to get work experience in some courses, or simply for pleasure. These students are turning over soil to prepare a flower bed for a children's park.*

Self-check

1. What is meant by "deforestation"?
2. Why did the first European settlers in North America cut down the forests?
3. (a) When a forest is cut down, what changes occur in the environment?
   (b) How might the changes you listed in (a) affect the growth of a tree seedling?
4. When a forest is cleared, what may happen to the area over many years?
5. What can be done to ensure reforestation?
6. List six possible jobs you might seek if you like to work with plants.

Chapter Objectives

NOW THAT YOU HAVE COMPLETED THIS CHAPTER, CAN YOU DO THE FOLLOWING?	FOR REVIEW, TURN TO SECTION
1. Define ecosystem and name its two main parts.	16.1
2. Explain the relationship between the soil and plants.	16.1
3. Define climate, and list five factors that contribute to the climate.	16.1
4. Explain how a leaf adapts its structure to bright and dim light.	16.1
5. Explain the adaptations of a plant for life in a dry place and a wet place.	16.1
6. Explain the terms pH, acidic, basic, and neutral.	16.2
7. Describe how the pH of soil can affect growing plants.	16.2
8. Interpret the meanings of the numbers on containers of fertilizer, and explain why different fertilizers have different effects.	16.2
9. Explain the effects of deforestation on soil and on seedlings.	16.3
10. List six jobs that involve working with plants.	16.3

Words to Know

ecosystem
adaptation
pH scale
acidic
basic
neutral
nitrogen
phosphorus
potassium
deforestation

Tying It Together

1. (a) What is an ecosystem?
   (b) What are the two main parts of an ecosystem?
2. (a) How does the soil affect the growth of plants?
   (b) How does humus improve soil?
3. In your notebook, identify each of the following soil types as either clay, sand, or silt.
   (a) It contains small particles with small spaces between them. Water drains away very slowly.
   (b) It contains medium-sized particles.
   (c) It contains large particles with large spaces between them. Water drains away very quickly.
4. How does climate affect the growth of plants?
5. If you were to use a microscope to compare cross sections of leaves of geraniums that grew in a sunny garden and in a house, what would you expect to find?
6. What special adaptations of leaf structure would you expect to see if you were to examine (using a microscope) cross sections of leaves of plants that grew
   (a) in the desert?
   (b) in a pond?
7. What is an acid?
8. What is a base?

9. What is a substance that is neither an acid nor a base? What pH does such a substance have?
10. What happens to the pH of soils when you add
    (a) peat moss?
    (b) limestone that has been ground to a powder?
11. (a) What are the three most important elements that plants must obtain from the soil?
    (b) What does each of these elements do for a plant?
12. (a) A fertilizer is rated 10–30–15. What percentage of each of the essential elements does it contain?
    (b) For what would you use the fertilizer described in (a)?
13. (a) What happens to the soil of a forest when the trees are removed?
    (b) How do the changes you described in (a) affect the growth of tree seedlings?

Applying Your Knowledge

1. Suppose you are going to plant a garden in which you hope to grow flowers or vegetables. What can you do to the soil to improve its ability to hold water and air so that the plants will grow well?
2. Figure 16.14 shows the leaf structure of a pine needle as seen in cross section under a microscope. List four adaptations of this leaf for conserving water. Why would pines need to conserve water?
3. Figure 16.15 shows the leaf structure of a pond weed as seen in cross section under a microscope. List four adaptations of this leaf for its life under water.

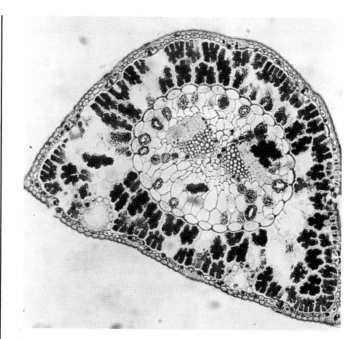

Figure 16.14 *Cross section of a pine needle*

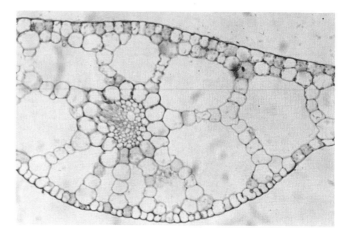

Figure 16.15 *Cross section of a leaf of pond weed*

Projects for Investigation

1. Prairies are areas where there is not enough rainfall to support a forest. Find out how grasses are adapted to live in places where there may be no rain for a long time. When the rain comes, how do grasses make use of the water quickly before it drains away?
2. Growing plants without soil in solutions that provide all the essential minerals is called *hydroponics*. Look up hydroponics in your library, and plan an indoor hydroponic garden. Could you succeed in producing tomatoes and lettuce during the Canadian winter? Figure 16.16 shows the growing of cucumbers hydroponically.

Coleus, ivy, *Philodendron, Tradescantia,* and *Zebrina* are foliage plants that do well in water gardens.

3. Recall what you learned about acid precipitation in Chapter 10, and then do research on the effects of acid precipitation on plants, soils, and lakes. In which parts of Canada does acid precipitation have its greatest impact on the environment? Why are some areas less affected than others?
4. Here is a question for group or class debate. What do you think people would have to do to reduce or prevent acid precipitation? Who should be responsible for these changes? Are they practical? Who would have to pay for them?

Figure 16.16 *This hydroponic garden was successful in producing cucumbers indoors during a Canadian winter.*

Unit Five: The Importance of Green Plants

MATCH

In your notebook, write the letters (a) to (g). Beside each letter, write the number of the word in the right column that corresponds to each description in the left column.

(a)	controls the opening and closing of a stoma	1. root hair
(b)	carries on photosynthesis	2. cuticle
(c)	carries sugars to the roots for storage	3. guard cell
(d)	absorbs water from the soil	4. stoma
(e)	prevents water loss	5. variegated
(f)	carries water to all parts of the plant	6. xylem vessel
(g)	allows gases to enter or leave	7. phloem
		8. fibres
		9. mesophyll

MULTIPLE CHOICE

In your notebook, write the numbers 1 to 10. Beside each number, write the letter of the best choice.

1. In Canada, most of the wood pulp for making paper is obtained from
   (a) white pine
   (b) black spruce
   (c) blue spruce
   (d) white cedar.

2. Many natural fibres, obtained from plants, are made into cloth. Which of the following groups contain only natural plant fibres?
   (a) cotton, linen, and jute
   (b) polyester, cotton, and wool
   (c) sisal, rayon, and acrilan
   (d) nylon, cotton, and angora

3. Destructive distillation is the process of
   (a) turning wood into pulp
   (b) heating wood to obtain such products as methanol and tar
   (c) recycling paper to make new paper
   (d) separating the fibres from flax to make linen.

4. Why do natural insecticides have an advantage over some synthetic insecticides?
   (a) They cost less.
   (b) They can be obtained from most plants.
   (c) They are more powerful.
   (d) They can be degraded by decay organisms.

5. Some plants contain substances that are useful drugs. North American native peoples chewed on willow twigs when they
   (a) were hungry
   (b) had malaria
   (c) wanted to repel insects
   (d) needed a pain reliever.

6. You are going to test a substance to see if it contains starch. You will know it contains starch if you
   (a) add a drop of iodine solution and the substance turns blue-black
   (b) add a drop of limewater and the limewater turns milky white
   (c) bring a burning splint close to the substance and hear a "pop"
   (d) bring a glowing splint near the substance and see the splint burst into flame.

7. A coleus plant with green and white leaves is placed in the dark for 48 h. As soon as you take it from the dark, you test the leaf for starch. The reason for doing this is to show that
   (a) light is necessary for photosynthesis
   (b) chlorophyll is necessary for photosynthesis
   (c) carbon dioxide is necessary for photosynthesis
   (d) plants in the dark use up starch.

8. The word equation for photosynthesis may be written carbon dioxide + water + energy ⟶ starch + _____ . What word is missing?
   (a) hydrogen
   (b) nitrogen
   (c) oxygen
   (d) iodine

9. You are given a plant that has very thick leaves covered with a shiny, waxy cuticle. When you look at a section of one of the leaves through a microscope, you see that there are very few stomata. The mesophyll is very compact, with no air spaces. What environment is this plant adapted for?
   (a) a pond
   (b) a forest
   (c) a desert
   (d) the tundra
10. Green plants are important because
   (a) they produce the foods that every living thing needs to survive
   (b) they supply oxygen to the atmosphere and use up carbon dioxide
   (c) they can add beauty to our surroundings.
   (d) All of the above are correct.

TRUE/FALSE

Write the numbers 1 to 10 in your notebook. Beside each number, write T if the statement is true and F if the statement is false. For each false statement, rewrite the statement so it becomes true.

1. Many plants store chemical energy in the form of starch.
2. Photosynthesis is the process of getting energy from stored foods.
3. In photosynthesis, the reactants are carbon dioxide and water.
4. The products of photosynthesis are starch or sugar and oxygen.
5. Water that plants need is absorbed from the air by transpiration.
6. Carbon dioxide enters the leaves and green stems through stomata.
7. Oxygen produced in photosynthesis is carried to the soil by the xylem.
8. Most of the organisms on earth get their energy from photosynthesis.
9. Photosynthesis takes place whenever light, carbon dioxide, and oxygen occur together.
10. Plants in the dark use carbon dioxide and water to make starch.

FOR DISCUSSION

Read the paragraphs and answer the questions that follow.

There is a limit to the amount of fossil fuels in the world. In your lifetime, that limit may be reached. But energy from the sun continues to reach the earth every day. And every day, plants trap some of that energy. If we could find plants that grow quickly and produce a lot of material (biomass), perhaps we could use them as an energy source.

Forestry scientists have found such plants. They are poplar trees. Poplars grow quite quickly. Scientists at the forestry research station at Maple, Ontario have been working with varieties of poplar from around the world. They have crossed different varieties to produce faster-growing hybrids. A hybrid is the offspring of different kinds of parents of the same species. After the first year, an average hybrid poplar will grow 1.5 m to 2.5 m taller, and 1.8 cm to 2.5 cm thicker every year.

The wood of the hybrid poplars is a good source of fuel. It can also be used for lumber, plywood, chipboard, or pulp for making paper. The poplar leaves can be fed to cattle, supplying them with energy that will become human food.

1. Why should we be concerned about finding alternatives to fossil fuels for energy?
2. Why is energy from the sun a good alternative to energy from fossil fuels?
3. State two different ways that energy can be obtained from wood.
4. What were scientists looking for when they produced hybrid poplars?
5. Humans can use hybrid poplars as fuel. In what other ways can these trees be used?

267

Food and Energy

Did you know that if you live to be seventy years old, you will spend the equivalent of more than four years of your lifetime eating! The food you eat provides you with all the nutrients you need to keep you healthy and active.

In your great-great grandparents' time, most people grew enough food to feed their own families. Today, a relatively small percentage of the world's population is involved in growing enough food to feed everyone. The path of food from field to table starts even before the seeds are in the ground. It continues through planting, growth, harvesting, processing, and distribution to your table. In this unit, you will investigate the importance of food and the steps involved in bringing it to your table.

Food: Sources, Energy, Nutrients

Key Ideas

- Nutrients in food supply us with energy and with materials to build and repair cells.
- Plant parts that are used as food include roots, stems, leaves, buds, flowers, fruits, and seeds.
- A healthy and nutritious diet includes foods from each of the four food groups.

Think back to this morning. What did you have for breakfast? Did you gulp down a glass of milk and rush through a bowl of cereal? Perhaps you had crisp bacon and a fried egg. Were you so late for school that you only had time for a quick muffin on the run? Regardless of what you had for breakfast, your body is already using the food that you ate this morning. In this chapter, you will learn how your body uses the food that you eat each day.

Why We Need Food

A **nutrient** is a substance that your body can use for any of its daily activities. Some of the nutrients found in the food you eat are used as fuel for the body (Figure 17.1).

Figure 17.1 *The running teenager uses food as fuel to produce energy. The barbecue uses gas as fuel to produce energy. For what purpose will the energy be used in each case?*

Many nutrients do not give us energy. They are necessary for repairing any parts of the body that are worn out and need to be replaced. This keeps your body in good working order. The nutrients in your food provide the materials for these repairs.

For the first 20 years of your life, your body grows in size. Many nutrients are used for this purpose.

The nutrients that you need can be divided into six major groups: carbohydrates, fats (lipids), protein, water, vitamins, and minerals.

Carbohydrates

At least half of the energy you need each day comes from **carbohydrates**. Potatoes, breads, and grains like wheat and rice contain large amounts of carbohydrates (Figure 17.2).

Sugars are the simplest type of carbohydrate. *Starch* is more complex but it is changed into sugar in your body before being used.

Most of the carbohydrates that you eat each day cannot be stored but must be used right away. Any carbohydrate that is not used within several hours of being eaten is changed into *fat*. Fat can then be stored in your body for later use.

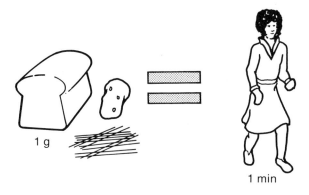

Figure 17.2 *One gram of carbohydrates provides you with enough energy to walk for about 1 min.*

Fats

Fats, also known as *lipids*, are found in butter, margarine, cooking oils, cheese, and nuts. They are important to have in the body for many reasons. Fats provide energy for daily activities. In fact, one gram of fat provides you with twice the energy of one gram of carbohydrates (Figure 17.3). Your body does not use fat as easily as carbohydrates. As a result, fat can be stored in various parts of your body. This fat insulates your inner organs to prevent them from getting cold. It also forms bulges that make it difficult to fit into last year's clothes!

Fats are an important part of cell membranes. If these membranes are damaged, fats are needed for quick repair. Some fats are necessary for the production of hormones. A **hormone** is a chemical messenger found within the body that is able to turn specific bodily functions on or off.

A particular type of fat known as **cholesterol** is an important part of all body cells. If too much cholesterol is present in the blood, it may begin to build up on the inside of the blood vessels, causing a condition known as **atherosclerosis**. As a result, the vessels are narrowed. This makes it difficult for blood to flow through them. If cholesterol continues to build up in the blood vessels, a heart attack or stroke may result.

Cholesterol is found in such foods as butter, egg yolks, and animal fat. It can also be made by the liver. In order to reduce the amount of cholesterol found in the blood stream, you should replace animal fat with fat from vegetable sources. Vegetable fat includes margarine and vegetable oil.

Protein

Your body uses most of the **protein** in your diet to build, maintain, and repair the cells of your body including skin, nails, hair, muscles, and nerves. Protein can also be used by the body to produce energy. One gram of protein provides about the same amount of energy as one gram of carbohydrates (Figure 17.4). Only 10% of all the energy needed by the body comes from protein.

Proteins have other functions within the body. Some are used as hormones to control the function of other tissues or organs. Other proteins act as **enzymes** which help to digest large food molecules. Fish, eggs, lean meat, nuts, and milk are all good sources of protein.

Figure 17.3 *One gram of fat provides you with enough energy to walk for about 2 min.*

Figure 17.4 *One gram of protein provides you with enough energy to walk for about 1 min.*

Water

Did you know that 70% of your body mass is water? Water is a very important part of your diet although it is not used as a source of energy. You cannot live very long without water. Even those people who go on hunger strikes would not survive more than a week without it.

Your body uses water for many things. *Perspiration* is mostly water. As this water evaporates from your skin, it removes heat from your body and helps to cool you off (Figure 17.5). All the chemical reactions that take place inside you would stop if water was not present. Water also helps your kidneys get rid of harmful waste materials.

You lose about 2.5 L of water each day as perspiration, water vapour in your breath, and urine. It is necessary for you to replace this volume each day. Don't worry, you don't have to drink 2.5 L of water! Many solid foods contain a great deal of water. Cucumbers and head lettuce are about 96% water. Even fried chicken is about 58% water. An average teenager needs to drink only five or six glasses of water a day to meet the body's needs.

Vitamins

Vitamins, like water, do not provide you with energy but are very important nutrients. There are many different vitamins and each has its own function (Table 17.1). Some of them can be made by the body. Most vitamins, however, must be obtained from the food you eat. The body needs only small amounts of each vitamin in order to function.

Vitamins take part in many chemical reactions that occur in your body. Problems may result if your diet is lacking in a particular vitamin. If, for example, you do not get enough vitamin C, you may develop *scurvy*. Your teeth would become soft and fall out. Your gums would bleed and you might die. This disease was common among early sailors during long voyages at sea. It was found that eating citrus fruits such as oranges, lemons, and limes prevented scurvy. These fruits are high in vitamin C.

Table 17.1 *Some Important Vitamins*

VITAMIN	FUNCTION	SOURCES
A	Necessary for maintaining healthy skin and hair, helps resist infection	Milk, liver, carrots
C	Helps maintain gums and teeth	Citrus fruit
D	Necessary for maintaining strong bones and teeth	Made by the skin in sunlight, fish oil
B	Helps the body to use carbohydrates	Whole grain cereals
K	Helps blood to clot	Green vegetables

Figure 17.5 *Remember how hot it was last summer?*

Minerals

When you think of **minerals**, you probably think of substances that are found in the ground and mined for industrial purposes—things like iron, zinc, copper, and so on. These minerals, and many more, are all needed to maintain a healthy body. A typical teenager with a body mass of 50 kg contains about 4 kg of minerals such as calcium, iron, iodine, and many others.

Luckily, it is not necessary to mine these minerals and eat them "raw"! Minerals are found in most of the foods you eat. Look at Table 17.2 to see how minerals help you to stay healthy.

Table 17.2 *Some Important Minerals in Your Diet*

MINERALS	FUNCTION	SOURCES
Calcium	Helps form bones and teeth, aids blood clotting, helps muscles function	Milk, cheese
Iron	Allows red blood cells to carry oxygen	Liver, green vegetables
Sodium	Controls water level in cells and blood, helps nerves to function	Table salt, most foods
Iodine	Controls activity of thyroid gland	Seafood, iodized salt

Food Analysis—Testing for Nutrients

It is possible to determine whether various nutrients are present in food samples using simple chemical tests. In the following experiments, you will test various food samples to determine if certain nutrients are present.

In each experiment, you will use water as a control because it does not contain any other nutrient and should give a negative result. You will test a substance known to contain the nutrient to see a positive result. You will then test three other samples for the presence of the nutrient. Your teacher may ask you to bring in some samples from home for class testing.

Ideas and Applications

Many years ago, it was observed that people living in mountainous, inland areas often developed a large swelling in their neck known as a goiter. It was found that these people were not getting enough iodine in their diet. Iodine is found in most seafood, which was lacking in the diet of these people. To overcome this problem, iodine was added to table salt and the problem was solved. Even today, if you read the label on a box of salt, you will see that it has been "iodized" to make sure that everyone obtains enough iodine to prevent goiter.

Activity 17A

Problem

Which of the following foods contain sugars?

Materials

water
glucose solution
milk
maple syrup
honey
5 Diastix reagent strips
5 clean test tubes
test tube rack

Procedure

1. Make a chart in your notebook similar to Table 17.3.
2. Line up five clean test tubes in the test tube rack.
3. Put water into the first test tube to a depth of 2 cm.
4. Add a sample of each test material to the remaining test tubes (Figure 17.6). Make sure you identify the contents of each test tube.
5. Dip the coloured end of a Diastix reagent strip into the water. Wait for 30 s. Observe any colour change of the Diastix.
6. Discard the Diastix.
7. Repeat steps 5 and 6 for all materials. A new Diastix must be used for each test.
8. Record your results in the observation chart in your notebook.

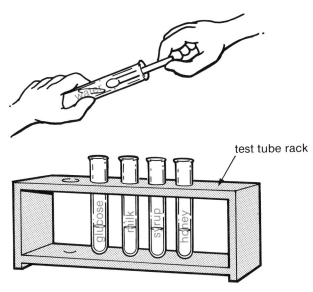

Figure 17.6 *Set-up for Activity 17A*

Observations

(See Table 17.3.)

Questions

1. Which of the substances that you tested contain sugar?
2. Examine the list from question 1. What do all of the substances containing sugar have in common?

Table 17.3 *Results of Sugar Test*

SUBSTANCE	ORIGINAL COLOUR OF DIASTIX	COLOUR OF DIASTIX AFTER DIPPING	SUGAR PRESENT? (YES/NO)
Water			no
Glucose			yes
Milk			
Maple syrup			
Honey			

Activity 17B

Problem

Which of the following foods contain starch?

Materials

water
starch solution
bread
macaroni
egg white
iodine solution
5 clean test tubes
test tube rack

Procedure

1. Make a chart in your notebook similar to Table 17.4.
2. Line up five test tubes in the rack.
3. Put water into the first test tube to a depth of 2 cm.
4. Put starch solution into the second test tube to a depth of 2 cm.
5. Add each of the other materials to the remaining three test tubes (one sample per test tube).
6. Recall the CAUTION about the use of iodine from Units 4 and 5. Add five or six drops of iodine solution to each test tube.
7. Observe any colour change in the iodine solution.

Observations

(See Table 17.4.)

Questions

1. Which of the substances that you tested contained starch?
2. What other foods might you expect to contain starch? Explain why.

Table 17.4 *Results of Starch Test*

SUBSTANCE	ORIGINAL COLOUR OF IODINE	COLOUR OF IODINE AFTER ADDING TO SAMPLE	STARCH PRESENT? (YES/NO)
Water			no
Starch			yes
Bread			
Macaroni			
Egg white			

SAMPLE ONLY

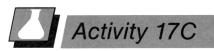

Activity 17C

Problem

Which of the following foods contain fat?

Materials

water
corn oil
margarine
mustard
peanut butter
brown paper
pencil

Procedure

1. Make a chart in your notebook similar to Table 17.5.
2. Draw five circles about 4 cm in diameter on the brown paper. Below each circle, print the name of each of the five substances to be tested (Figure 17.7).
3. Put a small drop of water in the middle of the circle marked "water."
4. Put a small drop of corn oil in the middle of the circle marked "corn oil."
5. Rub a small amount of margarine, mustard, and peanut butter in the middle of the circles marked "margarine," "mustard," and "peanut butter."

6. Leave the brown paper overnight to dry.
7. The next day, hold the brown paper up to the light. Observe whether or not light is able to pass through any of the five spots.
8. Record your observations in the chart in your notebook.

Figure 17.7 *Set-up for Activity 17C*

Observations

(See Table 17.5.)

Questions

1. Which of the substances that you tested contained fat?
2. What do all of the substances containing fat have in common?
3. What other foods might you expect to contain fat? Explain why.

Table 17.5 *Results of Fat Test*

SUBSTANCE	WAS LIGHT ABLE TO PASS THROUGH? (YES/NO)	FAT PRESENT? (YES/NO)
Water		no
Corn oil		yes
Margarine		
Mustard		
Peanut butter		

Activity 17D

Problem

Which of the following substances contain protein?

Materials

water
fingernail clippings
egg white
lard
cream cheese
nitric acid
5 test tubes

Procedure

1. Make a chart in your notebook similar to Table 17.6.
2. Line up five clean test tubes in the test tube rack.
3. Put water into the first test tube to a depth of 2 cm.
4. Add a sample of each test material to the remaining test tubes. Make sure that you identify the contents of each test tube.
5. Add three drops of nitric acid to each test tube (Figure 17.8).

> **CAUTION** Be very careful because nitric acid "burns" skin, clothing, books, and desks.

6. Observe any colour change in the nitric acid where it is in contact with the sample.

7. Record your results in the observation chart in your notebook.
8. Discard the contents of the test tubes as instructed by your teacher.

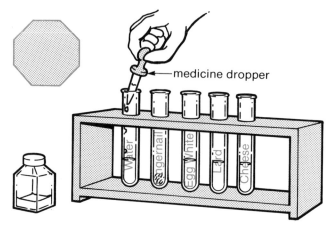

Figure 17.8 *Make sure that you do not drop any nitric acid on your skin or clothing.*

Observations

(See Table 17.6)

Questions

1. Which of the substances that you tested contained protein?
2. What other foods do you think might contain protein?
3. Besides fingernails, what other parts of the body might contain protein?

Table 17.6 *Results of Protein Test*

SUBSTANCE	ORIGINAL COLOUR OF NITRIC ACID	COLOUR OF NITRIC ACID AFTER ADDING TO FOOD SAMPLE	PROTEIN PRESENT? (YES/NO)
Water			no
Fingernail			yes
Egg white			
Lard			
Cream cheese			

Self-check

1. (a) List the six major groups of nutrients.
   (b) State the function of each of the nutrients listed in (a).
2. What do carbohydrates, lipids, and proteins all have in common?
3. What happens to carbohydrates not used by your body within a few hours after you eat them?
4. Why does your body need water?
5. What do you think is the best way to make sure that you get all the vitamins and minerals that your body needs to stay healthy?

Ideas and Applications

Carbohydrates are the source of most of our food: we eat starch-containing plants, or feed these same plants to animals. The animals convert the starch into meat or fat which we also eat. We clothe ourselves with cellulose in the form of cotton, linen, and other natural fibres. We build houses and furniture from cellulose in the form of wood. Thus, carbohydrates provide us with many of the basic necessities of life: food, clothing, shelter.

Energy Content of Food

You are going to be playing first string in the big basketball game after school. Your coach tells you to eat a good lunch. You are so worried about the game that you would rather go to the library and study the drill book (Figure 17.9). What should you do?

Figure 17.9 *Which activity is the best way of preparing for the game?*

Your coach gave you good advice. The food you eat is the fuel that your body will burn for energy. You are going to need energy to walk, run, and play in the basketball game.

Energy is the ability to do work. Energy can be found in many different forms (Figure 17.10). The energy found in food is called **potential chemical energy**. It is "potential" because the energy is being stored in the food, waiting to be used. It is called "chemical" because the energy can only be released through chemical reactions in the body. It is called "energy" because it can be used to do work. The potential chemical energy found in food can be changed into other forms of energy so that your body will be able to use it to do various types of work.

When you go camping, you use fuel like wood to produce heat to boil a pot of water. When you eat, your body uses fuel like apples, bread, and meat. This provides the energy for you to swim, hike, and run. It can also provide heat. This heat is used to keep your body temperature about 37°C all year around.

Imagine that you are given three food samples. Will all of these foods produce the same amount of energy? Could you determine which one provides the most energy? The following activity will help you answer these questions.

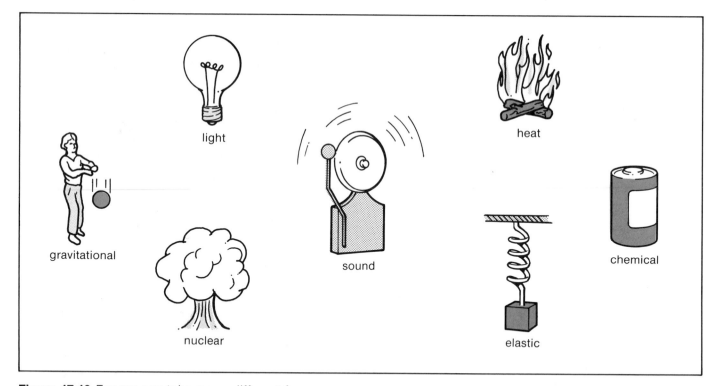

Figure 17.10 *Energy can take many different forms.*

280

Activity 17E

Problem

How can you determine how much energy is contained in a peanut?

Materials

peanut
pin
cork
balance
test tube holder
Pyrex test tube
graduated cylinder
10 mL of water
thermometer
match
safety goggles

> **CAUTION** Safety goggles must be worn whenever you are using a flame.

Procedure

1. In this activity, you must make up the instructions yourself. Keep in mind the following facts:
   (a) Food is just like any other fuel. It can be burned to produce energy.
   (b) Energy from a burning fuel can be used to heat water.
   (c) The more heat you apply, the hotter the water will become.
   (d) You may wish to repeat your experiment using a different kind of food. Make sure your experiment is controlled so you will be able to compare the results of future experiments with the original one.

2. When you have finished designing the procedure, show it to your teacher. When it has been approved, proceed with the experiment.
3. If you have time, repeat the experiment using a different type of food.

Observations

1. Make an observation chart comparing all conditions at the beginning of the experiment with all conditions at the end of the experiment.

Questions

1. What was the energy in the peanut used to do?
2. Was any energy from the peanut wasted in this experiment? If so, how was it wasted?
3. How could you improve the experiment to prevent energy wastage?
4. Suppose that you were able to test several different types of food. How would you know which one contained the most energy?
5. If you actually did test more than one food, which one contained more energy?

Food Energy Comparison

When you eat, your body digests the food and releases the energy for you to use. The amount of energy available from food is measured in **kilojoules (kJ)**. The amount of energy needed to lift a 50 kg teenager up over your head is about 1 kJ. An average slice of cucumber provides you with about 4 kJ of energy—enough energy to lift four teenagers over your head (or lift one teenager four times).

Look closely at Table 17.7. It lists several different kinds of foods and shows how much of each food is equal to 100 g. If you ate a 100 g portion of each food, see how much energy you would get! Compare the length of time that you could ride a bicycle if you ate 100 g of the foods listed.

Just as natural gas or electricity provides the energy to heat your home or run appliances, food provides the energy for your body to run and swim. If you could use food to supply the energy needed to run the appliances in your home, you would be able to run your refrigerator for nearly 6 min or keep a 150 W (watt) light bulb lit for nearly 40 min using the energy contained in one chocolate brownie!

Self-check

1. Define energy.
2. What unit is used to measure the amount of energy contained in food?
3. Using the information from Table 17.7, determine which food contains
   (a) the most energy
   (b) the least energy.
4. Using the information from Table 17.7, calculate the amount of energy contained in a bacon and tomato sandwich consisting of five slices of bacon, four slices of tomato, two slices of bread and 15 mL of mayonnaise. (Note: 15 mL of mayonnaise contains about 400 kJ of energy.)

Table 17.7 *Food Energy Comparison*

FOOD	AMOUNT EQUAL TO 100 g	ENERGY IN 100 g OF FOOD (kJ)	BIKE-RIDING TIME (min)
Chocolate brownie	5 brownies	1995 kJ	67
Bacon	5 slices	1260 kJ	42
Bread	4 slices	1092 kJ	36
Pork chop	1 medium	1092 kJ	36
Ice cream	1 large bowl	802 kJ	27
Peanuts	250 mL	583 kJ	19
Potato (baked)	1 medium	378 kJ	13
Cottage cheese	1 serving	357 kJ	12
Carrots (raw)	2 average	168 kJ	6
Strawberries	1 serving	151 kJ	5
Peach	1 small	130 kJ	4
Tomato	4 slices	93 kJ	3

Plants as Food

In Chapter 14, you found that plants are used for many different purposes in today's world. Serving as a food source for ourselves and animals is one of their most important uses.

Among the food plants, all of the parts of the plant (stems, roots, leaves, flowers, and fruit) are used as food. Some parts of some food plants, however, are poisonous and must not be eaten.

Plant Stems as Food

Asparagus is one of the first locally grown fresh vegetables to be seen in supermarkets in the spring (Figure 17.11). Because it loses nutrients very quickly when stored, it is best to buy it as soon after harvesting as possible. Today, you can enjoy asparagus fresh, frozen, or canned. With an energy value of 10 kJ per spear, it is a delicious addition to any meal.

Figure 17.11 *The asparagus you eat is a stem.*

Believe it or not, the potato that is served so often at mealtime is the modified stem of the potato plant (Figure 17.12). The tuber, as this specialized stem is called, grows underground.

Figure 17.12 *Here is a field of potato plants. The part of the plant that you eat is a tuber, which grows underground.*

Potatoes were brought from South America to Europe around 1550. It was the Irish, however, who made the potato a popular food crop in the seventeenth century. When a fungus disease wiped out the Irish potato crop in 1845, hundreds of thousands of Irish people migrated to America so that they would not starve.

Today, the potato is one of the most widely grown vegetables in the world. It is grown on every continent except Antarctica. As a result, it is eaten in a variety of ways. How many can you think of? The average potato contains 378 kJ of energy. It is a good source of vitamin C.

Plant Leaves as Food

The leaves of many plants are used for food. The cabbage is only one example of a "leafy" vegetable (Figure 17.13). The leaves of cabbage grow around each other to form a "head" that can have a mass of from 1 to 3 kg. It can be used cold to make coleslaw or served hot.

Figure 17.13 *A head of cabbage is made up of the leaves of the cabbage plant.*

Plant Flowers as Food

Most people think of flowers as the pretty, coloured parts of the plant used to attract insects. The cauliflower is one flower that is commonly eaten for food (Figure 17.14). The white variety, which is the more popular, is somewhat difficult to grow as it tends to get "sunburnt." When ripe, the heads are harvested and can be frozen, canned, pickled, or served hot at mealtime. Raw cauliflower makes a good snack.

Figure 17.14 *The cauliflower you eat is the flower of the plant.*

Plant Fruit as Food

Most people think of fruit as the sweet, juicy part of the plant that you put in a lunchbag or slice up over ice cream—things like oranges and strawberries. These are both examples of fruit. From a scientific point of view, the **fruit** is the ripened ovary of a flower that contains the seeds.

This means that peppers, cucumbers, squash, and peas, as well as watermelons and apples are all fruits. Tomatoes, because they contain seeds, are also classified as fruits. The tomato is nutritious and contains about 150 kJ of energy.

Like potatoes, tomatoes came from Peru and were brought to Europe by Spanish priests. Here, they were known as "golden apples" and even "love apples." Despite these nicknames, people in North America were not anxious to eat them. They thought tomatoes were poisonous. In fact, tomato stems and leaves are poisonous but the fruit is not. Once people realized this, they began planting them in their gardens (Figure 17.15). They are easy to grow and will produce fruit in as few as 45 d. Each tomato plant will produce 4.5 to 7.0 kg of tomatoes in one season. Tomatoes can be used in salads, soups, and sauces. Commercially, they can be canned or used to make paste, ketchup, or juice.

Plant Roots as Food

Carrots are a good source of vitamins, especially vitamin A. Each average-sized carrot provides 84 kJ of energy. Carrots are available throughout most of the year. This vegetable can be bought fresh, frozen, or even canned.

Carrots have many uses. Besides being served with roast beef and potatoes, carrots can be used to make soups, stews, muffins, and even cake. Next time you reach for a bag of potato chips, think again—fresh carrot sticks make a healthy and nutritious snack.

Ideas and Applications

The B and C vitamins are water soluble. This means that they can dissolve in cooking water and be lost from food before we eat it. One way to prevent this loss of vitamins is to steam vegetables instead of boiling them. Steaming vegetables has another advantage over boiling them. The vegetables have a brighter colour and a firmer, crisper texture.

Figure 17.15 *This tomato plant is full of ripe fruit.*

Problem

From what part of the plant do the following foods originate?

Materials

various plants from the produce section of the grocery store (will vary with the season)

Procedure

1. Make a chart in your notebook similar to Table 17.8. In the column under "Plant Name," list the various plant foods that your teacher has displayed at the front of the room.
2. Take one sample back to your desk.
3. Examine the food sample. Clearly describe its colour, texture, size, shape, odour, and any other physical characteristic. Do not taste any samples.
4. Return the food sample to the front and obtain a second sample.
5. Repeat steps 2–4 for all remaining samples.
6. Wash your hands when you have finished examining all samples.
7. When all observations have been completed, compare the descriptions of the various foods and try to identify the part of the plant from which each came. Is it a root, stem, leaf, flower, or fruit?

Observations

Table 17.8 *Description of Food Samples*

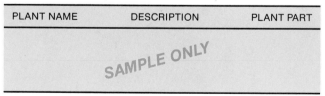

PLANT NAME	DESCRIPTION	PLANT PART
SAMPLE ONLY		

Questions

1. Make a list of all plant foods from the activity that were
   (a) stems
   (b) roots
   (c) leaves
   (d) flowers
   (e) fruits.

2. What characteristics did you use to determine if the plant was
   (a) a stem?   (d) a flower?
   (b) a root?   (e) a fruit?
   (c) a leaf?

Ideas and Applications

More and more people are becoming interested in eating wild plants—and with good reason! Some kinds of wild plants are very nutritious and tasty. But if you plan to search for "wild edibles," remember that the fields and meadows in which such plants grow are also full of poisonous plants that may look like the safe ones. Unless you are with someone who has had a lot of experience collecting wild edible plants, do not eat plants you find in a field. If eaten, some plants might cause nausea, vomiting, and even death.

Self-check

1. List three ways that plants can be prepared before they are eaten.
2. Classify each of the following foods as roots, stems, leaves, flowers, or fruits.
   (a) apple   (g) corn
   (b) bean   (h) lettuce
   (c) beet   (i) onion
   (d) broccoli   (j) parsnip
   (e) brussel sprouts (k) rhubarb
   (f) celery   (l) spinach
3. Based on your knowledge from Unit Five, explain why most stem and leaf vegetables are green.

Eat Right and Stay Healthy

Everyone knows that you have to "eat right" if you want to stay healthy. But how do you know what is right? It is necessary to eat both the right type and the right amount of food. The Department of Health and Welfare has prepared a food guide to make meal planning easier (Table 17.9). The numbers in parentheses indicate the number of kilojoules of energy per serving. Foods are divided into four basic food groups: dairy products, breads and cereals, fruits and vegetables, and meat and meat alternatives.

Table 17.9 *Canada's Food Guide*

DAIRY (3 servings a day)	BREADS AND CEREALS (3–5 servings a day)
1 glass of milk (520 kJ) 28 g cheddar cheese (480 kJ) 250 mL of yogurt (600 kJ)	1 slice brown bread (330 kJ) 1 slice white bread (340 kJ) 1 bran muffin (400 kJ) 1 plate of spaghetti (330 kJ)

FRUITS AND VEGETABLES (4–5 servings a day)	MEAT AND ALTERNATIVES (2 servings a day)
6 cherry tomatoes (100 kJ) 1 salad (200 kJ) ½ banana (230 kJ) 8 carrot sticks (160 kJ) 1 apple (290 kJ) glass of orange juice (500 kJ)	125 mL cottage cheese (360 kJ) 1 chicken breast (500 kJ) 125 g of ham (1300 kJ) 2 boiled eggs (740 kJ) 50 mL peanut butter (1100 kJ)

The Right Amount

The food you eat supplies you with the nutrients for maintaining your body in good condition. It also provides you with energy. If you do not eat enough in a day to supply your energy needs, your body will begin to use up fat that is stored in the body. If you eat too much food, your body will store the energy as fat. To maintain your body mass, you should eat just enough food to provide you with the energy you need to complete the day's activities.

A 50 kg teenage boy between the ages of 13 and 15 requires about 11 700 kJ of food energy each day. An average girl of the same mass and age needs only about 8 200 kJ of food energy each day. If you are more active, you will need to eat more. If you are less active, you will need to eat less. It is all a matter of counting your kilojoules!

Challenge

Why do you think boys need a greater energy supply than girls of the same age and mass?

Problem

How can you plan a healthy lunch?

Materials

Table 17.9 on page 287

Procedure

1. Examine the foods shown in Table 17.9. You will be using these foods to make your imaginary lunch.
2. Choose at least one item from the dairy group.
3. Choose at least one item from the meat and alternatives group.
4. Choose at least two items from the fruit and vegetable group.
5. Choose at least one item from the bread and cereal group.
6. Calculate the number of kilojoules of energy in your lunch. The total energy value of your lunch must be between 2800 kJ and 3200 kJ.
7. If you wish, you may add in some of the items shown below. Make sure that the total does not exceed 3200 kJ.

jam (enough for a sandwich)	340 kJ
candy bar	890 kJ
soft drink	480 kJ
diet soft drink	2 kJ
chocolate chip cookie	210 kJ
butter (2 pats)	300 kJ
salad dressing (for one salad)	80 kJ

Observations

1. List the items you have selected for your lunch and show how you calculated the total number of kilojoules in this lunch.

Questions

1. Which food group generally contains the highest number of kilojoules per serving?
2. A lunch consisting of eight cookies, one soft drink, and one candy bar has 3050 kJ. This contains the right number of kilojoules. Why is it not a good lunch?
3. To which food group or groups would each of the following items belong?
   (a) baked potato
   (b) milkshake
   (c) fillet of fish
   (d) apple juice
   (e) corn flakes

Self-check

1. Name the four basic food groups.
2. How many servings of meat or meat alternative should you eat each day?
3. Suppose that you want to lose weight. For breakfast, you have a glass of orange juice. For lunch you have a large salad. For dinner, you have a slice of bread and a diet soda.
   (a) Will you lose weight? Why or why not?
   (b) Is this a healthy diet? Why or why not?
4. How will exercising help you to lose weight?

Chapter Objectives

NOW THAT YOU HAVE COMPLETED THIS CHAPTER, CAN YOU DO THE FOLLOWING?	FOR REVIEW TURN TO SECTION
1. List the reasons why your body needs food.	17.1
2. State the function of carbohydrates, fats, proteins, vitamins, minerals, and water.	17.1
3. Determine if sugars, starches, fats, or proteins are present in a food sample.	17.1
4. Compare the energy content of various foods.	17.2
5. Observe and describe examples of plants that are used for food.	17.3
6. Identify the parts of plants used for food, and give examples of food from each part.	17.3
7. Classify foods based on the part of the plant from which they are derived.	17.3
8. List, with examples, the four basic food groups based on Canada's Food Guide.	17.4
9. List the number of servings from each food group that you should eat every day according to Canada's Food Guide.	17.4
10. Plan a healthy and nutritious meal.	17.4

Words To Know

nutrient
carbohydrate
fat
hormone
cholesterol
atherosclerosis
protein

enzyme
vitamin
mineral
energy
potential chemical energy
kilojoule (kJ)
fruit

Tying It Together

1. Which of the following nutrients can be used as an energy source—carbohydrates, fats, minerals, proteins, vitamins, water?
2. What might happen to you if the following were true?
   (a) Your diet is lacking in vitamin C.
   (b) Your diet is lacking in iodine.
   (c) Your diet contains too much cholesterol.
   (d) Your daily diet contains more kilojoules of energy than you use up each day.
3. List three reasons why your body needs water.
4. List three minerals necessary for good health.
5. In your notebook, match the nutrients from the first column with their tests in the second column.

NUTRIENT	POSITIVE TEST
(a) Starch	(i) Brown paper becomes translucent.
(b) Fat	(ii) Diastix turns brown.
(c) Sugar	(iii) Nitric acid turns black.
(d) Protein	(iv) Iodine turns black/purple.

6. Why is food often said to be "fuel for the body"?
7. State the part of the plant (root, stem, leaf, flower, fruit) from which each of the following comes:
   (a) cucumber
   (b) turnip
   (c) cauliflower
   (d) radish
   (e) sugar cane
   (f) blueberries

8. List three different sections in a grocery store where you might find food products derived from plants.
9. List, with examples, the four basic food groups based on Canada's Food Guide.
10. Using the information from Table 17.9, calculate the total number of kilojoules of energy you would get by eating a meal consisting of a glass of milk, a chicken breast, four carrot sticks, two bran muffins, and an apple.

Applying Your Knowledge

1. A student took an unknown sample of food and got the following results:
   The Diastix stayed blue.
   The iodine turned black.
   The nitric acid turned black.
   The brown paper did not allow light to pass through.
   What can you conclude from these results?
2. *Anorexia nervosa* is a disease that usually affects young girls. People who suffer from this disease are convinced they are overweight, no matter how slim they really are. They diet so drastically that they eat almost nothing. Explain why this is such a serious disease.
3. Examine Table 17.10 which summarizes the energy value of some common fast foods.

Table 17.10 *Energy Value of Some Fast Foods*

FOOD	ENERGY VALUE
Big Mac	2340 kJ
French fries	966 kJ
Onion rings	1260 kJ
Pizza (half of 35 cm)	3780 kJ
Milkshake	1428 kJ
Ice cream cone	966 kJ
Fudge sundae	2436 kJ
Three-piece chicken dinner	4135 kJ

Select a typical fast food meal that you might have. Calculate the total number of kilojoules in that meal. Compare this to the meal you designed in Activity 17G. Describe the advantages and disadvantages of these two meals.

Projects for Investigation

1. Choose one of the following diseases that result from vitamin deficiency. Find out which vitamin is lacking, what the symptoms of the disease are, and how it can be treated.

   Vitamin Deficiency Diseases: rickets, beriberi, pellegra, night blindness, pernicious anemia

2. Try to find out something interesting about each food that you studied in Activity 17F. This could include its uses, nutrient content, or how it is grown. Your family, the library, a seed catalogue, or a good cookbook will help.
3. Is fluoride added to drinking water in your area? Do research to investigate the value of fluoride. What is the recommended daily intake of fluoride? Officials at your town or city hall may be able to help you determine why fluoride is or is not added to your local drinking water.
4. Some people cannot drink milk and in some cases cannot eat any dairy products. They suffer from a condition called "lactose intolerance." Find out the causes of this condition. How can people who have lactose intolerance obtain sufficient amounts of calcium if they cannot drink milk or eat dairy products?
5. There are many myths and theories about the relationship between either specific nutrients or overall diet and the occurrence of acne. Visit the office of a dermatologist (a doctor who specializes in skin diseases), and ask for some information on this topic. Review any pamphlets or other material the doctor can provide. Summarize your findings.

Starting and Growing Food Plants

Key Ideas

- New plants are started from seeds or cuttings.
- New plants need the right temperature, moisture, and soil.
- Fertilizers help supply plants with necessary minerals.
- Viruses, bacteria, fungi, insects, and other invertebrates use plants as food, causing plant diseases.

If you have grown vegetables at home, you know that nothing tastes as good as the ones picked fresh from your own garden. To an outside observer, growing plants for food seems like an easy task—plant the seeds in the ground, water them once in a while, and pick the crop when it is ready. Ask a farmer or a serious vegetable gardener and you will find out that it is not quite this easy. In this chapter, you will find out just how complex farming is.

How Do Plants Start?

Seeds

Seeds come in different sizes. Some seeds, like those of radishes, are so small that they look like specks of dust. Other seeds, like those of the coconut palm, can weigh up to 7 kg. Regardless of their size, all seeds have the same basic structure (Figure 18.1).

A typical seed contains a tiny plant-like structure called an **embryo**. When the seed sprouts, the embryo will become the plant. Stored food is also found in the seed. This will supply the embryo plant with nourishment until it sprouts and begins to make its own food. The embryo plant and the food supply are surrounded by a hard **seed coat** that protects them and prevents sprouting until the environment is favourable for growth.

Dormancy refers to the resting stage of a seed during which it remains alive but does not sprout. The seed is respiring at a very slow rate. Seeds can remain in a period of dormancy for a long time. Most vegetable seeds can remain dormant for three to five years if they are stored in a cool, dry place. Some weed seeds can remain dormant for up to 100 years and will still grow when conditions are suitable.

Good seeds are clean and free from disease. They should also have a high viability rate. A seed is **viable** if it will sprout when supplied with moisture at the proper temperature. If the seeds will not grow, they are of no use to the gardener or farmer. The **viability rate** of seed refers to the percentage of seeds that will sprout when planted.

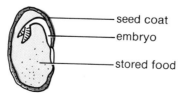

Figure 18.1
Parts of a seed

seed coat
embryo
stored food

Seed developers perform tests on samples of their seeds to determine the viability rate. In this activity, you will perform a similar test to determine the viability rate of packaged seeds from the store.

Problem

How do the conditions of germination affect the viability rate of seeds?

Materials

seeds (radish, tomato, lettuce, beans, etc.)
3 strips of white paper towelling (each 3 sheets long)
mist sprayer
pen

Procedure

1. In your notebook, make a chart similar to Table 18.1.
2. Obtain 60 seeds (all of one type) from your teacher.
3. Lay out the three strips of towelling. See Figure 18.2 (a).
4. Fold over the square on the right-hand side of each strip. On this square, print in pen your name, the date, and the type of seed you will be using. See Figure 18.2 (b). Also print the conditions under which the seeds will be germinated: one set will be in a cool place, one set will be at room temperature, and the third will be in a warm place.
5. Unfold the towels again, and moisten each towel using water from the mist sprayer. See Figure 18.2 (c).
6. Place 20 seeds on the centre towel of each set. Separate the seeds from each other as much as possible. See Figure 18.2 (d).
7. Fold each set of towels. See Figure 18.2 (e). The strips of towelling should look like the one shown in Figure 18.2 (f).
8. Place the towels in the appropriate places as directed by your teacher. Do not disturb the seeds for the remainder of the activity.
9. Check daily to make sure that the towels stay moist. Spray them with water if necessary.

Count the number of seeds that have germinated and record this information in the chart in your notebook (Table 18.1).

10. Repeat this procedure for ten days.
11. In your notebook, make a table similar to Table 18.2. Enter the results from other members of the class in this table at the end of ten days.

CAUTION Wash your hands after handling the seeds. Some seeds have been treated with poisonous chemicals called fungicides to prevent the growth of fungus.

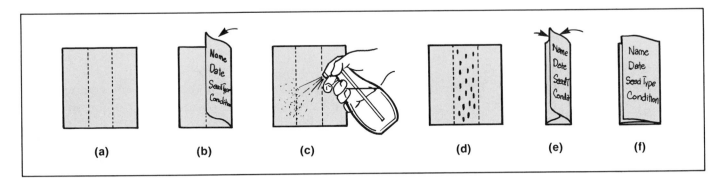

Figure 18.2 *Set up paper towels: (a) lay the towels flat, (b) fold over one square and write on it, (c) unfold again and moisten towels, (d) place 20 seeds on centre of towel, (e) fold both sides in, and (f) be sure that the writing can be seen.*

Observations

Table 18.1 *Germination of _____ Seeds*

CONDITION	NUMBER OF SEEDS GERMINATED									
	Day 1	Day 2	Day 3	Day 4	Day 5	Day 6	Day 7	Day 8	Day 9	Day 10
Cool temperature										
Room temperature										
Warm temperature										

SAMPLE ONLY

Table 18.2 *Class Results*

SEED TYPE	CONDITION	NUMBER OF SEEDS GERMINATED BY DAY 10 (OUT OF 20)
	Cool temperature	
	Room temperature	
	Warm temperature	
	Cool temperature	
	Room temperature	
	Warm temperature	

SAMPLE ONLY

Questions

1. Which condition resulted in the earliest germination?
2. Which condition resulted in the germination of the largest number of seeds?
3. Using the data from Table 18.2, calculate the viability rate of each type of seed under each condition.
4. Which temperature was best for germination?
5. Did all seed types germinate best under the same conditions? Explain.
6. Of the seeds studied in this experiment, which ones would you plant earliest in the spring? Why?
7. The seeds germinated at room temperature were placed in a box and not left out in the open. Why?
8. Besides temperature, what other conditions might affect germination?

Challenge

Design and complete an experiment to test the effect of various conditions on seed germination. For example, how could you determine whether the amount of moisture present will affect seed germination? Before performing the experiment, show your proposed procedure to your teacher for approval.

Seed Planting

To grow vegetables in Canada, you must plant seeds as early as possible in the spring to use the best growing conditions and the longest period free from frost. Frost kills most vegetable plants. Some seeds, like peas, continue to grow in cool weather but are still killed by frost.

Most seeds are buried at a depth that is two to three times the diameter of the seed (Figure 18.3). If seeds are planted too close to the surface, they will dry out before sprouting. If they are planted too deeply, all of the stored food will be used up before the seedling reaches the surface.

It is important to plant seeds in loose soil so that the seedlings will be able to push their way easily to the surface. As well, the roots will be able to grow downward with little difficulty. Once planted, the seeds must be kept moist. The seeds must absorb water in order to soften the seed coat. The softened seed coat will allow the embryo plant to break through and begin to grow.

The backyard gardener often starts plants indoors in small pots or egg cartons while snow is still on the ground so that seeds can start to grow inside and be well sprouted when outdoor conditions are warm. When the outside soil is warm enough, the seedlings are transplanted to the garden.

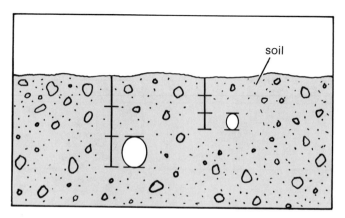

Figure 18.3 *Plant seeds at the right depth.*

Growing Plants without Seeds

Vegetative propagation is the process of producing plants without using seeds. This is a form of *asexual* reproduction. There are several different ways to do this. These methods provide quick results. Seedless fruits and vegetables can be produced using these methods.

Cuttings

Many plants can be easily grown from cuttings. A **cutting** is a leaf, stem, or root that has been removed from a growing plant. The cutting may be treated with a chemical to help new roots to grow. It is then put into either a container of water or directly into soil. Soon the rest of the plant will grow. This method is easy and highly successful.

Underground Stems

A **bulb** is a short, underground stem surrounded by leafy layers containing food. Inside these layers are one or more small bulbs (Figure 18.4). These small bulbs can be separated from the original one and then planted elsewhere to create a new plant. An onion is a good example of a bulb.

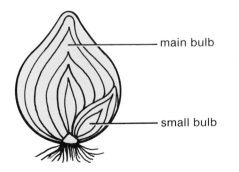

Figure 18.4 *The small bulb can be removed from the main bulb and used to start a new plant.*

Tubers are another type of underground stem. In this case, the stem becomes very big and stores large quantities of starch. A potato is an example of a tuber. If it is cut into smaller pieces, each containing an *eye*, each piece will grow into a new potato plant (Figure 18.5).

Figure 18.5 *When a chunk of potato containing an eye is planted, a new potato plant will grow. Potatoes that are sold in supermarkets have usually been treated with chemicals that slow down the growth of the eyes. These should not be used for planting. Farmers usually keep their own seed potatoes or buy them from a seed company.*

295

Runners

Some plants, like the strawberry, send out stems that crawl along the ground. These are known as **runners**. Where the runner is in contact with the ground, new roots can form (Figure 18.6). Once the roots develop, you can cut the original stem and move the new plant to another location. A healthy strawberry plant can produce as many as ten new plants from each runner.

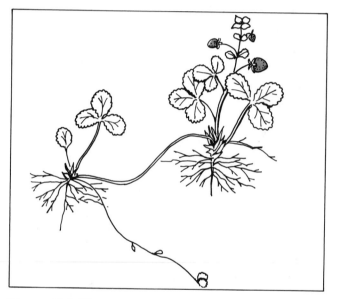

Figure 18.6 *The strawberry plant is a popular homegrown fruit. It reproduces by runners.*

Layering

Layering is used to produce new plants from such shrubs as raspberries. Where a branch droops over and touches the ground, new roots may develop. You can encourage root growth by heaping some soil on the part that is touching the ground (Figure 18.7). It is best to do this in the late summer so that roots will develop in the fall. In the spring, you can cut the old stem to separate the plants. The new one is then ready to be moved.

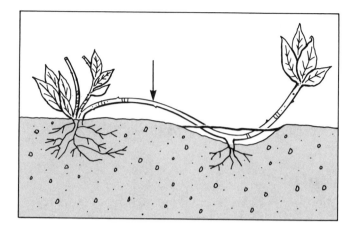

Figure 18.7 *Layering—If a cut is made at the arrow, two new plants will result.*

Grafting

Grafting is a very common technique used by fruit farmers. In this method, two stems are joined together and grow as one (Figure 18.8). The place where the joining took place is carefully sealed so that infection is prevented. It is important that the conducting tissues (xylem and phloem) of the two stems be closely lined up.

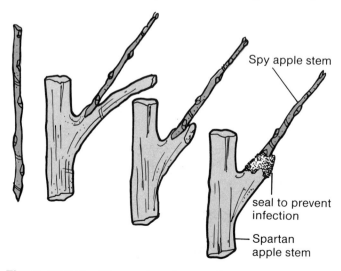

Spy apple stem

seal to prevent infection

Spartan apple stem

Figure 18.8 *Grafting a stem from a Spy apple onto that of a Spartan tree*

When this is done, the two stems will grow together, as a layer of cells (the cambium) between them continues to produce new xylem and new phloem by mitosis.

Grafting is done for many purposes. A branch from a frail tree with delicious fruit can be grafted onto a hardy tree with poor fruit. The result will be a hardy tree with delicious fruit.

Grafting can also be done to grow several different varieties of fruit on one tree, providing the trees are closely related. It would be impossible to grow apples on an orange tree or bananas on a plum tree, but one apple tree could be made to produce several different types of apples.

Self-check

1. (a) List the three main parts of a seed.
   (b) What is the function of each part?
2. What might happen to the germinating plant if
   (a) the seeds are planted too deep in the soil?
   (b) the seeds are planted too close to the surface?
3. Why must seeds be planted in loose soil?
4. (a) List four methods of vegetative propagation.
   (b) Give an example of a plant that reproduces by each method.
5. Name and describe two types of underground stems. Include an example of each.
6. Give two reasons why grafting might be used by fruit farmers.

Requirements For Growth

Once the seeds have sprouted or the cuttings have taken root, just sit back and wait to harvest the crop. Right? Wrong! Most food producing plants require lots of care in order to produce the largest possible crop.

Light

Plants require light for photosynthesis. As you learned in Unit Five, photosynthesis is the process in which carbon dioxide and water are used to produce food and oxygen. This is accomplished by green leaves using light energy. If there is not enough light, the plant will not be able to produce enough food for itself.

The amount of light required for growth depends on the individual plant. Most vegetables require a minimum of five to six hours of sunlight each day, although increasing this will result in more vigorous growth. Lettuce and pumpkin plants are said to be *shade tolerant* because they can survive with less light. Some plants such as ferns have adapted to low light conditions like those found on the forest floor. It is important to know the light requirements of your plants before you place them in the garden.

Challenge

Investigate the effect of light on plants at home. Grow some seedlings in small styrofoam cups. In the morning, place them near a window. By the end of the day, look to see if there has been a change in the direction of growth. After a few days, turn the seedlings around and watch what happens.

Activity 18B

Problem

How well do plants grow under various light conditions?

Materials

pea or sunflower seeds
sterilized, weed-free potting soil
4 large styrofoam cups
water
4 clear plastic bags
4 twist ties for the plastic bags
fluorescent lights
large tray

Procedure

1. Make a chart in your notebook similar to Table 18.3. You will need to make spaces for observations over many days.
2. Fill all of the cups to within 4 cm of the top with potting soil.
3. Obtain eight seeds (all of one kind) from your teacher.
4. Put two seeds in each cup.
5. Top with soil so that the seed is at a depth equal to 2–3 times the diameter of the seed. (Refer to Figure 18.3 if necessary.)
6. Moisten the soil with water.
7. Put each cup in a plastic bag and secure with a twist tie as shown in Figure 18.9.
8. Place cups in a warm place until the seeds begin to germinate.
9. When seeds have germinated and the seedlings are visible, remove the plastic bag.
10. Place one cup on each of the trays marked with the following conditions:
    (a) plants will be 1–2 cm from light
    (b) plants will be 8–10 cm from light
    (c) plants will be 20–30 cm from light
    (d) plants will be in the dark
11. The teacher will place the trays the correct distance from the light.
12. Water as necessary and adjust the lights on the plants daily so plants stay at the indicated distance from the light.

13. Compare the growth of the seedlings every other day. Record your observations in the chart in your notebook. Include the height, number of leaves present, colour, and appearance of seedlings.

> CAUTION Wash your hands after handling the soil and seeds.

Questions

1. What factors must be controlled during the growth of the seedlings?
2. Under which condition did the seedlings
   (a) become the tallest?
   (b) look the healthiest?
   (c) develop the most leaves?
   (d) have the poorest colour of leaves (yellow)?
3. Is the tallest plant necessarily the best? Explain.

Figure 18.9 *Set-up for Activity 18B*

Observations

Table 18.3 *Seedling Appearance Under Various Light Conditions*

DAY	OBSERVATION	CONDITIONS			
		1–2 cm From Light	8–10 cm From Light	20–30 cm From Light	Dark
1	Height Number of leaves Colour Comments				
3	Height Number of leaves Colour Comments				
Etc.					

SAMPLE ONLY

Water

Water is necessary for photosynthesis and for carrying materials through plants. The roots of most plants are responsible for absorbing water. If there is not enough water, the plants may wilt and die.

It is best to water crops thoroughly. When the water soaks into the ground, the roots grow deep and the plant will be well rooted.

Too much water is just as bad as too little. If the soil is too wet, the roots cannot get oxygen and are easily infected by disease organisms. As the excess water drains away, it takes away large amounts of nutrients from the soil.

Soil

Soil is important to plant growth for many reasons. Roots grow into the soil so that the plant is well anchored. Roots absorb from the soil the water a plant needs. Soil contains many minerals and nutrients that plants need to grow and stay healthy. To review the main types of soil, see Figure 16.3 in Chapter 16.

Minerals

Your body requires vitamins and minerals in order to stay healthy. Plants also need minerals so they can stay healthy. *Potassium, calcium, magnesium*, and *phosphorus* are examples of elements found in the soil. Their compounds dissolve in water and are then absorbed by the roots of plants.

Temperature

It would be unusual to see orange trees growing along the streets of Sudbury. It would be just as unusual to see blueberries growing on a tropical beach. Why? The temperature affects seed germination and blooming time. The temperature also regulates the length of the growing season.

Most farm crops grow well between 5°C and 40°C. Plants like broccoli, radishes, and peas do better in cooler temperatures within this range and can be planted in early spring. Cucumbers and eggplants grow best in the warmer range and should be planted later in the season when there is no risk of frost.

Self-check

1. Why must plants receive light each day?
2. How might water drainage affect plant growth?
3. In what temperature range will most food plants grow?

Ways To Improve Growth

Healthy crops just don't happen by good luck. They happen because of good management. There are many ways to improve the growth of growing plants.

Irrigation

Crops need water throughout the growing season. Rain provides a large amount of this water. In some areas, there is not enough rain during the growing season to keep the plants well supplied. Even fields in high rainfall areas may experience periodic dry spells. **Irrigation** refers to methods used to increase artificially the amount of water that reaches the plants. Home gardeners usually rely on a watering can, garden hose, or sprinkler system to water plants when there is not enough rainfall. Farmers would find it difficult to water large fields in this way. They have devised many different ways to irrigate their crops. Most of these methods are quite expensive.

An irrigation system must have a source of water. Nearby lakes, streams, or reservoirs are good possibilities. Wells or underground springs can also be used.

How will the water reach the fields from the source? Ditches can be dug along which the water will flow if the fields are downhill from the source. If the water must go uphill, a pump must be installed to move the water against gravity.

The flow of water must be controlled so that the fields are only watered when needed. Large sprinkler systems are used to imitate the rain by sending a gentle mist of water over the fields. See Figure 18.10 (a). Although this is expensive, it can be used on any soil and any crop at any time. Fertilizer can be added to the water so that the plants receive nutrients at the same time.

A less efficient method of irrigation occurs when fields are simply flooded with water from the irrigation source. See Figure 18.10 (b). There is very little control over water flow.

Figure 18.10

(a) *Sprinkler irrigation*

(b) *Natural flooding*

(c) *Drip irrigation*

Fruit tree farmers often rely on the drip method. Instead of sprinkling or flooding a large area, hoses deliver water to individual trees at a very slow rate to maintain soil moisture. See Figure 18.10 (c). This reduces waste and is very efficient for this type of crop.

The last thing that must be considered is how to drain off excess water from the fields. Usually, the water runs off into ditches that lead to nearby lakes or streams.

Fertilization

Fertilizers are added to the soil when there are not enough nutrients present. Even though the soil may be very rich when the seeds are planted, the plants remove many nutrients as they grow. Water also washes away many nutrients from the soil. These must all be replaced. To review the types and uses of commercial fertilizers, turn to Section 16.2 in Chapter 16.

Animal Manure

Animal manure contains large amounts of organic material. It also acts as a fertilizer. If you were to analyze cow manure, you would find that it would be a 10–5–10 type fertilizer. Which nutrient is found in the lowest amount in cow manure? It is important to let the manure *decompose*, or rot, before applying it to the crops, especially near planting time. If it is too fresh, it will harm the plants. Fresh manure can be added to the soil in the fall so that it will rot over the winter and be ready for spring planting.

Farmers must be careful that manure does not enter the water run-off from the fields. This water may drain into lakes and rivers. The micro-organisms contained in the manure could multiply in the water. If humans drink this water, the micro-organisms could cause disease.

Mulching

You may have seen farmers spreading straw or hay on the surface of the fields. This is known as **mulching**. Mulching keeps the roots of the plants cool, prevents the ground from drying out, and prevents weed growth. Grass clippings, gravel, wood chips, pine needles, sawdust, peat moss, and nut shells can all be used as mulch.

Treatment with Hormones

A relatively new area of plant improvement is hormone treatment. **Hormones** are chemical substances that control many activities in a living organism.

Plants produce hormones naturally. Many of these can now be produced in the lab and used to improve garden plants. One such hormone, indoleacetic acid, can be used to increase the rooting ability of plant cuttings. This hormone also helps the fruit of the plant to enlarge without producing seeds.

Ethylene, another hormone, causes fruit to ripen. Ripe apples give off large amounts of this hormone. If a few ripe apples are placed in a bag with the green tomatoes harvested at the end of the season, the tomatoes will ripen quickly.

Bananas are often shipped while they are still green. Adding ethylene to the containers will cause them to ripen by the time they reach the supermarket.

Self-check

1. (a) Why is the sprinkler method of irrigation quite often the best way to irrigate fields?
   (b) Why might some farmers use other methods of irrigation?
2. List two ways mulching helps plants to grow better.
3. Describe the effects that hormones can have on plants.

Plant Pests

No matter how well you water and nourish the crops, there is always the danger of pest attack. Native pests are usually controlled by other organisms. Pests that are brought in from other countries do considerably more damage. The pest population continues to grow because its controlling agent is left behind.

Pests can take many forms. Many are so small that they can only be seen with the aid of the microscope. Larger pests, such as insects and other animals, can also be a problem. Even other plants can be a nuisance. Once a pest attacks the garden, something must be done or the crop may be ruined.

Micro-organisms

Viruses and bacteria are very tiny, but they are able to cause diseases. Viruses often attack tomatoes, tobacco, and corn. When viruses attack a plant, the leaves may develop yellow patches and black splotches, or they may turn dark and curl. Unusual growths may appear. The plant may eventually die.

There is no chemical that will control a viral infection. The only treatment is to cut off and burn the infected part of the plant. To prevent the spread of the disease, look for and remove any insects that might carry the virus.

Bacteria are found everywhere. Many bacteria are useful. They help change nutrients in the soil into forms that the plant can use. Others help decompose organic material in the soil and in compost heaps. Many bacteria are harmful. They use animals and plants as a source of food, causing disease. The potato is often attacked by a bacterium that causes common *potato scab*. The tuber develops brown spots. It becomes weak and can then be infected by other harmful organisms. The size of the potato crop in an infected field will be much smaller than usual. Luckily, the bacteria are killed by winter temperatures.

Fungi

Fungi (singular: fungus) are organisms that do not have chlorophyll. Because they cannot make their own food, they must obtain their nourishment from other organisms. Mushrooms and moulds are common types of fungi.

Fungi are responsible for many plant diseases. They cause *apple scab, downy mildew* of grapes, corn and potato *blights*, and *rusts* and *smuts* of grain.

Wheat is often attacked by a fungus that causes *wheat rust*. Yellow spots appear on the leaves and release thousands of powdery *spores* that are blown onto other wheat plants. The leaves then turn brown. The grain is weakened and easily attacked by other pests. Because of this, many farmers use only disease-resistant seeds.

Damping off is one of the most common problems in young seedlings that are grown indoors. The seeds sprout and grow two or three leaves and then rot at soil level. Damping off is caused by a fungus that lives in the soil. One way to avoid damping off is to sterilize the soil first. Alternatively, you can soak the soil with a fungicide (fungus killer).

Insects

There are more species of insects than of any other animal in the world. They are found almost everywhere on the earth. Some insects spread plant disease by carrying micro-organisms from one field to another. Many insects eat plants as food. If these insects find their way into a home garden or into a farmer's field, they can cause a great deal of damage.

Cutworms, which are the larvae of moths, come out at night and chew through the stems of tomato or bean plants at ground level. This completely kills the plant. If you inspect the garden daily, you can find the cutworms just under the soil surface and crush them.

Tomato hornworms are green caterpillars that live on tomato plants (see Figure 18.11). Hornworms are large enough that they can be picked off by hand or cut in half with scissors. This is not true of all insect pests. Many cannot be seen until it is too late and the damage has been done.

Figure 18.11 *The tomato hornworm can grow up to 8 cm in length and can eat all of the leaves on a small tomato plant in one night.*

Weeds

A weed is any plant that grows where it is not wanted. Weeds compete with desired plants because they absorb nutrients and water from the soil. They also provide a home for many harmful insects that will attack plants.

Farmers and gardeners control weeds by hoeing them out so they will dry and die. Mulching will also prevent weeds from growing. *Herbicides* (weed killers) can also be used with care. There is always a danger that the chemicals will get on the crops and kill them as well.

Problem

How do pests affect plants?

Materials

pictures of healthy and diseased plants (Figures 17.13, 17.14, 17.15, 18.12, Unit Six Opener)

Procedure

1. Make a chart in your notebook similar to Table 18.4.
2. Examine the picture of a healthy tomato plant (Figure 17.15). Describe it.
3. Examine the picture of the diseased tomato plant (Figure 18.12(a) and (b)). Describe it.
4. Repeat steps 2 and 3 for the remaining plants in Figure 18.12.
5. Enter all information in the observation chart in your notebook.

Observations

Table 18.4 *Comparison of Healthy and Infected Plants*

PLANT NAME	DESCRIPTION OF HEALTHY PLANT	DESCRIPTION OF INFECTED PLANT

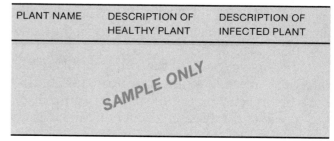

Questions

1. What parts of the plant are affected by disease or pests?
2. (a) What are some of the symptoms of infected plants?
   (b) What is the most common symptom of infected plants?
3. Do you think that each infected plant will die? Why or why not?

Figure 18.12 *Examine these pictures of diseased plants. How are they different from healthy plants?*

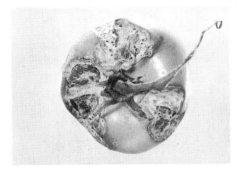

(a) *Tomato hornworm injury*

(b) *Tomato plant suffering from an unidentified disease*

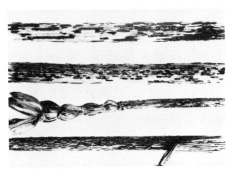

(c) *Stems of grain plant infected with stem rust*

(d) *Maggot damage to cauliflower*

(e) *Insect damage to cabbage*

(f) *Leafhopper damage to grape leaf*

(g) *Leaf miner damage*

(h) *Southern corn leaf blight*

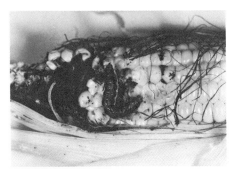

(i) *Insect injury to corn*

Self-check

1. How can plants that have been infected by viruses be treated?
2. How can bacteria
   (a) harm plants and (b) help plants?
3. List two diseases of plants that are caused by fungi.
4. How do tomato hornworms harm tomato plants?
5. List two reasons why weeds are not welcome in gardens.

Pest Management

What should you do when you are confronted by an army of pests that are attacking and destroying the plants that you have looked after so carefully? You have two basic choices. You can use chemical pesticides or a natural approach.

Pesticides

Pesticides are chemicals that are used to kill pests. They are usually quite effective providing they are used as directed. Read the label carefully to see what pest it kills, what plants it can be used on, the best time to use it, the proper amount to be used, and the correct way to apply it. People using pesticides on home gardens must use ones specifically marked for domestic use.

There are many dangers associated with the use of pesticides. Helpful organisms may be killed by the pesticide. Most pesticides are poisonous to animals and people. It is important to follow the *no spray interval* as described on the pesticide container. This tells you how long after spraying you must wait until it is safe to harvest the plants. If you eat the plants too soon after using a pesticide, you could get very sick.

Pesticides are easily washed away by rain and may enter lakes and streams in the run-off. This could contribute to water pollution.

Because of these difficulties, pesticides should not be applied to the garden "just in case" there are pests. They should only be used when actual damage occurs.

Challenge

Examine the label of a pesticide container. Make a list of the important information found on the label. Compare your findings with those of others in the class.

Natural Control

There are many ways to control pests in the garden without using chemicals. The choice of method depends on the type of pest you want to eliminate.

(a) Helpful Insects—Many insects already in the garden will destroy pests for you. For example, ladybugs will eat hundreds of harmful aphids from rosebushes, raspberries, or currants.

Remember the tomato hornworm that is able to eat all of the leaves of a tomato plant in one night? (Refer to Figure 18.11.) The hornworm is kept under control by a particular species of wasp. The wasp lays its eggs right inside the tissues of the caterpillar. When the eggs hatch, the wasp larvae eat the hornworm alive.

(b) Other Animals—Frogs, toads, spiders, and snakes eat potentially harmful insects. They are useful in controlling such pest populations.

(c) Crop Rotation—The changing of crops in a field from year to year is known as **crop rotation**. This technique is used to control the spread of disease. Suppose that a farmer grows wheat in the same field year after year. One year, the wheat becomes infected with wheat rust. Even though the wheat is harvested, the spores of the rust fungi remain in the soil until the spring. If the farmer plants wheat again, the fungi will attack the crop and continue to grow and spread. If, however, the farmer plants carrots in the field in the spring, the rust fungi will not have a suitable food source and they will die.

(d) Hormones—Insects produce hormones that control their growth and development. Scientists are currently trying to produce these hormones in the lab. If they are successful, these hormones could be applied to the crops where the insects live to stop the insect from becoming mature and reproducing. This would prevent further crop damage.

(e) Water—If insects are visible on the leaves of the crop, they can be sprayed off with a stream of water. This, and all watering, should be done in the morning or early afternoon so that the leaves can dry before the sun goes down. Viruses and bacteria grow more quickly on wet leaves.

(f) Seedling Collars—If you are worried about seedlings being attacked by slugs or cutworms, make a collar from an old milk carton. Place it around the seedling and push it a few centimetres into the ground (Figure 18.13). This forms a barricade to prevent pests from reaching the seedling.

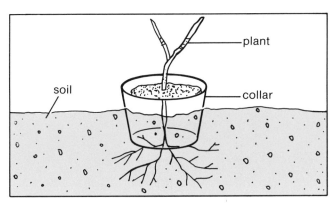

Figure 18.13 *The collar acts like a wall to keep pests out.*

Self-check

1. What kind of information is found on the label of a pesticide?
2. What is meant by the "no spray interval"? Why is it important?
3. Describe three methods of controlling garden pests without the use of chemicals.

Chapter Objectives

NOW THAT YOU HAVE COMPLETED THIS CHAPTER, CAN YOU DO THE FOLLOWING?	FOR REVIEW TURN TO SECTION
1. List the three parts of a seed and the function of each.	18.1
2. Describe how plants can be started.	18.1
3. List five methods of vegetative propagation.	18.1
4. List the five factors necessary for plant growth.	18.2
5. Explain how the addition of fertilizer and organic matter to soil helps to improve it.	18.3
6. Describe the importance of mulching to a home gardener.	18.3
7. List five types of garden pests.	18.4
8. Describe the symptoms of potato scab and wheat rust.	18.4
9. List reasons why pesticides should not be used in all situations.	18.5
10. List five methods of natural pest control.	18.5

Words to Know

embryo
seed coat
dormancy
viable
viability rate
vegetative propagation
cutting
bulb
tuber
runner
layering
grafting
irrigation

fertilizer
mulching
fungi (singular: fungus)
herbicide
pesticide
crop rotation

Tying It Together

1. Draw and label the parts of a seed.
2. Explain why each of the following is a necessary part of growing plants from seed:
   (a) Plant seeds in loose soil.
   (b) Plant seeds at a depth that is 2–3 times the diameter of the seed.
   (c) Keep the soil moist.
   (d) Plant seeds after the last frost if you are planting seeds outdoors.
   (e) Sterilize the soil if you are planting seeds indoors.
3. What is meant by the viability rate of seeds?
4. Read each of the following descriptions and identify the method of vegetative propagation it describes:
   (a) an underground stem containing a large supply of starch
   (b) a horizontal stem that grows along the ground and sends down roots
   (c) a method used to propagate shrubs like raspberry
   (d) a short underground stem surrounded by leafy layers of food
   (e) a method used to improve trees in orchards by joining two stems
   (f) part of a plant (stem, leaf, or root) that is removed from one plant and put into water or soil to start another plant
5. Water is important to plant growth. Answer the following questions about plants and water:
   (a) Why must plants have water in order to grow?
   (b) Why should you not overwater plants?
   (c) Why should you not let plants dry out?
   (d) During what part of the day should you water plants? Why?
6. (a) What is irrigation?
   (b) List three methods of irrigation.
   (c) For each method you listed in (b), explain one advantage that it has over the other methods.
7. List three reasons why you should mulch a garden.
8. (a) What are hormones?
   (b) How can green tomatoes be made to ripen after they have been picked? Explain how this works.
9. What precautions should you take when using a pesticide?
10. List three alternatives to chemical pesticides that can be used to control garden pests.

Applying Your Knowledge

1. A young scientist spent many years developing a delicious-tasting seedless watermelon. It was so good that all her friends wanted to grow some in their backyards. She could not give them seeds to plant since her watermelons did not have any. How could she make sure her friends were able to grow these delicious seedless watermelons?
2. A grade 10 student went to the local bait store and bought 20 large earthworms. He took them home and scattered them in the garden. The worms soon burrowed into the soil and out of sight. He thought this would improve the soil in his vegetable patch. Was he correct? Why or why not?

Projects for Investigation

1. Tent caterpillars are a problem in many orchards in Ontario. These caterpillars live in large numbers in "tents" in the branches of trees. One tent full of caterpillars can easily eat all the leaves from a tree and will then move on and attack other trees. Figure 18.14 summarizes the life cycle of the tent caterpillar.

   In order to control these caterpillars, scientists are developing a hormone that will cause the eggs to hatch in late summer or fall instead of spring. How would this prevent the trees from being damaged by the caterpillar?

2. Farmers plant "seed potatoes" in their fields. How are seed potatoes produced commercially? Try to sprout some potatoes yourself. Under what conditions do potatoes sprout best?
3. Composting is the process where vegetable peelings, coffee grounds, egg shells, garden clippings, and similar substances are decomposed. This rotted material is then added to the soil to increase its humus content. Find out how to make a compost pile. How long does it take for decomposition to occur?

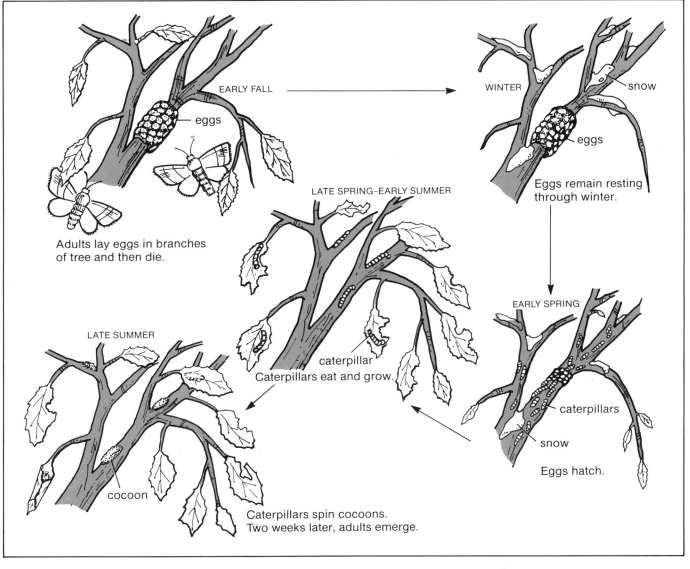

Figure 18.14 *Life cycle of the tent caterpillar*

From Field to Table: Processing Food

Key Ideas

- Much energy is needed to produce, transport, process, and package food.
- Meat products require much more energy than plant products.
- Substances may be added to foods to improve their appearance or flavour, or to make them last longer.

Cows and tomatoes are raised on farms. These raw materials are used to make the hamburger and ketchup that you might eat for lunch. The food-processing industry is responsible for changing the plants and animals from the field into the food products that you put on your table. In this chapter, you will find out how this is done.

Problem

What is the best use for good farming land?

Materials

Procedure

Read this story about the Wilkinsons, and then answer the questions that follow.

Mr. and Mrs. Wilkinson bought a 100 ha (hectare) farm in 1954. At that time, the soil was quite fertile and suitable for crop farming. A nearby river provided a source of irrigation. There were about 170 frost-free days each year. The nearest town was located 15 km away. The population of the town in 1954 was 9000 and provided a good market for the farm produce.

The Wilkinsons have managed the farm well since then. By means of crop rotation, sensible fertilization, and continuous composting, they have maintained the fertility of the soil. Over the years, they have acquired some farm machinery which is now beginning to break down regularly.

The population of the nearby town has increased to 60 000. Because of this, the city limits are only 4 km away. A major food-processing plant has become established in the town. It employs many people from the community. This plant buys a large quantity of the produce from the farm.

The Wilkinsons found it discouraging to note that out of every dollar Canadians spend on food, only 38 cents goes to the farmer, while 62 cents goes to the processors, wholesalers, distributors, and retailers.

The Wilkinsons wish to retire soon. Their children live in another province and have no wish to take over the farm. They plan to put the farm up for sale.

When the farm was put on the market, they received a number of offers. The best one came from a large developer who wanted to build homes on the land (Figure 19.1). The Wilkinsons could retire in comfort if they sold their land to this developer. Only one offer was made by anyone interested in

Figure 19.1 *Is this the best solution?*

maintaining the property as a working farm. This was from a young couple from out west. Since they were just starting out, they could not afford to pay nearly as much as the developer for the land. Either Mr. or Mrs. Wilkinson would have to work part-time if they decided to sell the farm to the young couple.

Only 11% of all land in Canada is suitable for agricultural purposes. The remainder consists of tundra, rock, lakes, or land that is too cold or too dry to be cultivated. As a result, the municipality does not want to rezone the land for commercial or residential building. They want to maintain the land as agricultural farmland.

Many people are involved in this issue. What is the best solution?

Observations

1. What characteristics make the Wilkinsons' property an ideal location for a farm?
2. What characteristics make the Wilkinsons' property an ideal location for a residential community?

Questions

1. Would you consider farming as a possible career? List reasons you might have both for and against a farming career.
2. Choose one of the following people/groups to represent in a discussion. Present your opinions to the class and try to come up with the best possible solution to the Wilkinsons' dilemma.
   (a) Mr. and Mrs. Wilkinson
   (b) the developer
   (c) the young farm couple
   (d) town politicians
   (e) residents of the town who work at the food-processing plant
   (f) residents of the town who work for the development company

Farming Concerns

There are many different types of farms in operation in Canada. Figure 19.2 shows the amount of income of each type of farm in Canada.

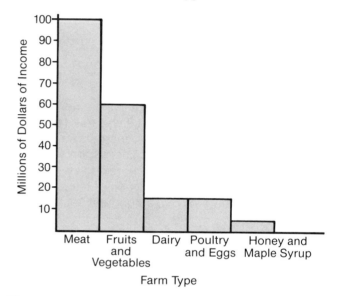

Figure 19.2 *Income of various types of farms in Canada (1984).*

Each type of farm has its own problems. Crop farmers are greatly affected by weather conditions but do not have to worry about veterinary fees. Beef and dairy farmers are more concerned about animal feed than fertilizers. Regardless of what type of farm is in operation, there are a great many costs that must be considered in order to produce a good crop or animal herd. Figure 19.3 shows how the average farm dollar is spent on Canadian farms.

How money is spent is a major consideration on the farm. It affects the price of the products sold to you, the consumer. It also determines the amount of profit a farmer can make. How energy is used is also very important. Three percent of all energy used in Canada goes toward farming. Fertilizer is the most energy-expensive item on the farm.

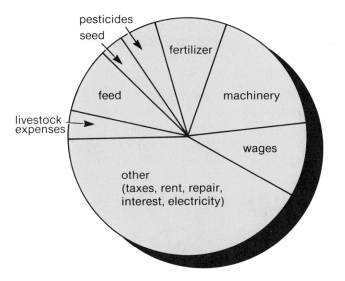

Figure 19.3 *How the dollar is spent on the farm*

The basic unit for measuring energy is the **joule**. The amount of energy contained in food is measured in **kilojoules** (1 kilojoule = 1000 joules). Because the amount of energy used on a farm is so large, it is measured in **megajoules** (1 megajoule = 1000 kilojoules = 1 000 000 joules). The abbreviation for megajoule is MJ.

Two crops that are grown to feed animals are grain corn and soybeans. All animals, including ourselves, need to eat in order to obtain energy for daily activities. If a farmer planted 1 ha of land with grain corn, the resulting crop would provide 80 000 MJ of energy for the animals who eat it. If the same field was planted with soybeans, the resulting crop would provide only 37 193 MJ of energy. As a farmer, which crop would you plant if the amount of energy it provides is your main concern?

The answer to this question is not as easy as it seems. Each step in the production of a crop requires energy. The energy cost at each step depends on the crop being grown and the environmental conditions. Some crops require a great deal of fertilization. Others need extensive irrigation. Still others make use of machinery

that consumes energy in large amounts. Table 19.1 compares the amount of energy used in the growing of grain corn and soybeans. It takes almost four times as much energy to produce 1 ha of grain corn as it does to produce 1 ha of soybeans.

In comparing the two crops, it can be seen that fields of grain corn must be heavily fertilized. Because large amounts of energy are needed to manufacture fertilizers, their use greatly increases the energy cost of grain corn. In addition, grain corn must be dried. This is also an energy-consuming process which is not needed in the production of soybeans. Now which crop would you grow?

Table 19.1 *Energy Inputs for the Production of 1 ha of Grain and 1 ha of Soybeans.*

	ENERGY NEEDED TO PRODUCE GRAIN CORN (MJ)	ENERGY NEEDED TO PRODUCE SOYBEANS (MJ)
Land preparation	1 348	1 480
Planting	364	364
Fertilizer/application	14 865	2 626
Herbicide/application	1 962	1 449
Harvesting	892	842
Drying	6 062	0
Hauling to storage	186	106
Total Energy Input	25 679 MJ	6 867 MJ

Self-check

1. List three things that a crop farmer must pay for that an animal farmer does not.
2. List three things that an animal farmer must pay for that a crop farmer does not.
3. As a farmer, would you grow corn or soybeans? Defend your answer.

Processing Plants for Food

Corn and tomatoes can go directly from the field to your table. They can also be processed into different forms before being eaten. The following section describes some of the ways these two food plants can be processed.

Corn Flakes

Corn is a very popular crop in Canada. Some is eaten fresh from the cob. You can also buy corn frozen or tinned. It can be processed into corn meal or used to make corn flakes. The average Canadian eats nearly 72 kg of corn each year in one form or another.

The path from corn seed to breakfast cereal is a long and complicated one (Figure 19.4). Corn in Canada is grown from special seed called *hybrid seed*. Farmers purchase hybrid seed every year from seed companies. This, of course, costs money. The seed is planted, cultivated, and harvested by machine. These machines need energy in order to work. In between planting and harvesting, the crops must be fertilized. Pesticides may be used to control insects and fungi. These chemicals require energy to produce and apply and therefore increase the cost of raising the crop. On average, it takes about 200 kJ of energy to grow enough corn to make one serving of corn flakes. That is about the same amount of energy as you would get from burning 15 mL of gasoline.

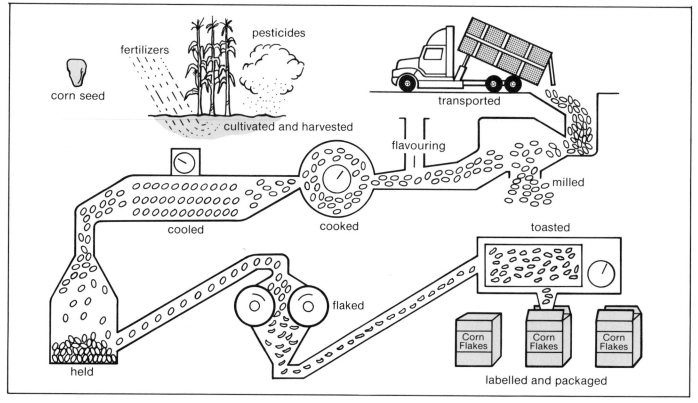

Figure 19.4 *How corn flakes are made*

Once harvested, the corn is transported by truck or rail to the processing plant. Again, it takes money and energy to do this. At the processors, the grain is *milled* to remove the unwanted portions of the corn and ground to the right size. Flavouring is added and then the corn is cooked. It must then be cooled, dried, and allowed to set for several hours before further processing. All of these steps use up more energy and cost more money. The mixture is then pressed between large rollers, flaked, and toasted to produce the cereal that you pour into your bowl each morning. Once cooled, the cereal is packaged and labelled. The cost in energy from milling to packaging is about 680 kJ per serving.

You can't buy boxes of cereal at the processing plant. They must be shipped by truck or rail to warehouses and then sent to retail stores across the country. This requires more money and energy. Once at the store, the boxes of cereal are placed on the shelf for you to purchase. The cashier rings up your order and you can take the cereal home to enjoy. The shipping and retailing of the cereal adds another 350 kJ of energy to the "energy price" of your cereal.

A little arithmetic shows that about 1230 kJ of energy were used to produce one serving of corn flakes from start to finish. When you eat this cereal, you get only about 300 kJ of energy from it. Is this a wise use of energy?

Ketchup

Remember the hamburger and fries you ate last week? Did you put ketchup on them? Did you think about how tomatoes are grown in the field and processed into ketchup in a bottle?

Even before planting, market analysts and growers get together to decide how many hectares of land will be planted with tomatoes. The type of tomato to be grown is selected. The seeds are planted in greenhouses to get a head

Figure 19.5 *Tomato seedlings are planted in greenhouses so that they can begin to germinate even though it is still too cold outside. They grow into tall, mature plants inside the greenhouses.*

start in growing. The seedlings may be transplanted into fields once they have become established or they may continue to grow in greenhouses (Figure 19.5).

During the warm months of summer, the plants are sprayed with water and fertilized. Weeds are removed so they have lots of room to grow. After many days of sunlight and rain, the tomatoes ripen. They are picked by hand and put into baskets. Those that are still green remain on the vine and will be picked at a later date. The ripe tomatoes are then dumped into a truck that takes them to the processors. See Figure 19.6 on the next page.

The truck stops. Water carries the tomatoes down a long chute into the processing plant. This water slide removes dirt and gets the tomatoes inside without being bruised.

Once inside, the tomatoes are examined by government inspectors. They are sorted by size and washed again. Now the tough part. Boiling water is poured on the tomatoes and the stem, skin, and seeds are removed. All that remains is the *pulp*.

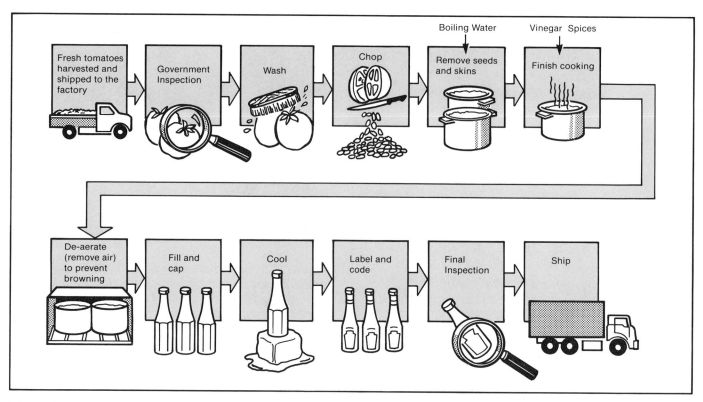

Figure 19.6 *How ketchup is made*

The pulp is heated to over 93°C. This prevents chemicals inside the pulp from spoiling the tomato. Vinegar and spices are added for flavour. The result? Ketchup! Any air that may be present is removed to improve the appearance and prevent micro-organisms from spoiling the product. The ketchup is poured into sterilized glass bottles. The bottles are sealed, cooled, labelled, and inspected once again. If they pass inspection, the bottles are sent to retailing stores where they can be purchased and put on tables across the country.

Of all tomatoes grown in Canada, 29% are used to make tomato juice, 25% are eaten fresh, 25% become ketchup, and the remaining 21% are used for other purposes.

Self-check

1. What is the most "energy expensive" step in the production of corn flakes?
2. What is the purpose of milling?
3. Why are processed foods such as breakfast cereals said to be uneconomical sources of energy?
4. Why are tomato seeds planted in greenhouses?
5. Why is water used to transport tomatoes from the truck to the processing plant?
6. What is the purpose of pulping tomatoes?
7. Seventy-nine percent of tomatoes are eaten fresh, in juice, or in ketchup. Twenty-one percent of tomatoes are used for other purposes. What might some of these other uses be?

From Field to Table — Meat Production

Did you know that the average Canadian ate nearly 73 kg of meat in the last year? Of that, over half was likely beef. Those hamburgers, steaks, and roasts really add up.

Farmers who raise beef must first establish a good herd. Animals used for breeding are carefully selected so they will produce good calves to maintain the herd.

It may take up to three years for a calf to reach a mass of 500 kg. At this point, it can be slaughtered. The cost of raising a calf to this point works out to about $1.85/kg. Feed makes up nearly 85% of this cost. Veterinary costs, labour, and shelter make up the remainder.

Once the animal has reached its full size, it is sold to a meat packer. Here, it will be slaughtered. The head, feet, hooves, hide, and innards are removed. These can be processed into *by-products*. The hides are used to manufacture leather. Hooves can be used to make glue. Many internal organs are used to produce medicines. Any remaining parts can be used to make fertilizer. These inedible parts make up 200 kg. Only 300 kg is left for processing into meat products. As a result, the value of the meat has increased to $3.08/kg.

The carcass of the cow must now be cut into portions suitable for sale to wholesalers. These cuts include the hip, sirloin, flank, rib, chuck, and shank. These large pieces are bought by butchers who trim and cut them into roasts, steaks, and ribs. Some is made into ground beef.

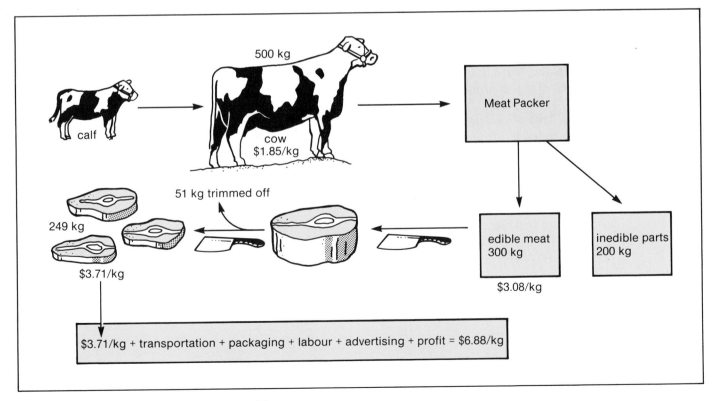

Figure 19.7 *How a roast reaches your table*

Approximately 51 kg of material is trimmed away. The meat now costs $3.71/kg.

The costs given so far reflect the price of the meat alone. Other costs, like transportation, packaging, labour, advertising, and profits for all those along the way bring the total average price of the beef to $6.88/kg. This varies greatly according to the cut of meat purchased and the amount of processing that occurred along the way. Figure 19.7 on the previous page summarizes the entire process.

An average serving of beef provides approximately 1600 kJ of energy when eaten. It takes 120 000 kJ of energy to produce one serving of beef from start to finish. Compare this to the energy inputs and outputs of plant products. Which is more energy efficient—plant or meat processing?

Self-check

1. Not all of the steer is used as meat. What happens to the parts of the steer that are not eaten?
2. Examine Table 19.2. This table compares the amount of energy needed to produce one serving of fresh corn and meat with the amount of energy each provides when eaten.
   (a) Which food takes more energy to produce?
   (b) Which food provides more energy when eaten?
   (c) Explain why people in Third World countries use more land for raising plants than they do for raising animals.

Table 19.2 *Energy Input and Output of Corn and Meat*

	AMOUNT OF ENERGY TO PRODUCE ONE SERVING	AMOUNT OF ENERGY PROVIDED BY ONE SERVING
Corn	100 kJ	300 kJ
Meat	120 000 kJ	1 600 kJ

The Final Product

You are having guests for dinner tonight. You must go to the store and buy a roast, potatoes, frozen corn, tomatoes for a salad, and a can of sliced peaches for the meal. Confidently, you push the shopping cart down the aisle of the grocery store or move from stall to stall at the local farmers' market. Before you select anything, you stop and think (Figure 19.8). Even though you know what a good roast looks like, can you be sure that the meat will be tender? Were the tomatoes grown using pesticides that are dangerous to people? Are the potatoes old or diseased? Is the corn discoloured or dirty? Maybe the peaches are bruised and damaged. All of a sudden, you are not so sure anymore. How will you choose your purchases so that you can be sure they are high quality items?

Figure 19.8 *How does this shopper know how to select his purchases?*

318

All food products sold in Canada are thoroughly examined by trained *inspectors* to make sure that the growers and processors are following government regulations. Microbiologists and chemists perform a variety of tests on food material at various stages of processing to make sure that it is safe for you to eat. Only products that meet government standards can be sold to consumers. Food packers or processors must then grade their products to indicate the quality of the item they are selling.

Meat

Meat is graded according to the age and quality of the cow when slaughtered. Young meat is usually more tender than that from older cows. The grade is stamped on the meat using harmless, edible dyes (Figure 19.9). Top quality, young beef is stamped *Canada Grade A* in red.

Good quality beef from young cows is marked *Canada Grade B* in blue. *Canada Grade C* appears in brown letters on middle-aged beef. The toughest beef from mature cows is stamped *Canada Grade D* in black. Regardless of the grade, all meat is safe for you to eat.

Canned and Frozen Food

Because the consumer cannot examine the contents of canned or frozen foods directly, a great deal of information can be found on the package (Figure 19.10). The grading of canned and frozen fruits and vegetables is very thorough. The product is rated as to the flavour, aroma, and colour of the food. The tenderness and level of maturity are also noted. The contents must be of uniform size and shape. The appearance of any liquid is also studied. Of course, there must be no defects or foreign material present.

Figure 19.9 *All meat in Canada is graded to indicate its level of quality.*

Figure 19.10 *What information is contained on the label of canned foods?*

Based on these criteria, frozen and canned fruits and vegetables are graded in order of quality as *Canada Fancy* (best quality), *Canada Choice, Canada Standard*, and *Canada Substandard*. The last one is generally not sold in supermarkets. All grades are suitable for eating. Canada Fancy is usually more expensive than Canada Choice and Canada Standard. If you are going to serve sliced peaches and ice cream to your guests, you might choose Canada Fancy so that it will look the best. If you are going to purée the peaches for a sauce, then you could easily use Canada Standard and save money.

Investigating Fresh Produce

Apples are graded as *Canada Extra Fancy, Canada Fancy,* and *Canada Commercial*. But there is more to an apple than just its grade. There are many differences between varieties of apples, even if they all have the same grade. In the following activity, you will determine some of the differences between varieties of apples.

Activity 19B

Problem
How do varieties of apples differ?

Materials
different varieties of apples—Spy, McIntosh, Spartan, Delicious, etc. marked with their cost
knife
paper towel
balance
overflow can
graduated cylinder
cutting board

Procedure
1. Make a chart in your notebook similar to Table 19.3.
2. Take one apple to your desk. Note the variety and cost.
3. Examine the apple closely and record its colour, shininess, shape, skin texture, and odour.
4. Using the balance, determine the mass of the apple in grams.
5. To determine the volume of your apple, first fill the overflow can to the top.
6. Carefully lower the apple into the can and collect the water that flows out of the can.
7. Using the graduated cylinder, measure the amount of water forced from the can in millilitres. This volume of water has the same volume as the apple. Record this volume in your chart.
8. Pour the water down the sink.
9. Dry the apple well with a paper towel. Does it look any different from the initial examination?
10. Your teacher will leave one apple of each variety on a shelf in the room. Compare the appearance of each apple after
    (a) one week
    (b) one month
    (c) two months.
11. The class will now move to the cafeteria. Your teacher will cut each apple into slices. One student should count the number of seeds contained in each apple and report it to the class.

12. Obtain a slice of one variety of apple.
13. Describe the cut surface of the slice.
14. Remove a seed from the apple slice and describe it.
15. Take a small bite of the apple slice. Describe the tastiness and crispiness of the apple.
16. After 10 min, reexamine the cut surface of the apple. Describe any changes that have occurred.
17. Complete the table to summarize the results of the tests.
18. Repeat steps 12-17 for each variety of apple.

Observations

(See Table 19.3)

Questions

1. What are three major differences between the different varieties of apples?
2. Next to taste, what is the most important characteristic of an apple that will be used for
   (a) eating raw?
   (b) making pie?
   (c) making applesauce?
   (d) baking?
   (e) making fruit salad?
   (f) storing for winter eating?
3. What might account for differences in the cost of each apple?
4. What factors might contribute to unfairness or error in your survey?
5. What would you have to do to determine which apple is the juiciest?

Table 19.3 *Comparing Different Varieties of Apples*

	SPY	McINTOSH	SPARTAN	DELICIOUS
Cost				
Colour				
Shininess				
Shape				
Skin texture				
Odour				
Mass				
Volume				
Changes after washing and drying				
Seeds: Description				
Number				
Cut surface: Immediately				
After 10 min				
Taste				
Crispiness				
Appearance after one week				
Appearance after one month				
Appearance after two months				

Food Additives

There are many problems facing the farmers and food processors of today. These same issues affect you as consumers. The solutions are not always simple. Read through the following information on one such issue and discuss the use of additives with other members of the class.

Using Food Additives

A **food additive** is something that is added to a food product to either improve its appearance, act as a preservative to increase its shelf life, or add flavour.

There is a great deal of controversy surrounding the use of food additives. Many people prefer food to look and taste as nature intended without artificial colour or flavour. If you examine the list of ingredients contained in many processed foods, it may be difficult to determine exactly what you are eating. One or more of the additives may cause *allergic* reactions in some people.

The Health Protection Branch of Health and Welfare Canada sets the standards for the use of food additives. A great deal of research is done on each and every food additive to make sure that it is safe to use. The amount of additive that can safely be added to food is also determined. Unfortunately, the effect of some additives on people cannot be seen until after a long period of time. As a result, the Health Protection Branch must change the laws regarding food additives as more and more information becomes available.

Some of the more common food additives are described in Table 19.4.

Table 19.4 *Common Food Additives*

ADDITIVE	PURPOSE	WHERE USED
Potassium nitrate Sodium nitrate	Prevents growth of micro-organisms	Meat, canned goods
Calcium salts	Prevents softening of tissues	Tomatoes
Calcium proprionate Sodium proprionate	Prevents growth of mould	Bread
Colouring agents	Adds colour to make food more appealing to the eye	Puddings, candies
Emulsifiers	Prevents separation of components of product	Coffee cream
Stabilizers	Preserves foods during freezing	Frozen pizza
Sugar	Prevents growth of bacteria	Jams
BHA	Keeps products from becoming stale	Cereal
Natural flavouring	Changes or improves flavour	Ice cream, soft drinks, cake mixes
Artificial flavouring	Changes or improves flavour	Ice cream, soft drinks, cake mixes
Aspartame	Sweetens food	Diet foods
Vitamins	Improves nutritional value of food	"Fortified" cereal

Careers in the Food Industry

Food production is becoming more and more complex. Gone are the days when a farmer simply grew food crops in the back fields, cultivated and harvested them by hand, and sold them at a roadside stand. Today, the production and processing of food from field to table involves hundreds of different people with a variety of skills.

Challenge

Fungi are organisms that are able to use bread as a food source. When they invade bread, they cause it to spoil. Preservatives are added to prevent fungal growth and increase the shelf life of bread.

Design an experiment to test the effectiveness of preservatives used in different types of bread. You may wish to include homemade breads, bakery breads, and grocery store breads. Include a control for your experiment. When you have designed the procedure, show it to your teacher for approval. When the experiment is complete, share your results with the rest of the class.

Challenge

List, in order, the many people involved in getting French fried potatoes from the field to your table.

 **Activity 19C**

Problem

How do various people help process food as it moves along from field to table?

Materials

Procedure

1. Choose one of the occupations listed in Table 19.5 or one suggested by your teacher.
2. Go to the library and find out as much as possible about the role of this person in the food industry.
3. If possible, interview someone who is currently performing this job.
4. Prepare a poster summarizing the role of this person in the production and processing of food.

Table 19.5 *People Involved in the Food Industry*

Accountant	Home economist
Advertiser	Inspector
Artist	Market analyst
Bank manager	Nutritionist
Butcher	Sales person
Chemist	Store manager
Consumer surveyor	Trucker
Ecologist	Veterinarian
Geneticist	Weather forecaster

Questions

1. Could the person you selected be involved in producing and processing any food or is the job limited to one particular food product? Explain your answer.
2. What training would a person need in order to perform the occupation you selected?

The Future of the Holland Marsh

Why should you care about the Holland Marsh? For one thing, many of the vegetables you eat each year are grown there. But there's another reason why you should care. This valuable crop land is literally disappearing.

Near Bradford, Ontario, there is a group of low, flat basins. For a long time, these were covered by waters draining into Lake Simcoe. For thousands of years, plant material from these marshes settled under water but did not completely decay. Instead, it turned into peat.

In 1930, a series of dikes and canals were completed. Water was pumped out of the Holland Marsh, making available 2900 ha of rich brown muck for farming. This peat soil is ideal for holding plants and supplying them with water and air. But it is very acidic, and it has few available minerals. When the peat is very dry, the wind can blow it away. Scientists at the Muck Research Station of the Ministry of Agriculture and Food have found that the soil is disappearing. They have learned how to slow this loss of soil by controlling the water table. The *water*

The Holland Marsh in Bradford, Ontario

Location of Holland Marsh

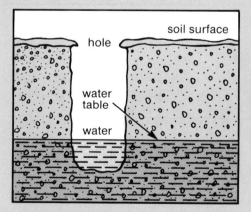

The water table

A section through the marsh at a drainage ditch shows the depth of the peat above the clay.

table is the upper level of the part of the soil where all the spaces are filled with water. If you dig a hole below the water table, the hole fills with water to the level of the water table.

Scientists have showed Holland Marsh farmers how to keep the water table at a suitable distance (from 30 cm to 100 cm, depending on the crop) below the surface of the ground. Drainage pipes are buried in the ground. They lead to cisterns (storage tanks) where the level of water is controlled by the height of a pipe system. In dry periods, water from the canals surrounding the marsh is pumped into the same underground pipes, raising the water table.

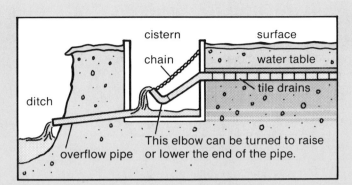

The system for controlling the height of the water table

These lettuce plants are getting water and fertilizer from overhead sprinklers.

Harvesting celery plants

Plants with shallow roots can be irrigated with overhead sprinklers. This irrigation system lets farmers harvest, for example, three crops of celery each year.

Used this way, the peat soil of the Holland Marsh may last for another 50 a. In a typical year, the marshes around Bradford produce about 71 000 kg of carrots, 42 600 kg of onions, 148 000 bunches of celery, 35 600 heads of lettuce, and smaller amounts of potatoes, parsnips, cabbage, beets, cauliflowers, and many other kinds of vegetables. Where will this much food be produced when the peat is all gone?

Chapter Objectives

Words To Know

joule
kilojoule
megajoule
food additive

Tying It Together

1. List four different types of Canadian farms.
2. When farmers are growing crops, on what do they spend
   (a) the least money?
   (b) the most money?
3. In your notebook, list in order the following stages of corn flake production. Do not write in this textbook.
   (a) The cooked mixture is pressed, flaked, and toasted.
   (b) The ears of corn are harvested.
   (c) The corn flakes are put into boxes.
   (d) The corn mixture is cooked.
   (e) Flavourings are added.
   (f) The corn is milled.
   (g) Seeds are planted.
   (h) The corn is transported to a processing plant.
   (i) The crop is cultivated and fertilized.
   (j) The corn mixture is dried.

4. Which of the following tomato parts are not used to make ketchup?
   (a) skin
   (b) seeds
   (c) pulp
   (d) juice
   (e) leaves
   (f) stem
5. Name four processed food products that are made from tomatoes.
6. What is the most expensive part of raising cattle?
7. Why does it cost $1.85/kg to raise a cow but we pay $6.88/kg at the store?
8. What is the highest grade of
   (a) meat?
   (b) canned food?
   (c) fresh produce?
9. Name some of the qualities a food inspector looks at when grading food products.
10. Name two varieties of apples and list four differences between them.
11. List three uses of food additives.
12. Describe two careers associated with the food-processing industry.

Applying Your Knowledge

1. Give two reasons why tomatoes are so expensive to harvest.
2. Why does it take more energy to produce one serving of meat than it does to produce one serving of vegetables?
3. Suppose that you had a bag of corn and a chicken as food. Consider the following alternatives:
   (a) Eat the chicken and then eat the corn.
   (b) Eat the corn and then eat the chicken.
   (c) Feed the corn to the chicken and then eat the chicken.

Would you get the same amount of energy from each alternative? If not, which alternative would provide you with
   (a) the least amount of energy?
   (b) the greatest amount of energy?

Projects for Investigation

1. Many fruits such as strawberries and peaches can be processed into jam. Consult a cookbook to see how this is done. If possible, try to make some strawberry or peach jam. Describe the changes that occur to the fruit at each stage of the processing. Calculate the cost of the final product. Try to include labour costs in the calculation.
2. What characteristics are used to grade apples and other fresh produce? Your library or the produce manager at the local grocery store may be able to help you.
3. Investigate the "pros and cons" of at least one food additive from Table 19.4. Keep the following questions in the mind as you look for information.
   (a) Do you think food additives should or should not be used in the food industry? Why or why not?
   (b) What kinds of tests would you want to do on a food additive before permitting it to be used by the food industry?
   (c) Should additives whose only purpose is to make the food "look nice" be permitted? Why or why not?
   (d) What would be the effect on imported goods if additives were not allowed?
   (e) What characteristics do you think would make a good food additive?

Unit Six: Food And Energy

MATCH

In your notebook, write the letters (a) to (f). Beside each letter, write the number of the word in the right column that corresponds to each description in the left column.

(a) an underground stem containing a large supply of starch

(b) a horizontal stem that grows along the ground and sends down roots

(c) a method used to propagate shrubs such as raspberry

(d) a short underground stem surrounded by leafy layers of food

(e) a method used to improve trees in orchards by joining two stems together

(f) part of a plant that is removed from one plant and put in water or soil to start another plant

1. runner
2. grafting
3. tuber
4. cutting
5. clone
6. bulb
7. diffusion
8. layering

MULTIPLE CHOICE

In your notebook, write the numbers 1 to 10. Beside each number, write the letter of the best choice.

1. Which nutrient provides you with at least half of the energy you need daily?
   (a) fat
   (b) protein
   (c) carbohydrates
   (d) vitamins

2. Which of the following is *not* a function of proteins?
   (a) help in repairing damaged cells
   (b) insulate the body from the cold
   (c) provide the body with energy
   (d) act as enzymes

3. Which food group is missing from a lunch consisting of a ham and lettuce sandwich, two peanut butter cookies, an apple, and a can of soda pop?
   (a) dairy
   (b) meat and meat alternatives
   (c) breads and cereals
   (d) fruits and vegetables

4. Why is drip irrigation often used on commercial farms?
   (a) It is most like natural rain.
   (b) It can be used over large areas.
   (c) It is expensive.
   (d) It delivers water at a slow and steady rate.

5. Which of the following techniques helps to reduce the growth of weeds?
   (a) fertilizing
   (b) composting
   (c) mulching
   (d) irrigation

6. What is the best way to control a viral infection of a crop?
   (a) Treat it with insecticide.
   (b) Cut off and burn infected parts of the plants.
   (c) Introduce ladybugs into the field.
   (d) Spray the crops with water.

7. Which type of farm in Canada results in the largest income?
   (a) meat farms
   (b) fruit and vegetable farms
   (c) poultry and egg farms
   (d) dairy farms

8. Why is corn milled when making corn flakes?
   (a) Milling adds flavour.
   (b) Milling keeps the cereal crisp.
   (c) Milling grinds the corn to the right size.
   (d) Milling increases the nutritional value of the cereal.

9. Why is tomato pulp heated when making ketchup?
   (a) to remove dirt and small particles
   (b) to prevent the pulp from spoiling
   (c) to add flavour
   (d) to prevent bruising and discoloration of the tomatoes

10. It costs about $1.85/kg to raise a calf to maturity. What is the most expensive consideration in this process?
    (a) veterinary costs
    (b) feed
    (c) shelter
    (d) labour for upkeep

TRUE/FALSE

Write the numbers 1 to 10 in your notebook. Beside each number, write T if the statement is true and F if the statement is false. For each false statement, rewrite it so that it becomes true.

1. One gram of fat will provide twice as much energy as one gram of carbohydrates when processed by the body.
2. Perspiration helps to cool the body.
3. Boys and girls need the same number of kilojoules of energy each day.
4. Seeds should be planted in loose soil.
5. All plants need at least seven hours of direct sunlight daily.
6. Manure should be fresh when added to the soil.
7. Outdoor plants should be watered in the late evening after the sun goes down.
8. Fertilizers are the most "energy expensive" item in crop farming.
9. It is more energy efficient to raise plants than animals for food.
10. *Canada Choice* is the highest grade of canned vegetables in Canada.

FOR DISCUSSION

Read the paragraphs and answer the questions that follow.

The effects of air pollution on humans are well known—dry and irritated eyes, coughing, skin problems, and lung disease. Air pollution also affects plants—especially the leaves of plants.

Recall that sulphur dioxide results from the burning of petroleum and other industrial processes. This air pollutant enters the leaves of plants through the stomata. The leaves soon become blotchy. After long exposure to sulphur dioxide, the entire leaf may turn yellow and drop off. The plant may eventually die.

Ozone, another pollutant, is found in car exhaust. Grapes, lettuce, potatoes, corn, and tomatoes can be easily damaged by ozone. The leaves of the plants become spotted and develop a yellowish colour. High humidity and damp conditions increase the effect of ozone on plants.

Pollutants such as soot and dust are not absorbed into the leaf. Instead, they settle on the leaf and prevent photosynthesis. This coating also prevents pesticides from doing their job. Thus, insects and other pests might destroy the crop.

The Ministry of the Environment is working hard to identify causes of air pollution and find ways to reduce it. This will result in healthier people and healthier plants.

1. What part of the plant is most affected by air pollution?
2. Where does sulphur dioxide come from?
3. How do pollutants get inside the leaves?
4. What are the symptoms of sulphur dioxide damage?
5. List three plants that may be affected by ozone.
6. What are the symptoms of ozone damage?
7. What conditions affect the amount of damage done to plants by ozone?
8. Give two reasons why soot and dust are harmful to plants.
9. Which branch of the government is concerned with reducing air pollution?

The Science of Sports and Leisure

Does science play a part in any of the activities you do in your spare time? It does if you bowl or play baseball, tennis, or golf or if you have a hobby such as photography. In this unit, you will find out how some simple scientific principles can increase your skill in and enjoyment of ball sports and photography.

20

Sports and Motion

Key Ideas

- Scientific terms and ideas help to explain the kinds of motion that occur in sports.
- Forces such as air resistance, gravity, lift, and drag affect motion in sports.
- Spin affects motion whenever a ball is rolled, bounced, or thrown.
- Sports research based on scientific knowledge can benefit both athletes and non-athletes.

Why do tennis players put topspin on their shots? Why do bowlers practise to perfect their sidespin, while golfers try to eliminate it? What makes a ball go really fast? What is the "hidden opponent" in every ball game? How does scientific research help coaches train faster athletes?

This chapter will help you answer these questions—and you won't have to become a scientist or an athlete to understand the answers.

Understanding Speed and Acceleration

Speed tells how fast an object is moving at any particular point on its path. In the laboratory, speed can be measured automatically. In sports, however, speed at a single point is not always meaningful. For example, marathoners such as those in Figure 20.1 may run at 20 km/h downhill, but slow to 15 km/h uphill. Occasionally, a runner may even stand still. Speed also varies in ball sports such as baseball, football, and tennis. In cases like these, the following formula may be used to calculate **average speed**.

$$\text{average speed} = \frac{\text{total distance travelled}}{\text{total time taken}}$$

Figure 20.1 *In a race, determining speed is not always so easy.*

SAMPLE PROBLEM

A track team is practising a "4 x 100 m" relay race. The best individual times are

first runner: 14 s   second runner: 12 s
third runner: 13 s   fourth runner: 11 s

What is the best average speed this team can expect?

SOLUTION

Total distance = 4 x 100 m = 400 m
Total time = 14 s + 12 s + 13 s + 11 s = 50 s
$$\text{Average speed} = \frac{\text{total distance}}{\text{total time}} = \frac{400 \text{ m}}{50 \text{ s}} = 8 \text{ m/s}$$

Units for Speed

The highway sign in Figure 20.2 shows speed in units of km/h. This unit is useful for cars, which often travel for hours. Relay runners, however,

Figure 20.2 *How does the maximum allowable speed differ from average speed?*

race for just a few minutes. Their coaches might prefer to express speed in m/min. In fact, any convenient distance and time units may be combined to form speed units (Table 20.1.).

Table 20.1 *Average Speeds in Sports and Space*

USUAL UNITS	SPEEDING OBJECT	SPEED IN m/s
25 km/min	Rocket taking off for the moon	417 m/s
272 km/h	Racing car in Indianapolis 500	76 m/s
45 m/s	Baseball in Blue Jays game	45 m/s
1600 m/min	Horse in Queen's Plate	27 m/s
10 m/s	Olympic athlete in 100 m dash	10 m/s
360 m/min	Olympic athlete in 5000 m event	6 m/s

In this activity, you will work in groups of three to determine average speed experimentally. The questions for both Part 1 and Part 2 may be found after Part 2.

PART 1

Problem

How can the average speed of a moving ball be determined?

Materials

"racetrack" (for example, 30 cm plastic ruler)
"ball" of suitable size (for example, marble)
watch or clock with second hand

Procedure

1. Copy Table 20.2 into your notebook.
2. Set up ball and racetrack as in Figure 20.3.
3. Divide up the following jobs and carry them out. They may require some practice.

   Timer: Watch the clock closely, and give the player a start signal. After exactly 10 s, give a stop signal.
   Player: At the start signal, roll the ball so it travels smoothly back and forth without pausing or jumping the track (Figure 20.3). Keep the ball moving until you hear the stop signal.
   Recorder: Count the number of complete trips made by the moving ball. At the stop signal, the ball may be part-way along the track. If so, note the ball's exact position, and measure the length of the partial trip.

4. Copy the data collected on your table. Complete the calculations as shown.

Figure 20.3 *One possible race-track arrangement for Activity 20A*

Observations

Table 20.2

Total time	10 s
Distance of complete trips	A
Distance of any partial trip	B
Total distance	A + B
Average speed = $\dfrac{\text{total distance} = (A + B)}{\text{total time } 10 \text{ s}} = ?$	

PART 2

Problem

How can the average speed of a relay team be determined?

Procedure

(Note: This is a suggested procedure. Before beginning, plan exactly what you will do, and draw up a suitable data table.)

Form groups of six. You will need four runners, a timer, and a recorder. Each runner will make a return trip across the gym floor or the parking lot. The recorder will measure the length of the course. The timer will measure how many seconds the team takes to complete the course.

Observations

Complete the data table you designed.

Questions

1. (a) When a ball is rolled back and forth along a track, at what point does the ball move fastest? Explain.
   (b) Does the ball ever stop moving entirely? Explain.
2. (a) Which measurement—distance or time—is most likely to be inaccurate? Explain.
   (b) Is this more of a problem in Part 1 or Part 2? Explain why.
3. (a) How could your method of measurement be improved to eliminate any inaccuracy?
   (b) Is there some source of inaccuracy that you cannot eliminate? Explain.
4. Why do you think you were asked to keep the ball in motion for 10 s? Why not just measure the time for a single trip?

Force and Acceleration

Before a ball can start moving, a force must set it in motion. In bowling, for example, a forward force is needed to speed the ball up. Then the force of friction slows the ball down. To curve instead of travelling straight, the ball must also have sidespin (Figure 20.4). The bowler causes sidespin by exerting an extra sideways force.

These three kinds of change in motion (speeding up, slowing down, and changing direction) are all examples of acceleration. In science, **acceleration** tells how fast an object is changing its speed or direction. To avoid confusion, slowing down is sometimes called *negative acceleration*.

Acceleration always requires a force. A **force** is any push or pull that can cause acceleration. Not every force does cause acceleration, however. For example, you can push very hard against a locked door without moving it at all. But whenever acceleration occurs, there is always a force involved.

Both acceleration and force are very important in sports. For example, track coaches study films to learn how runners can accelerate out of the starting blocks more quickly. Researchers use computers to learn how discus throwers can use force more effectively.

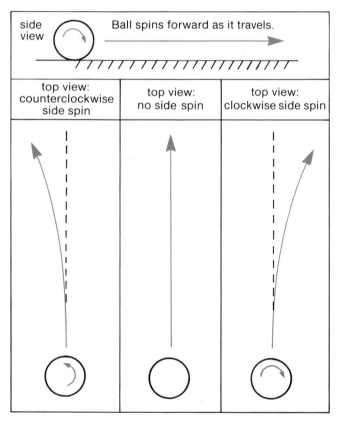

Figure 20.4 *As the ball rolls, it spins forward like a car wheel. To curl left or right, the ball must also spin around like a top.*

Challenge

Two grade 9 students made the following measurements to determine the average speed of a relay runner:

    distance travelled by runner   = 150 m
    time taken   = 20 s

(a) Which would have more effect on the answer, an error of 1 m in the distance or an error of 1 s in the time? To find out, repeat the above calculation four times:
  (i) with the distance 1 m greater
  (ii) with the distance 1 m smaller
  (iii) with the time 1 s greater
  (iv) with the time 1 s less

(b) With the measuring tools available in your school, which is more likely, an error of 1 m or an error of 1 s? Which do you think is more important in determining speed, the instrument used to measure distance or the instrument used to measure time? Explain.

Self-check

1. (a) Define speed.
   (b) Define average speed.
   (c) Use an example from sport to explain the difference between speed and average speed.
2. (a) Define acceleration.
   (b) Define force.
   (c) Use an example from sport to explain how acceleration depends on force.
3. (a) A motorist drove on a highway with a speed limit of 80 km/h. The car travelled a distance of 160 km in exactly 2 h. What was the motorist's average speed?
   (b) A highway patrol stopped the car. The police officer said that a radar gun had measured the car's speed at 100 km/h. The motorist got a ticket for speeding. Explain how the motorist could have exceeded the speed limit without going more than 80 km in 1 h.
4. (a) Examine Figure 20.5. For each picture, decide whether or not acceleration is taking place, and explain how you know.
   (b) For each picture that shows acceleration, identify the source of the force that is causing it.
5. (a) A wheelchair athlete travelled 40 000 km in a little over 2 a. Do you think the athlete travelled exactly the same distance every day? Explain.
   (b) If the trip lasted 800 d, calculate the athlete's average speed, in km/d.
6. During a baseball warm-up, a pitcher and catcher stand 20 m apart. The team manager clocks them at 8 s for 10 throws. (From pitcher to catcher = 1 throw.) Calculate the average speed of the ball (in m/s).

Figure 20.5

Vertical Motion: Falling and Bouncing

Two changes in motion take place at the edge of a waterfall. First, the water's direction changes from horizontal to vertical. Second, the water speeds up as it falls downward. See Figure 20.5 (e). Both changes are signs of acceleration. You probably realize that the force causing this acceleration is gravity.

Acceleration Due to Gravity

Earth's **gravity** is a force that pulls objects towards the planet's surface. To learn how gravity affects vertical motion in sports, researchers study experiments such as the one in Figure 20.6. It shows a ball being dropped from the roof of a tall building.

Gravity accelerates the ball so it travels farther and faster each second. The ball continues to speed up until it hits the ground. This is because gravity is a continual force. It continues to pull even after the ball lands. At that point, however, the ball cannot fall any farther.

Gravity also affects vertical motion when a ball is thrown upward (Figure 20.7). As soon as the ball is released, it starts to slow down and before long comes to a stop. Then, it changes direction and accelerates downward.

Researchers who design safety equipment must allow for gravity. For example, both high jumpers and pole vaulters use landing pits. High jumpers fall about 2 m before they land (Figure 20.8). Pole vaulters drop about three times as far. Because of acceleration due to gravity, falling three times as far means landing nine times as fast! Thus, pole vaulters need much more cushioning to land safely.

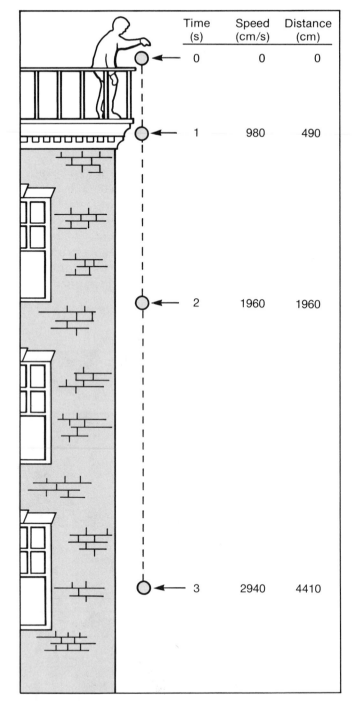

Time (s)	Speed (cm/s)	Distance (cm)
0	0	0
1	980	490
2	1960	1960
3	2940	4410

Figure 20.6 *The same speed and distance values are obtained every time this experiment is repeated.*

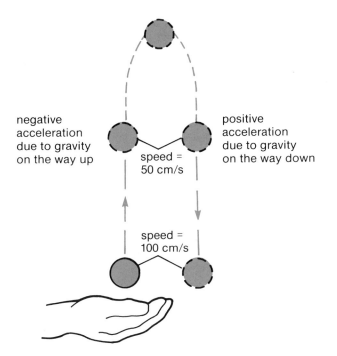

negative
acceleration
due to gravity
on the way up

positive
acceleration
due to gravity
on the way down

speed =
50 cm/s

speed =
100 cm/s

Figure 20.7 *This example of vertical motion shows all three forms of acceleration. Gravity first slows down the ball, then changes its direction, and finally speeds it up.*

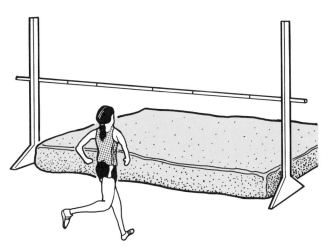

Figure 20.8 *A foam pit allows the accelerating high jumper to land safely.*

Spin, Bounce, and Rebound

You have already seen how spin affects the horizontal path of a bowling ball. Does spin affect vertical motion too? In Activity 20B, you and a partner will investigate this question by bouncing a ball and measuring its rebound. To do this, you will use the most elastic ball available.

In many sports, a ball is bounced off a floor, a bat, or a boot. Bouncing a ball flattens it temporarily on one side (Figure 20.9). The ball then springs back to its original shape. This "springiness" makes the ball rebound away from whatever flattened it. The scientific name for "springiness" is **elasticity**. The more elastic the ball, the faster and farther it rebounds.

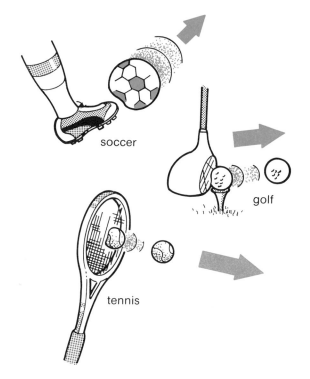

soccer

golf

tennis

Figure 20.9 *What happens when a ball bounces. The flattening effect has been greatly exaggerated here. It would not normally be visible.*

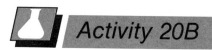

Activity 20B

PART 1

Problem

Which kind of ball is the most elastic?

Materials

metre stick
masking tape
assortment of balls
  (as shown in table, or others
  may be substituted)

Procedure

1. Copy Table 20.3. Choose one partner to drop the ball and one to make measurements.
2. Place tape on the wall 2 m from the floor. Standing at least 1 m from the wall, line up the top of the ball with the tape.
3. Do a test drop, so the recorder can practise measuring rebound height. Measure to the top of the ball (Figure 20.10).
4. For each ball, do three drops and measure the rebound. In your table, add the three rebound measurements. Divide the total by three. This value is average rebound. Enter it in the table.

Observations

(See Table 20.3)

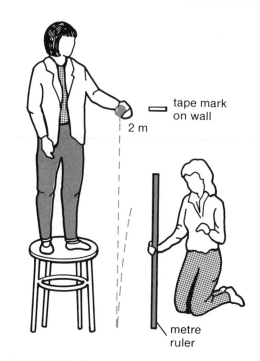

Figure 20.10 *Hold the ball 2 m high. If you have to stand on something, be sure it is steady and safe.*

Questions

1. (a) Which kind of ball is most elastic?
   (b) Which kind of ball normally travels long distances?
   (c) Explain any connection between (a) and (b).

Table 20.3

	HEIGHT OF REBOUND			
	Basketball	Golf Ball	Lacrosse Ball	Tennis Ball
A	cm			
B				
C				
A+B+C				
AVERAGE REBOUND	(A+B+C) ÷ 3 =	(A+B+C) ÷ 3 =	(A+B+C) ÷ 3 =	(A+B+C) ÷ 3 =

2. Suppose each kind of ball was hit with the same amount of force. Which would go farthest? Explain.
3. (a) Which is more elastic, a golf ball or a tennis ball?
   (b) Why not use a small tennis ball to play golf?
   (c) Why not use a large golf ball to play tennis?
4. (a) Did the tennis ball travel at the same speed each time it was dropped? Explain how you know.
   (b) Is your answer to (a) true for all the other balls tested? Explain.
5. (a) Describe any source(s) of error in your measurements.
   (b) How could you measure more accurately?

PART 2

Problem

How does spin affect rebound?

Materials

metre stick
masking tape
most elastic ball from Part 1

Procedure

1. Practise putting topspin and backspin on the most elastic ball (Figure 20.11).
2. Copy Table 20.4 in your notebook allowing enough space for the sketches required in step 5. The coloured arrows show what path the ball will follow until it hits the floor.
3. Drop the ball from a height of 2m (a) with no spin, (b) with topspin, and (c) with backspin.
4. Stand 2 m from your partner. Mark the halfway point with tape. Aiming at the tape, bounce the ball toward your partner (a) with no spin, (b) with topspin, and (c) with backspin.
5. As you do each step, complete the sketch in your observation table. Use an arrow to show the rebound path of the ball after it bounces. In the same space, note any other changes in the ball's motion.

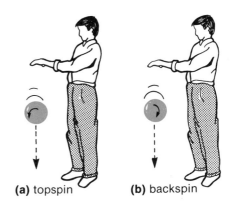

(a) topspin   **(b)** backspin

Figure 20.11 *For topspin (a), drop the ball with its top surface spinning away from you. For backspin (b), drop the ball with its top spinning back toward you.*

Observations

Table 20.4

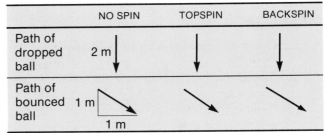

	NO SPIN	TOPSPIN	BACKSPIN
Path of dropped ball	2 m ↓	↓	↓
Path of bounced ball	1 m ↘ 1 m	↘	↘

Questions

1. (a) How does spin affect the rebound path of a dropped ball?
   (b) Is the effect the same for topspin as backspin?
2. Repeat question 1 for a ball bounced at an angle.
3. Does spin affect rebound the same way after a vertical drop as it does after an angled throw? Explain.
4. (a) Does a spinning ball rebound as quickly as a non-spinning ball?
   (b) Is your answer to (a) the same for topspin as backspin?
5. (a) Does a ball with topspin still have topspin after it bounces?
   (b) Repeat (a) for a ball with backspin.

How Spin Affects Rebound in Sports

No matter what game is being played, a ball with topspin rebounds faster than one with no spin. It also rebounds at a lower angle. Experienced players learn to allow for this effect.

Spin is used deliberately in games such as table tennis, cricket, and tennis. Tennis players often put backspin on bounce shots made near the net. The ball gets over the net safely but rebounds at a very steep angle. With enough backspin, the ball can actually rebound backwards, away from the opponent. Even an experienced opponent might miss such a return shot.

Self-check

1. (a) Define the following terms, and give an example of each: gravity and acceleration due to gravity. Explain how the two ideas are connected.
   (b) Explain the difference between gravity and acceleration due to gravity.
2. Describe how gravity affects vertical motion
   (a) in the upward direction
   (b) in the downward direction.
3. (a) Define elasticity, and give two examples to show that you understand how it affects ball sports.
   (b) Explain how elasticity affects vertical motion when a ball is dropped from a height of 2 m.
4. (a) Explain the meaning of topspin and backspin.
   (b) How does topspin affect vertical motion when a ball is dropped from a height of 2 m?
   (c) How does backspin affect vertical motion when a ball is dropped from a height of 2 m?

Sports Research

Today's Olympic medalists are the fastest and strongest ever, thanks to sports research. Sports research applies the principles of science to problems encountered in sport. Its main goals are as follows:

1. improving health and physical fitness
2. treating and preventing injuries
3. improving performances of athletes
4. improving the design of equipment

Most sports researchers specialize in a particular field of study. One such field is *sports physiology*, which studies how exercise affects the body. One obvious benefit of this research is that it results in a more scientific approach to coaching and training.

Non-athletes benefit, too. For example, X rays show that baseball pitchers have stronger, denser bones in their throwing arms. From this and other evidence, researchers conclude that exercise strengthens the bones as well as the muscles. Doctors have always recommended regular, vigorous exercise to keep the lungs, heart, and blood vessels in good condition. They now recommend it for everyone to help prevent bones from weakening.

A personal fitness program also plays a vital part in maintaining mental health. Research shows that physical activity eases stress and tension, relieves depression by promoting a sense of well being, and defuses feelings of anger and frustration. Exercise also suppresses the appetite, making it easier to maintain ideal body mass.

Sports medicine studies the cause of athletic injuries. It has introduced many completely new methods of treatment. For example, old-fashioned knee surgery used to keep athletes on the sidelines for months. Now, they recover in just a few days, thanks to arthroscopic surgery (Figure 20.12). Sports medicine helps athletes learn how to perform better without hurting themselves. It has also led to new rules aimed at preventing injury. For example, protective helmets must now be worn by all young hockey players. Most manufacturers employ sports researchers to help design better safety features.

Kinesiology studies how the body learns to run, jump, and throw. It can tell athletes how to change their swing, their grip, or their body position for the best result. This research shows that even the simplest motions are very complex. For example, a single golf swing involves a hundred different forces acting on a dozen different body parts, as well as the golf club!

The *physics of sports* studies equipment instead of the human body. Research topics here include the designing of accessories (such as balls, bats, and running tracks) and the predicting of flight paths (in events such as javelin, discus, and ball sports). In the next section, you will be able to try out some of the research methods used in the physics of sport.

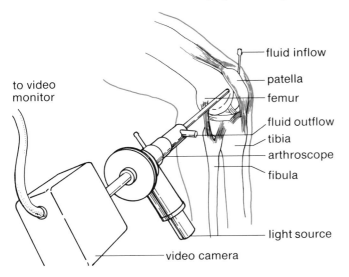

Figure 20.12 *Only a small incision is needed to insert the arthroscope so the joint can be inspected. Fluid may be injected to expand the joint slightly. If surgery is needed, tiny instruments are inserted through other small incisions.*

The Paths of Projectiles

A **projectile** is any object that can be thrown, fired, or shot. The path of a projectile is called a **trajectory**. Trajectories can be difficult to study because they are usually invisible.

In Activity 20C, you will study a visible trajectory.

Ideas and Applications

Osteoporosis is a disease characterized by brittle bones in old age. It sometimes affects men, but it is much more common in women. One in four women in Canada over the age of 65 suffers from osteoporosis. Osteoporosis often results in broken hips or other bones, and back and shoulder pain. There may also be some loss of height because of the deterioration and partial collapse of the spinal column. It used to be thought that the disease was caused by a diet low in calcium. Doctors still recommend a diet with sufficient calcium throughout life (not just in later years). But diet alone cannot increase bone density. Only the action of muscle pulling on bone will do that. Experiments have shown that exercise helps restore bone mass even in women who already have symptoms of osteoporosis. Medical researchers now believe that a regular exercise program throughout life will help to prevent this disease from occurring in old age.

Problem

What determines the shape of a water trajectory?

Materials

2 large cans, opened at one end and punched as in Figure 20.13

1 large can, opened at one end and left unpunched

Procedure

1. In your notebook, trace the can outlines from Figure 20.13.
2. Fill an unpunched can with water. Place a three-holed can at the edge of a sink. Plug the holes with your fingers.
3. Pour water from the unpunched can into the three-holed can. Unplug the holes. Quickly sketch the three water trajectories.
4. Repeat steps 1 and 2 with a one-holed can. Sketch the trajectory when the can is (a) nearly full and (b) nearly empty.

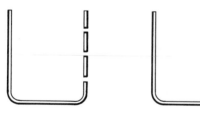

three-holed can   one-holed can

Figure 20.13 *Trace these outlines and use them when you sketch the water trajectories.*

Observations

Sketch the trajectories on the can outlines. Record any other observations on the same diagram.

Questions

1. The following questions refer to the one-holed can.
   (a) When does the water move fastest: when it leaves the can, at the bottom of the sink, or halfway between?
   (b) In what direction is the water moving at first?
   (c) Describe how it changes direction as it pours out. Try to explain why this change happens.
2. The following questions refer to the three-holed can.
   (a) Which trajectory is fastest? Try to explain why.
   (b) Which curves the most? Try to explain why.
   (c) Which is closest to horizontal? Try to explain why.
3. (a) Define acceleration.
   (b) What examples of acceleration did you observe?
   (c) What was causing the acceleration named in (b)?
4. (a) What affects trajectory shape the most? Explain.
   (b) What else affects trajectory shape? Explain.

How Gravity Affects Projectiles

In an orbiting space vehicle, projectiles act as if there were no gravity. Back here on earth, all unsupported objects are affected by gravity. Dropped footballs fall. Baseballs thrown upward come down again. Arrows aimed sideways curve downward.

The path of a projectile results from two forces: launching force and gravity. Because of gravity, no trajectory can ever be perfectly horizontal. Even if the launching force is horizontal, the trajectory will curve (Figure 20.14).

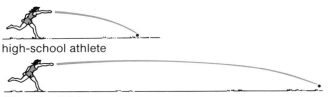

high-school athlete

Olympic athlete

Figure 20.14 *The ball that is launched with the greater force travels the farthest before it lands.*

In many ball sports, the goal is to send a ball as far forward as possible. This distance is limited by the way gravity affects falling time. Many stopwatch studies have been done in situations like Figure 20.14. When two balls are launched from the same height, their falling time is always the same. No matter how fast one ball is thrown sideways, both land at the same time. The faster ball travels farther, but does not spend any more time in the air. The only difference is that the faster ball can travel farther in the time available.

There are ways to increase a ball's air time, however. One is to launch it from a greater height (Figure 20.15). In most sports this is not practical. Another way is launch the ball at an angle. If the angle is too steep, the ball stays in the air a long time, but doesn't go very far forward. If the angle is too low, the ball hits the ground before it goes very far forward.

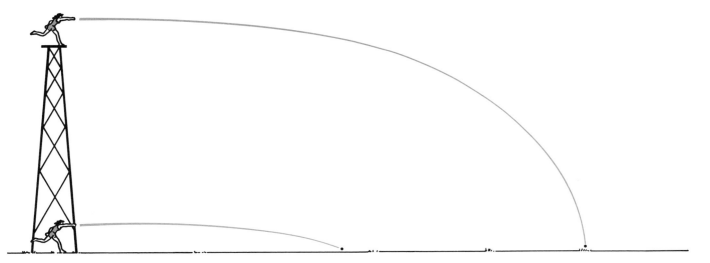

Figure 20.15 *When the launching force is the same, the ball launched higher up travels farther because it takes more time to fall.*

Sports researchers have done mathematical studies to find the best launching angle. These studies show that an angle of 45° gives the maximum forward distance (Figure 20.16).

So far, we have studied projectiles as if they were travelling through a vacuum. In the real world, they travel through air. Section 20.5 will consider how air affects the path of a projectile.

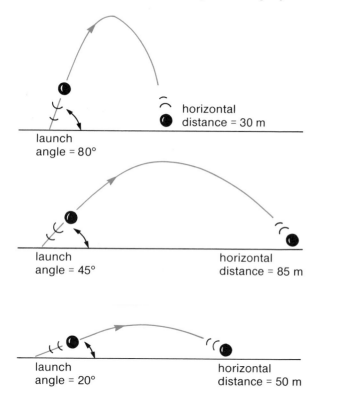

launch angle = 80°

horizontal distance = 30 m

launch angle = 45°

horizontal distance = 85 m

launch angle = 20°

horizontal distance = 50 m

Figure 20.16 *All three balls were launched with the same force.*

1. (a) What is sports research?
   (b) How has sports research helped athletes? Give two examples.
   (c) How has sports research helped non-athletes? Give two examples.
2. (a) Define projectile.
   (b) Give three examples of projectiles, and state how each one must be launched.
3. (a) What is the path of a projectile called?
   (b) What two forces affect the path of a projectile?
4. (a) What limits the amount of time a projectile can spend in the air?
   (b) What can an athlete do to keep a ball in the air for the greatest possible time?
   (c) What can an athlete do to send a ball the greatest possible forward distance?

Motion through Fluids

Materials that flow are called **fluids**. All gases and liquids are fluids. All fluids resist motion. **Air resistance** is caused when air molecules collide with a moving object (Figure 20.17). Each tiny collision reduces forward motion.

Figure 20.17 *Air resistance affects most sports because it resists forward motion. For example, a ball moving through a vacuum (black line) travels much farther than the same ball moving through air (coloured line).*

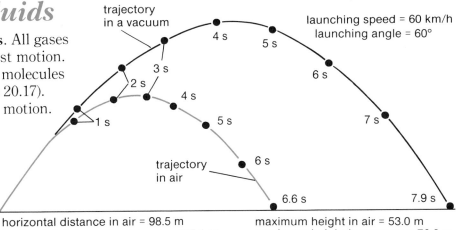

trajectory in a vacuum

launching speed = 60 km/h
launching angle = 60°

trajectory in air

horizontal distance in air = 98.5 m
horizontal distance in vacuum = 177.1 m

maximum height in air = 53.0 m
maximum height in vacuum = 76.8 m

The Drag Force

The braking force caused by air resistance is called **drag**. Its direction is always opposite to the moving object's direction. The amount of drag depends on the object's speed, size, surface, and shape (Figure 20.18).

In most sports, drag is a disadvantage because it makes athletes work harder. To reduce drag, sports equipment is often polished or specially shaped. This is called **streamlining**. See Figure 20.18 (c).

Figure 20.18

(d) *Shape. A shape that lets the airflow change direction gradually reduces drag.*

sharp changes in airflow

gradual changes in airflow

(a) *Speed. The faster an object moves, the greater the drag. Ten times as much speed means 100 times as much drag!*

(b) *Size. The larger the frontal area, the greater the drag. Hunching over reduces the frontal area that must cut through the air.*

(c) *Surface. The smoother the surface, the smaller the drag.*

Drag is not always a disadvantage, however. For example, it helps skydivers land safely (Figure 20.19). It also slows them down before the chutes open.

The first part of a jump is called "free fall." During this period, the divers fall faster and faster each second. But drag grows even faster than speed. For example, four times the speed gives sixteen times the drag. Once drag equals gravity, the divers stop accelerating. Their bodies continue to fall, but at a steady rate called **terminal speed** (Figure 20.20).

Because of drag, all objects falling through air reach a terminal speed eventually. (If the ball in Figure 20.6 could fall far enough, it would also reach terminal speed.) The actual value of terminal speed depends on the object's shape, size, and surface area. See Table 20.5.

Table 20.5 *Terminal Speeds*

Human body	190 km/h
Volleyball	56 km/h
Table-tennis ball	32 km/h
Raindrop	24 km/h

Figure 20.19 *Skydivers open their parachutes to increase drag.*

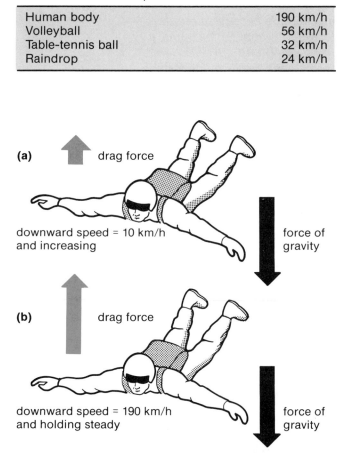

(a) drag force

downward speed = 10 km/h and increasing

force of gravity

(b) drag force

downward speed = 190 km/h and holding steady

force of gravity

Figure 20.20 *When drag becomes equal to gravity, objects falling in air reach a final steady rate called terminal speed.*

Air flowing around an object always causes drag. If the airflow is unequal, another effect may be seen.

Problem

How are movable objects affected by unequal airflow?

Materials

strip of light-weight typing paper (e.g., 20 cm x 5 cm)
2 small styrofoam balls with long threads attached
ringstand
bendable straw

Procedure

1. Read step 2. Predict the outcome by choosing (a), (b), or (c).
   (a) The paper will move up, toward the faster airstream.
   (b) The paper will move down, away from the faster airstream.
   (c) The paper will flap up and down about equally.
   Write down your prediction, and then carry out step 2.

2. Using both hands, hold the paper to your lower lip (Figure 20.21). Blow a narrow stream of air over the top of the paper. Start with a slow airstream, and gradually speed it up. Note how the paper moves, and record your observations.
3. Set up the apparatus shown in Figure 20.22.
4. Read step 5. Predict the outcome by choosing (a), (b), or (c).
   (a) Both balls will move toward the faster airstream.
   (b) Both balls will move away from the faster airstream.
   (c) Both balls will sway from side to side about equally.
   Write down your prediction, and then carry out step 5.
5. Through the straw, blow a strong horizontal airstream between the two styrofoam balls. Record your observations.

Observations

Observe the paper and the styrofoam balls carefully. Is there any sign of a motion you did not predict?

Figure 20.21 *How to hold the paper in step 2 of Activity 20D*

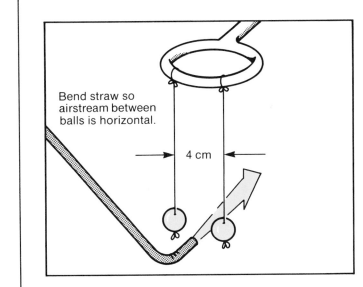

Bend straw so airstream between balls is horizontal.

4 cm

Figure 20.22 *Apparatus for step 5 of Activity 20D*

Questions

1. (a) What caused unequal airflow around the paper?
   (b) What caused unequal airflow around the styrofoam balls?
2. (a) Was the motion of the paper what you predicted? If not, how was it different?
   (b) Is lift a good name for the motion of the paper? Explain.
3. (a) Was the motion of the styrofoam balls what you predicted? If not, how was it different?
   (b) Is lift a good name for the motion of the styrofoam balls? Explain.
4. Which sentence does the best job of describing how unequal airflow affects movable objects? Explain why.
   (a) When airflow is unequal, a force acts to push objects toward the faster airstream.
   (b) When airflow is unequal, a force acts to push objects away from the faster airstream.
   (c) No noticeable effect can be seen when airflow is unequal.

Lift

The force caused by unequal airflow always pushes toward the faster airstream. If the object is pushed upward, the force is called **lift**. Lift is possible only when airflow on top is faster than airflow below. If airflow is equal, no lift results (Figure 20.23).

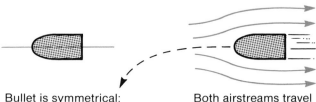

Bullet is symmetrical: Both sides of centre line match.

Both airstreams travel same distance and at the same speed.

Figure 20.23 *Because a bullet is symmetrical, it cuts the air into two equal streams. No lift results. As the bullet pushes forward, gravity pulls it downward, and it falls to the ground. Compare with Figure 20.24 (b).*

Lift is the force that enables hang-gliders to fly (Figure 20.24). When a glider heads into the wind, air flows around the wing. Its front edge breaks the air into two streams that meet again at the back. The shape of the wing forces air to move faster along the top. As a result, the whole glider is pushed up, toward the faster air stream. This same lift effect makes birds and airplanes fly. It also helps skiers jump farther. (Recall Figure 20.18 (b) and try to explain why.)

The force caused by unequal airflow does not always push upward. Its direction depends entirely on the faster airstream. For example, one styrofoam ball moved left while the other moved right. But both were pushed toward the faster airstream.

You may also have noticed that the styrofoam balls began to spin. Unequal airflow and spin often go together. Section 20.6 will examine how spin and unequal airflow affect motion in ball sports.

Figure 20.24 (a) *Lift results when the glider is pushed into the wind.*

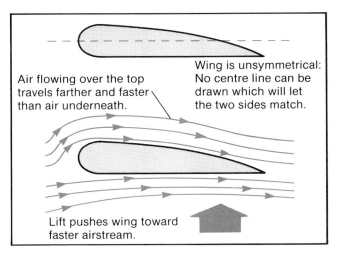

Air flowing over the top travels farther and faster than air underneath.

Wing is unsymmetrical: No centre line can be drawn which will let the two sides match.

Lift pushes wing toward faster airstream.

(b) *Lift is caused by the wing's unsymmetrical shape. The bulge forces air to flow farther and faster on top. This pushes the wing up, toward the faster airstream.*

Self-check

1. (a) Define air resistance, and explain what causes it.
   (b) Give two sports-related examples of air resistance.
2. (a) Explain the meaning of drag.
   (b) Name two sports in which drag is a disadvantage. Explain why.
3. (a) Define streamlining.
   (b) Give an example of streamlining in sports.
   (c) Give an example of streamlining in nature.
4. There are two ways to create unequal airflow.
   (a) What method did you use in Activity 20D?
   (b) What method is used in hang-gliding?
5. (a) Define lift, and explain what causes it.
   (b) Give a sports-related example of lift.
   (c) Give an example of lift in nature.
   (d) Give an example of lift in industry.

Spinning through a Fluid

Earlier in this chapter, you learned how spin affects the path of a rolling or bouncing ball. Spin also affects the path of a ball soaring through the air.

Problem

How does spin affect trajectories in table tennis? (Note: If space or equipment are limited, your teacher may arrange for a demonstration.)

Materials

table-tennis ball	pencil
bat	blue pen
masking tape	red pen

Procedure

1. Hit the ball with a straight follow through, so it flies forward without spinning. See Figure 20.25 (a). This is flight path A. Observe it, and mark where the ball lands.
2. Hit the ball with topspin. To do this, bring the bat up and over the ball as you hit. See Figure 20.25 (b). This is flight path B. Observe it, and mark where the ball lands.
3. Hit the ball with backspin. To do this, "slice" under the ball as you hit. See Figure 20.25 (c). This is flight path C. Observe it, and mark where the ball lands.

Observations

On a single diagram, sketch the three flight paths. Show path A in pencil, path B in blue, and path C in red.

Questions

1. (a) Describe the basic flight path of the non-spinning ball.
   (b) Explain why the path is not a straight line.
2. (a) How does topspin alter the basic flight path?
   (b) How does backspin alter the basic flight path?
3. Compare the three flight paths as follows:
   (a) What do they all have in common?
   (b) Which is longest? Which is shortest?
   (c) Which curves most sharply? Which curves least sharply?

Figure 20.25

(a) *A straight hit—the ball does not spin.*

(b) *Putting topspin on the ball*

(c) *Putting backspin on the ball*

How Spin Affects Trajectories

Figure 20.26 shows three trajectories. Ball A was thrown with no spin. Ball B was given a topspin, and Ball C was given a backspin.

Trajectory A curves downward because of gravity, but trajectory B curves even more sharply. Apparently, topspin causes an extra downward push on Ball B (one that adds to the effect of gravity). Trajectory C curves less sharply than A. Apparently, backspin causes an upward push on Ball C (one that counteracts the effect of gravity). This effect can be noticed in many ball sports, including baseball, golf, and tennis.

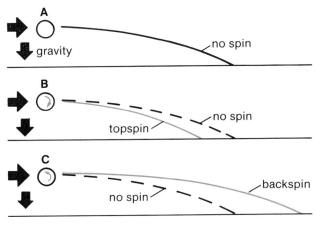

Figure 20.26

The Magnus Force

The way spin changes trajectories is called the Magnus effect, after the scientist who first noticed and studied it. This effect is caused by the Magnus force. The **Magnus force** acts only when objects spin through a fluid. It may push the trajectory down, up, left, or right, depending on spin direction.

353

MAGNUS FORCE WITH TOPSPIN

For a non-spinning ball, airflow is equal on all sides. See Figure 20.27 (a). But a topspinning ball drags some air around with it. See Figure 20.27 (b). On top of the ball, the dragged air opposes natural airflow. The result is unequal airflow. The top airstream moves slower, while the bottom airstream moves faster.

Figure 20.27

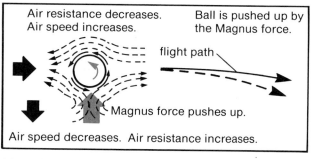

(a) *Airflow around a non-spinning ball*

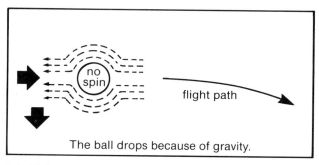

(b) *Airflow around a topspinning ball*

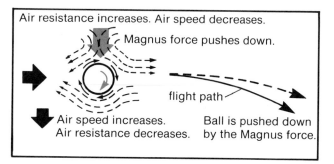

(c) *Airflow around a backspinning ball*

The force caused by this unequal airflow is known as the Magnus force. The Magnus force always pushes the ball toward the faster airstream. During topspin it pushes the ball down. As a result, the ball hits the ground sooner than it would with no spin.

MAGNUS FORCE WITH BACKSPIN

Figure 20.27 (c) shows the airflow around a ball with backspin. This time, the dragged air opposes natural airflow *under* the ball. The bottom airstream moves slower, while the top airstream moves faster.

Thus backspin creates a Magnus force that pushes the ball upward, toward the faster air stream. This extra upward force "lifts" the trajectory. The ball stays in the air longer than it would with no spin.

MAGNUS FORCE WITH SIDESPIN

Sidespin can make the ball curve left or right (Figure 20.28). The Magnus force always acts to push the ball toward the faster airstream. If the front of the ball spins left, its path will curve left. If the front of the ball spins right, its path will curve right.

NO SPIN MEANS NO MAGNUS FORCE

If a ball is thrown with no spin, airflow is equal on all sides. There is no faster airstream. Thus, no Magnus force is created.

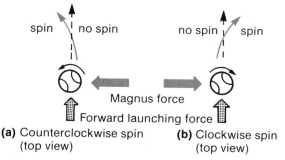

Figure 20.28 *Sidespin causes a Magnus force that pushes the ball left or right.*

How Spin Enhances Ball Sports

Baseball pitchers depend on spin to control the ball's trajectory. Knowing just how and when to put spin on a ball is important in many other sports as well. For example, topspin is used in many tennis shots. This is because topspin makes the ball drop sharply just after clearing the net. Opponents may have trouble reaching the ball in time.

Sidespin is often used to make the tennis ball curve left or right. Except for bounce shots, however, tennis players avoid backspin. A ball with backspin will stay in the air longer, giving the opponent more time to reach it.

Golfers also use topspin and backspin but try to avoid sidespin. A golf ball with backspin can travel much farther before it hits the ground. Topspin can be used to bring the ball down sooner, preventing the ball from going astray. But the slightest bit of sidespin will create a Magnus force the golfer does not want. If the front of the ball spins left, the Magnus force will push the ball's path to the left. Spin to the right will push the path to the right. Golfers call these problems hooking and slicing.

The Size of Magnus Force

The size of the Magnus force depends on five factors:
1. diameter (the larger its diameter, the more the ball curves)
2. mass (the smaller the mass, the more the ball curves)
3. roughness (the rougher its surface, the more the ball curves)
4. spin rate (if two identical balls are thrown with the same forward speed, the one that spins faster will curve more sharply)
5. forward speed (if two identical balls are given the same spin rate, the one that moves forward more slowly will curve more sharply)

The Magnus effect is very important in a game like golf. The golf ball has a small mass, can be made to spin rapidly, and has a rough surface. These points more than make up for its rapid forward speed and small diameter.

The Magnus force is also important in baseball but has very little effect in basketball. A basketball has a much slower forward speed than a baseball. It also has a larger diameter and a rougher surface. These three factors should make a spinning basketball curve sharply. However, two other factors have just the opposite effect. A basketball has much more mass than a baseball and also a much slower spin rate. As a result, the Magnus force on a basketball is too small to have much effect on its flight path.

In table tennis, however, the ball has a very small mass, and a rapid spin rate combined with a fairly slow forward speed. These more than make up for its smoothness and small diameter. The Magnus effect is so great that the curve can be seen with the naked eye, even over short distances. You may actually see a table tennis ball with backspin fly upward.

Self-check

1. Sketch the path of a baseball that is
   (a) thrown horizontally with no spin
   (b) thrown horizontally with topspin
   (c) thrown horizontally with backspin.
2. (a) What is Magnus force?
   (b) What causes Magnus force?
3. (a) What is the final effect of Magnus force on a ball with topspin?
   (b) What is the final effect of Magnus force on a ball with backspin?
4. In which of the following sports is Magnus force important? Explain why or why not.
   (a) golf   (d) tennis
   (b) hockey   (e) bowling
   (c) basketball   (f) football

Chapter Objectives

Words to Know

average speed
acceleration
force
gravity
elasticity
projectile
trajectory
fluid
air resistance
drag

streamlining
terminal speed
lift
Magnus force

Tying It Together

1. (a) Explain the difference between speed and average speed by giving an example of each.
    (b) Explain the difference between speed and acceleration by giving an example of each.
2. (a) A runner in the 100 m event finishes in a time of 15 s. What is the runner's average speed? (Correct to one decimal place.)
    (b) A runner in the 5000 m event finishes in a time of 750 s. What is the runner's average speed? (Correct to one decimal place.)

3. (a) Define force, and give three examples.
   (b) Define acceleration, and give three examples.
   (c) Explain how force and acceleration are related.
4. (a) A car crashes into a brick wall. Did acceleration occur? Explain.
   (b) A marathoner maintained a steady pace for 10 km. Did acceleration occur? Explain.
   (c) A skydiver jumped from a plane. Did acceleration occur? Explain.
5. (a) Define gravity.
   (b) Describe how earth's gravity affects motion in each of the following sports: bowling, sailing, hang-gliding, sky-diving, golf, and pole-vaulting.
6. (a) With a sketch, explain the meaning of elasticity.
   (b) Explain how elasticity affects motion in ball sports.
   (c) Which of the balls mentioned in this chapter is most elastic.
7. (a) List the goals of sports research.
   (b) Name four areas of speciality in sports research.
   (c) Briefly describe one of these specialities.
8. A centrefielder throws a baseball to home plate.
   (a) At what angle should the ball be launched to spend the longest time in the air?
   (b) Will this launch angle also give the greatest horizontal distance? Explain.
9. (a) Define trajectory.
   (b) Two identical one-holed cans containing water are placed on a sink. The water from can A travels a farther horizontal distance than the water from can B. Sketch both trajectories.
   (c) State two possible reasons for the difference in trajectory.
10. (a) Explain why all objects falling in air eventually reach a terminal speed.
    (b) Which has the faster terminal speed, a human body or a baseball?
    (c) Explain why there is no set value for terminal speed.

11. (a) With a simple sketch, show how lift and drag act on a hang-glider.
    (b) What do lift and drag have in common? (Hint: Both have the same basic cause.)
    (c) How are they different? (Hint: They have a different effect on moving objects.)
12. (a) Why would a speed-skater want to reduce drag?
    (b) List some ways this could be done.
13. (a) Why would a skydiver want to increase drag?
    (b) List some ways this could be done.
14. Use a sketch to show the trajectory of
    (a) a ball with no spin
    (b) a ball with backspin
    (c) a ball with topspin.
15. Use a sketch to show the forces acting on a topspinning ball as it travels forward through the air. Label each force and show its direction.
16. Explain in detail
    (a) the meaning of Magnus force
    (b) how it affects a ball with backspin
    (c) why it is more important in baseball than basketball.
17. (a) Is Magnus force important in a sport like bowling? Explain.
    (b) Is Magnus force important in a sport like tennis? Explain.

Applying Your Knowledge

1. Which ball would travel farther when bounced: one made of crushed paper or one made of Plasticine? Explain.
2. A car is stopped at a red light. When the light turns green, the car starts moving and reaches a speed of 60 km/h. The car continues at a constant speed of 60 km/h until it comes to another red light. It then comes to a stop.
   (a) On what parts of the trip was the car accelerating?
   (b) Name the force that caused each acceleration named in (a).

3. (a) Describe how the surfaces of baseballs, tennis balls, and golf balls have been deliberately roughened.
   (b) Explain why they are roughened. What difference would it make if they were smooth?
4. (a) Bicycle racers often choose to follow the leader, even when it would be easy to pass. Use a sketch to show how this gives an advantage to the following rider.
   (b) List two other methods used by bicycle racers to reduce the drag force, and explain how these methods work.
5. Suppose a computer is used to design the fastest land animal on earth. Draw a sketch to show what you think this animal would look like.
6. In a windstorm, the roof of a house may be "blown off," even though the wind cannot get under the roof. Explain how this is possible.
7. (a) What forces act on a Frisbee once it leaves your hand?
   (b) What effect does each of the forces in (a) have on the Frisbee's motion?

Projects for Investigation

1. Use your imagination plus what you learned in Activity 20A and Section 20.1 to collect data and calculate average speed for one of the following:
   (a) a bowling ball as it rolls down the alley
   (b) a student walking into school in the morning
   (c) a student walking out of school in the afternoon
   (d) a tennis ball or volleyball being batted back and forth
   (e) a basketball as it is being passed up and down the floor
2. Use the methods of Activity 20B to test different brands of golf balls. Find out whether the most expensive ball is the most elastic. Why would a golfer want the most elastic ball?

21

Light and Photography

Key Ideas

- Photography depends on the nature of light and light-sensitive materials.
- Various lens-containing instruments can be used to form images, alter their size, or project them onto a screen.
- A multi-step procedure involving camera, film, and chemicals is needed to record images on permanent photographs.
- Photography is an important part of the work, study, and recreation of scientists and non-scientists.

All of the pictures on the previous page were taken by amateur photographers. No matter how complex the picture or how much equipment is used, all photography is based on a few simple scientific principles. In this chapter, you will learn enough about these principles to increase your own enjoyment of photographs and photography.

Drawing Pictures with Light

Photography means "drawing with light." To take good photographs, you need to know some of the scientific facts about light.

1. Light is a form of energy that you can see. Some visible objects, such as the sun, give off their own light. Most visible objects, however, only reflect light from some other source.
2. Light travels very fast. At the speed of light, 300 000 km/s, you could circle the earth seven times in one second. Thus, photographers can usually ignore the time it takes for light to get from the scene to the camera.
3. Light travels outward in all directions. The farther light travels from its source, the dimmer it gets (Figure 21.1). This affects the way photographers use artificial lighting. For example, using flashbulbs to shoot distant scenes such as mountains is just a waste of money.
4. Light travels in straight lines. In science, straight lines with arrowheads are often used to show the path of light. The lines are called **light rays**. Drawings that use them are called **ray diagrams** (Figure 21.2).

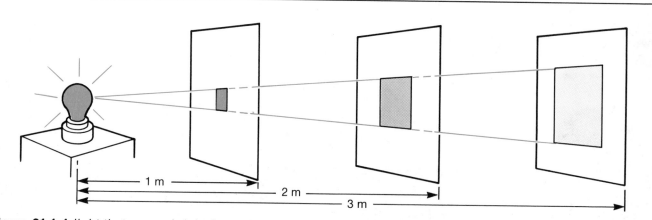

Figure 21.1 *A light that seems bright from 1 m away will be 1/4 as bright from 2 m and 1/9 as bright from 3 m.*

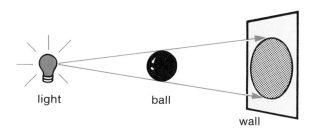

light   ball

wall

Figure 21.2 *This ray diagram predicts the shadow formed by a ball. (Note: Light is travelling away from the bulb in all directions, but rays are drawn only for light travelling toward the ball.)*

When Light Meets Matter

What happens when light strikes a surface? Depending on the surface material, light can be transmitted, reflected, or absorbed (Figure 21.3).

Transmission occurs when light passes through **transparent** materials, such as glass or clear plastic. **Opaque** materials, such as metal or black plastic, prevent transmission. Cameras have transparent parts to let light in, and opaque parts to keep it out.

Reflection occurs when light bounces off of a surface. Most of what we see or photograph is viewed by reflected light. Pale surfaces reflect most of the light that strikes them. Dark surfaces reflect very little.

Smooth, shiny surfaces reflect light in an orderly fashion, producing a "mirror image." See Figure 21.4 (a). An **image** looks like the original object because the reflected rays repeat the original pattern. Few surfaces are smooth enough to do this. Even white typing paper has microscopic projections. Thus, light rays reflected off paper jumble the original pattern, and no image is seen. See Figure 21.4 (b).

Absorption occurs when light energy becomes "trapped" inside the material. The trapped energy may simply warm up the material by making its molecules move faster.

In some cases, absorption of light energy causes a chemical reaction and may result in a colour change. This happens only with so-called **light-sensitive** materials.

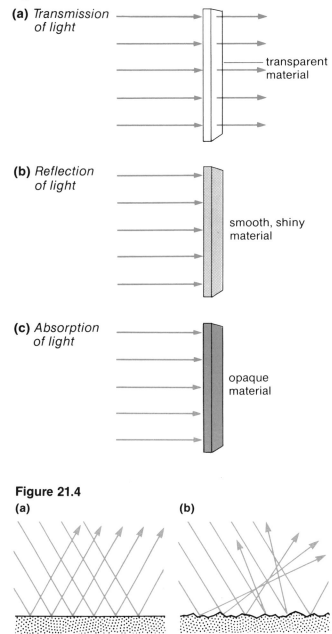

Figure 21.3

(a) *Transmission of light* — transparent material

(b) *Reflection of light* — smooth, shiny material

(c) *Absorption of light* — opaque material

Figure 21.4

(a)   **(b)**

All photography depends on the action of light-sensitive materials, especially certain silver compounds. These compounds are white to start with, but they break down when exposed to light. The result is very small bits of dark, metallic silver. The unexposed parts stay white. In this activity, you will use paper coated with silver compounds to record a pattern formed by light. You won't need a camera or darkroom. You will, however, have to work in dim light until step 3. Before you begin, make a copy of Table 21.1 in which to record your observations.

Problem

How can a picture be taken without a camera?

Materials

"photo sandwich" made of materials shown in
Figure 21.5
3 small objects that won't roll easily
clock or watch with second hand
scissors or paper cutter

Procedure

1. Your teacher will adjust the classroom lights to a dim setting. Place your "photo sandwich" on the worktable, cardboard side up. Lift the cardboard and slide it under the typing paper.
2. Arrange three objects on the print paper. Make sure this paper has its glossy side up.

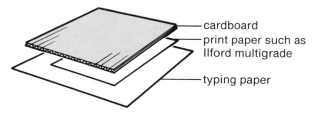

cardboard
print paper such as
Ilford multigrade

typing paper

Figure 21.5 *Arrangement of "photo sandwich" for Activity 21A*

3. When your teacher turns up the lights, compare the colour of the print paper to that of the typing paper.
4. Without moving any of the objects, hold the sandwich so that print paper is exposed to bright light. Note the time.
5. After two minutes of exposure, remove one object. Note the condition of the print paper, and record your observations in your copy of Table 21.1. After two more minutes, remove a second object. Again, record your observations. After a total of six minutes, remove the third object and record your observations.
6. Cut the print paper in two. Write your name on each part. Store one in a dark place and the other in a bright place, as your teacher directs. Next day, compare the two halves.

Observations

Table 21.1 *Data for Activity 21A*

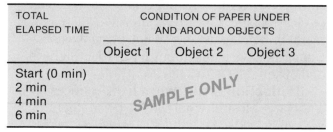

TOTAL ELAPSED TIME	CONDITION OF PAPER UNDER AND AROUND OBJECTS		
	Object 1	Object 2	Object 3
Start (0 min) 2 min 4 min 6 min			

Questions

1. Describe how the print paper changes when exposed to light.
2. How does exposure time affect the resulting picture?
3. (a) How good is the quality of your picture? For example, can other students use your picture to identify the objects used?
   (b) How does your picture compare to a photograph of the same objects?
   (c) Recall the definition of an image. Is your picture an image? Explain.

Outlines and Images

The pictures you made in Activity 21A are called **photograms**. They differ from photographs in two important ways.

1. Photograms are not permanent. The paler parts keep on darkening until the whole sheet is the same colour. Section 21.4 will explain how photograms can be preserved.

2. Photograms are not detailed. A photogram is only an outline of the original object, not an image (Figure 21.6). This is because a photogram collects only light that gets past the object. To make a true photograph, the print paper must be exposed to light from the object itself.

Simply allowing light from the object to fall onto print paper does not produce an image, however. This is because light travels from every point on the object to every point on the print paper. Figure 21.7 shows how the original pattern of light and dark gets lost in a jumble of rays. As a result, the paper darkens without producing an image.

To form an image with all the original detail, we need a device that will gather light from the object as well as keep the pattern of light and dark from the object sorted in its original order. The device most commonly used to do this is a camera. In Section 21.2, you will investigate the most basic kind of camera and learn how it produces an image.

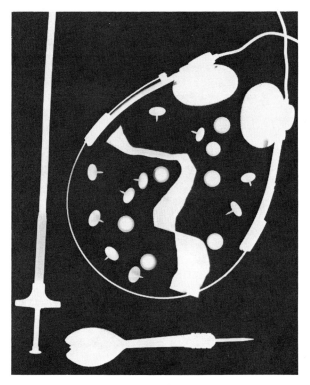

Figure 21.6 *How does this photogram differ from an image of the same objects?*

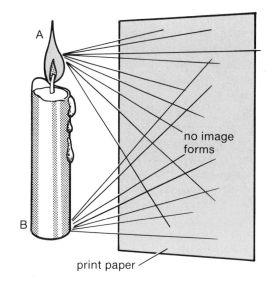

Figure 21.7 *Only a few of the light rays that could be drawn are shown here.*

Self-check

1. (a) What evidence is there that light travels in straight lines? (Give two examples of evidence from this chapter, and one from your own experience.)
   (b) Why is the fact that light travels in straight lines important in photography?
2. (a) Define image. Give an example to show that you understand the meaning of this term.
   (b) Explain why the picture produced by a photogram is not an image.
3. (a) There are three ways that light may act when it meets matter. Give the correct term for each way.
   (b) With a sketch, show the meaning of each term listed in (a).
4. (a) What is a light ray? Use a simple sketch to show that you understand the meaning of this term.
   (b) Does a light ray contain light? Explain.
5. In your notebook, make a ray diagram to predict the shadow that will form in Figure 21.8.
6. (a) Explain the meaning of "light-sensitive material."
   (b) Name a light-sensitive material that is not used in photography. Describe what happens when light strikes it.
   (c) What light-sensitive materials are used in photography? Describe what happens when light strikes them.

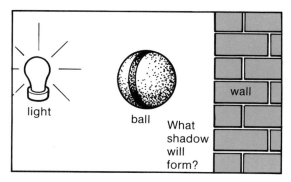

Figure 21.8 For Self-check question 5

Forming Images with a Pinhole

The strange-looking apparatus in Figure 21.9 is actually a camera. It is one type of pinhole camera. The simple **pinhole camera** shown in Figure 21.10 (a) consists of a sealed container with a small hole at one end. The hole can be left closed or opened to admit light. The container is otherwise lightproof.

The modern camera in Figure 21.10 (b) has the same basic features as this simple pinhole camera. The modern camera also has a light-sensitive surface inside. The simple pinhole camera shown here has only a screen. Thus, it does not take photographs. Its sole purpose is to form images. The part that makes this possible is the pinhole itself.

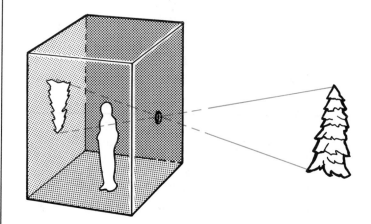

Figure 21.9 *What cameras looked like before light-sensitive compounds were discovered. This device, called a camera obscura, helped non-artists to draw accurately—in every way but one!*

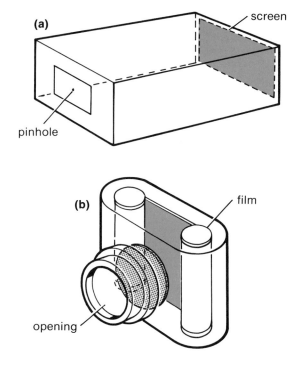

Figure 21.10 *Studying pinhole cameras will help you understand the modern cameras discussed in Section 21.4.*

How a Pinhole Forms an Image

Light spreads out in all directions from the tree in Figure 21.11. However, only those rays heading towards the pinhole can get inside the pinhole camera. For example, Ray A follows a straight-line path from the top of the tree to the hole. Once it passes through, Ray A continues along the same downward path, until it reaches the bottom of the screen.

Ray B starts out part way from the top of the tree. After it passes through the hole, it strikes the screen part way from the bottom. Ray C starts out near the bottom of the tree but strikes the screen near the top. Ray D travels from the very bottom of the tree to strike the very top of the screen. Many other rays could be drawn to pass through the pinhole. Together, they show how the pattern of light from the tree repeats itself on the screen to form an image.

The pinhole does two important things to prevent the jumbling of light rays from the tree. First, it blocks off most of the unwanted light. Only the light headed in the right direction can enter the camera. Second, the pinhole sorts the light that does enter into the original pattern. The only important difference is that the image is upside down!

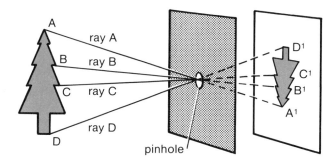

Figure 21.11 *This ray diagram shows why pinhole images are upside down.*

In this activity, you will use a pinhole camera under various conditions and look for factors that affect the quality of the image. (A factor is anything you can change that makes a difference.) You will work in dim light for steps 3, 4, and 6.

Problem

What factors affect the image produced by a pinhole camera?

Materials

pinhole camera (as in Figure 21.12 or similar)
brightly lit subject (e.g., candle on teacher's desk or slide projected on wall)
geometry compass (or similar pointed instrument)

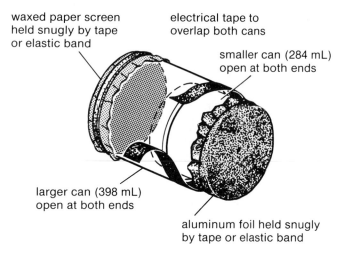

waxed paper screen held snugly by tape or elastic band

electrical tape to overlap both cans

smaller can (284 mL) open at both ends

larger can (398 mL) open at both ends

aluminum foil held snugly by tape or elastic band

Figure 21.12 *One type of pinhole camera*

Procedure

1. In your notebook, make an observation table similar to Table 21.2.
2. Your teacher will show how to make your pinhole camera. Adjust your camera to its maximum length, and tape the join securely. THE ROOM LIGHTS SHOULD NOW BE TURNED OFF.

3. In this step, you will observe the effect of changing the distance between camera and subject.
   (a) Move as far from the subject as possible. Aim the camera at the subject until an image appears on the screen. Observe the size, brightness, and clarity of the image formed by your camera.
   (b) Move halfway toward the subject. Note any change(s) from the image you saw in 3 (a).
4. In this step, you will observe the effect of changing the distance between the pinhole and its screen.
   (a) From your worktable, observe the size, brightness, and clarity of the image on your camera's screen.
   (b) Carefully strip off the tape, and set the camera at its shortest length. Retape the join.
   (c) From your table, observe the image formed by the shorter camera. Note any change(s) from the image you saw in 3 (a). THE ROOM LIGHTS MAY NOW BE TURNED ON.
5. Before performing step 6, make a prediction. How do you think enlarging the pinhole will affect the image? Record your prediction in the observation table. You will test it in step 6. Hint: Before making a prediction, look back at Figure 21.11, and ask yourself:
   (a) How will a larger pinhole affect the amount of light that enters the camera and reaches the screen?
   (b) How will a larger pinhole affect the way the light enters the camera and reaches the screen?
   (c) How will the size of the image be affected? (Will it become larger, smaller, or stay the same?)
   (d) How will the brightness of the image be affected? (Will it become brighter, dimmer, or stay the same?)
   (e) How will the clarity of the image be affected? (Will it become more detailed, less detailed, or stay the same? THE ROOM LIGHTS SHOULD NOW BE TURNED OFF.

6. In this step, you will observe the effect of changing pinhole size.
   (a) From your table, observe the image formed by the small pinhole.
   (b) Enlarge the pinhole by pushing the point of the compass part way into the aluminum foil.
   (c) Observe the image formed by the larger pinhole. Note any change(s) from the image you saw in 5 (a).

CAUTION Handle sharp, pointed objects carefully.

Observations

(See Table 21.2.)

Questions

1. (a) How is a pinhole image like the original subject?
   (b) How is a pinhole image different from the original subject?
2. Did any step in the procedure turn the image right side up? Explain.
3. Draw a ray diagram to show how the distance between camera and subject affects image size, brightness, and clarity
   (a) when the camera is far from the subject
   (b) when the camera is close to the subject.
4. Draw a ray diagram to show how the distance between pinhole and screen affects image size, brightness, and clarity
   (a) when the screen is far from the pinhole
   (b) when the screen is close to the pinhole.
5. (a) What did you predict would happen to the image when the pinhole was enlarged?
   (b) What effect did the larger pinhole actually have on image size, brightness, and clarity?
   (c) Draw a ray diagram to explain why this happens.
6. How would you change the pinhole camera if you wanted to use it to take photographs?

Table 21.2 *Data for Activity 21B*

EQUIPMENT/SETUP	PINHOLE CAMERA IMAGE QUALITY		
	Size	Brightness	Clarity
Long camera			
Short camera			
Camera close to subject			
Camera far from subject			
Small pinhole			
Large pinhole (predicted)			
Large pinhole (observed)			

SAMPLE ONLY

The Limits of Pinhole Images

Pinhole images can be made as large or as small as needed. They are easy to produce, and pinhole cameras are inexpensive. However, pinhole images have one major disadvantage. They are always quite dim. This is because so little light is used to form them. Unless the room is very dark and the subject is very bright, pinhole images may be difficult to see.

The dimness of pinhole images becomes a serious problem when pinhole cameras are used for photography. Can you guess what the problem is? In Section 21.3, you will learn more about this problem and how photographers solve it.

Self-check

1. Compare a modern camera to a pinhole camera.
   (a) How are they alike?
   (b) How are they different?
2. Simply holding a screen in front of a brightly lit object does not result in an image. With a diagram, explain why.
3. In order to project an image on a screen, you need an image-forming device.
   (a) What is the image-forming device in a pinhole camera?
   (b) What two important jobs does this part perform?
4. What major limit of pinhole cameras makes them unsuitable for use in photography?

21.3

Forming Images with a Convex Lens

The world's first photograph was taken by a pinhole camera. A light-sensitive plate was placed in front of the screen. Then, the open pinhole was aimed out a window, exposing the plate to light for eight hours. Unfortunately, the plate was not very sensitive. Thus, the resulting rooftop scene is far from clear (Figure 21.13).

Films have improved so much since then that clear, sharp pinhole photographs are now possible. Long exposure times are still needed, however. Otherwise, pinhole images are far too dim to affect the film.

For faster pictures, more light is needed to make a brighter image. Modern cameras do this by placing a *lens* in front of the opening.

A **lens** is a transparent device that alters the path of light by bending the beam. Which way the beam bends depends on the shape of the lens. There are two basic lens shapes: concave and convex.

Concave lenses are thin through the middle and thick at the edges. A concave lens causes a beam of light to diverge, or spread out (Figure 21.14). This decreases brightness and prevents image formation.

Convex lenses are thin at the edges and thick through the middle. A convex lens causes a beam of light to converge, or become narrower. This concentrates the light falling on a large area into a narrow beam (Figure 21.15).

Convex lenses increase brightness, which makes them useful in cameras. They are also used in other instruments that form images. The kind of image that forms depends on how the lens is used (Figure 21.16).

Figure 21.13 *This scene was photographed by the French scientist Nicéphore Niépce in 1826. A shiny pewter plate was coated with a kind of tar that hardened when exposed to light. The unexposed tar stayed soft and soluble. Niépce used oil to wash away the tar that had not been hardened (exposed). This left an image so permanent that it can still be seen today.*

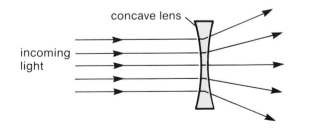

Figure 21.14 *Concave lenses bend the incoming light to make it spread out.*

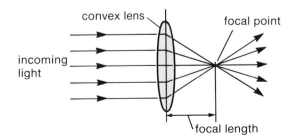

Figure 21.15 *Concave lenses are widely used in cameras.*

The place where a convex lens concentrates light most strongly is called the **focal point**. The distance between the focal point and the lens is called the **focal length**.

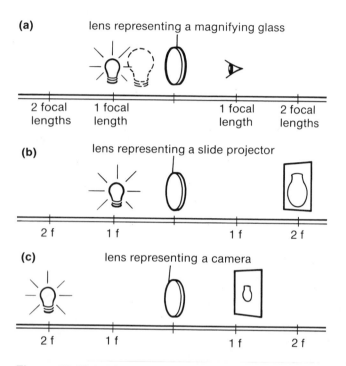

(a) lens representing a magnifying glass

| 2 focal lengths | 1 focal length | | 1 focal length | 2 focal lengths |

(b) lens representing a slide projector

2 f   1 f   1 f   2 f

(c) lens representing a camera

2 f   1 f   1 f   2 f

Figure 21.16 *(a) In a magnifying glass, the convex lens forms a larger, upright image. (b) In a slide projector, the convex lens forms a much larger, inverted image. (c) In a camera, the convex lens forms a much smaller, inverted image.*

Activity 21C

Problem

What factors affect the image produced by a convex lens?

Materials

metre stick
thin convex lens and mounting bracket
small screen (for example, white card 15 cm x 15 cm)
candle and mounting bracket
matches or burner flame
masking tape and marking pen

CAUTION Be careful of your hair and clothing when using an open flame.

Procedure

1. Make an observation table similar to Table 21.3.
2. In this step, you will locate the focal point of the lens. Darken the room, and cover all windows but one. Hold the lens up so it can gather light from the window. Move the screen back and forth until you see a bright, sharp image of the scene outside. The screen is now at the *focal point* of the lens. Observe what happens to the image when the screen is moved closer to the lens and then farther away.
3. In this step, you will measure the focal length of the lens. Mount the lens on a metre stick (Figure 21.17). Have one partner aim the lens-stick at the window. Have the other partner slide the screen along the stick to find a bright, sharp image. Subtract the metre stick reading under the lens from the reading under the screen. The result will be the distance between the screen and the lens. This distance is the focal length (f). Record it.
4. Set up the apparatus in Figure 21.18. Cover the marks on the metre stick with a strip of masking tape. Mark off multiples of the focal length from the centre of the metre stick in both directions: 1f, 2f, 3f.

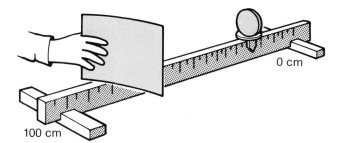

Figure 21.17 *Using a lens-stick to find focal length (f)*

0 cm

100 cm

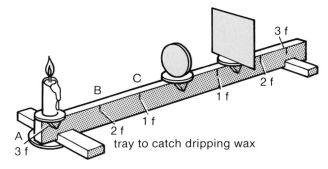

3 f

2 f

1 f

C

B

1 f

2 f

A

3 f

tray to catch dripping wax

Figure 21.18 *Use the focal length (f) of your lens (as measured in step 2) to mark the tape at 1 f intervals.*

5. In this step, you will observe how image and focal length are related.
   (a) Place the lighted candle at A, three focal lengths (3 f) from the lens. Slide the screen along the stick until you locate a bright, sharp image of the candle. Record your image observations in the table.

(b) Repeat (a) with the candle at B (2 f from the lens).
(c) Repeat (a) with the candle at C (1 f from the lens).
(d) Repeat (a) with the candle at "in-between" positions, as shown in the table.

Observations

(See Table 21.3.)

Questions

1. (a) What arrangement of candle, screen, and lens gives the largest possible image?
   (b) What arrangement of candle, screen, and lens gives the smallest possible image?
2. Compare the different arrangements in the procedure with Figure 21.16.
   (a) What arrangement was most like the magnifier?
   (b) What arrangement was most like the projector?
   (c) What arrangement was most like the camera?
3. How must slides be placed in a projector? Explain.
4. A detective wants to magnify a fingerprint. How should the magnifying lens be held to give the largest possible upright image?
5. The image formed by a convex lens depends on several factors. Name these factors, and explain the effect of each one.

Table 21.3

LOCATION OF CANDLE	NATURE OF IMAGE			LOCATION OF IMAGE
(How Many f's from Candle to Lens?)	Upright or Inverted?	Smaller or Larger?	Sharp or Blurred?	(How Many f's from Image to Lens?)
(A) 3 f				
2 f–3 f				
(B) 2 f				
1 f–2 f				
(C) 1 f				
Less than 1 f				

How a Convex Lens Forms an Image

In a modern camera, light-sensitive film is used instead of a screen. The lens is placed so a small, bright, inverted image falls on the film. The ray diagram in Figure 21.19 shows how this image is formed.

Light spreads out in all directions from every point on the person. Only a few of many possible light rays are drawn here. The two rays from A diverge (move apart) until they reach the convex lens. Then the lens makes them converge. It bends the rays so they meet again at a focal point. This forms an image of A on the bottom of the film.

Many other rays travel from A to the lens. All meet the same focal point. Thus, all help to form the image of A.

The rays from B also diverge until they reach the lens. The lens makes the two rays converge. They meet and form an image of B on the top of the film. Many other rays travel from B to the lens. All meet at the same focal point. All help to form the image of B.

In a similar way, the lens forms an image of each and every point on the person. All of these images form on the film, exactly one focal length from the lens. As a result, the pattern of light from the person repeats itself on the film. Except for being inverted, the light pattern of the image is identical to the original.

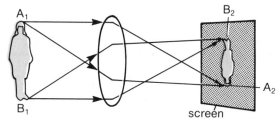

Figure 21.19 All rays from A_1 will be bent to meet again at A_2. All rays from B_1 are bent to meet at B_2. Rays from all points between A_1 and B_1 will meet at corresponding points between A_2 and B_2.

The image formed by a convex lens is much brighter than a pinhole image. This is because a pinhole image contains only the light falling on the small pinhole. But the image formed by a convex lens contains all the light falling on the surface area of the lens. The lens concentrates this large amount of light onto a small area of the film.

Focal Length and Photography

The thicker the lens, the closer is its focal point and the shorter its focal length (Figure 21.20). Knowing the focal length of a lens is important to photographers. There are two reasons for this.

1. The lens and film must be exactly 1 f apart for the brightest, sharpest image. For example, a lens with a long focal length must

Figure 21.20 Two convex lenses bending light rays. Which lens is thicker in the middle?

be placed a long way in front of the film. This is why some lenses are mounted on long tubes (Figure 21.21).

2. Focal length determines how wide a cone of light the lens can gather. Thick lenses have a short focal length because they can bend light so drastically. This means they can gather light from a wide cone (Figure 21.22). The image of a large scene can be crowded onto the film.

Thin lenses have a long focal length. They bend light less drastically. This means they gather light from a much narrower cone. The film gets only a small part of the scene in front. But, that image is spread all over the film. The result is a much more detailed picture. It looks as if the camera was much closer to the subject.

Figure 21.21 *Which tube contains the thickest convex lens? Which contains the thinnest?*

Figure 21.22

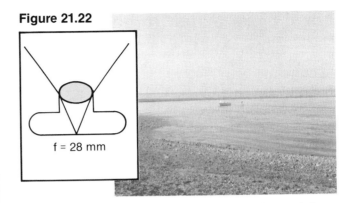

f = 28 mm

(a) *The very thick lens that took this "wide-angle" shot can focus light from a very wide cone. Thus, a very large part of the scene in front reaches the film.*

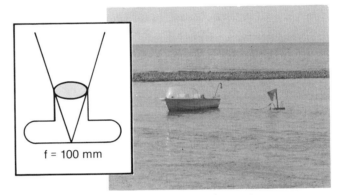

f = 100 mm

(b) *A thinner lens took this photograph. The cone of light it can gather and focus is more narrow. Thus, a much smaller part of the scene reaches the film.*

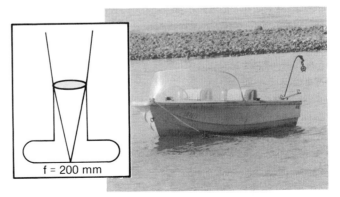

f = 200 mm

(c) *The very thin lens that took this "close-up" can only gather light from a very narrow cone. Thus, only a very small part of the scene in front reaches the film.*

Focal Length and Human Vision

Focal length is also important to human vision. The human eye is often called a "living camera." Figure 21.23 shows that eyes and cameras have certain basic features in common. In a healthy, normal eye, the lens and retina are about one focal length apart. The human lens is even self-adjusting, so the retina always receives a bright, sharp image.

However, a misshapen eyeball can cause vision problems. In some people, the eye is too short for the focal length of the lens, no matter how much the lens adjusts itself. In others, the eye is too long. Placing an extra lens in front of the eye can correct this kind of vision problem (Figure 21.24). Convex lenses are used if the eye is too short. Concave lenses help if the eye is too long.

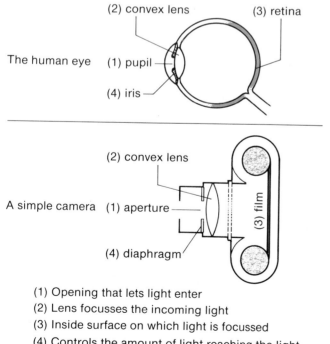

Figure 21.23 *The human eye is sometimes called a "living camera."*

The human eye
- (2) convex lens
- (3) retina
- (1) pupil
- (4) iris

A simple camera
- (2) convex lens
- (1) aperture
- (3) film
- (4) diaphragm

(1) Opening that lets light enter
(2) Lens focusses the incoming light
(3) Inside surface on which light is focussed
(4) Controls the amount of light reaching the light-sensitive surface

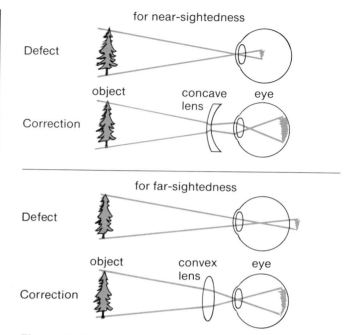

for near-sightedness
Defect
object   concave lens   eye
Correction

for far-sightedness
Defect
object   convex lens   eye
Correction

Figure 21.24 *How lenses correct vision defects*

Self-check

1. (a) What problem is there in using a pinhole camera to take photographs?
   (b) What device does a modern camera use to solve this problem?
   (c) How does the device named in (b) solve the problem in (a)?
2. (a) Name and describe two kinds of lenses.
   (b) Which kind is used to form images?
   (c) Why is the other kind not used to form images?
3. Compare the human eye to the modern camera. Tell how they are alike and how they differ.
4. (a) What is the meaning of focal point?
   (b) How would you locate the focal point of a convex lens?
   (c) Explain why you cannot locate a focal point for a concave lens.
5. (a) What is the meaning of focal distance?
   (b) What is the importance of focal distance in a camera?
   (c) What is the importance of focal distance in an eye?

Modern Photography

Photography has become an important part of our lives. Photography makes news more interesting and understandable. It lets scientists observe fast-moving events by "freezing" the details. Some professionals use photography to earn their living. Many amateurs find it a highly enjoyable hobby. And many people who never use a camera still enjoy collecting photographs.

People have always enjoyed pictures. Pinhole images were known as early as 1100 A.D., and lenses were used to focus light before that. Despite this, modern photography did not begin until around 1850. It took that long to work out a way to make a permanent image.

Forming a Permanent Image

Your eye cannot form a permanent image. The photograms you made in Activity 21A were not permanent, either. By themselves, image-forming devices and light-sensitive materials are not enough. Five steps go into making a permanent photograph (Figure 21.25).

1. Gathering light from an object and forming an image
2. Exposing light-sensitive film to a focussed image
3. Developing exposed film to produce a negative
4. Passing light through a negative onto print paper
5. Developing print paper to produce a positive print

Let's examine these steps in more detail.

1. A camera lens gathers light from the object to be photographed. As the light pattern passes through the lens, a bright, sharp image forms on the film at the back of the camera. The film is made of clear plastic, coated with *emulsion*. In photography, an emulsion is a thin paste that contains light-sensitive silver compounds.

Figure 21.25

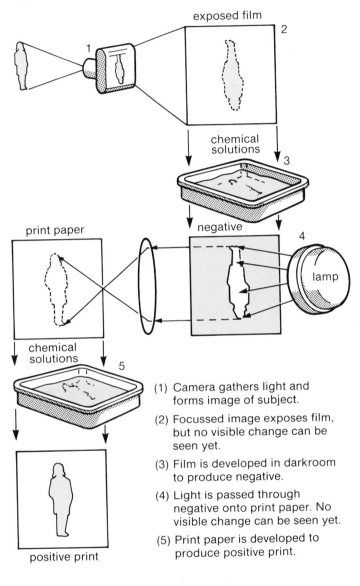

(1) Camera gathers light and forms image of subject.

(2) Focussed image exposes film, but no visible change can be seen yet.

(3) Film is developed in darkroom to produce negative.

(4) Light is passed through negative onto print paper. No visible change can be seen yet.

(5) Print paper is developed to produce positive print.

2. Light from the image exposes the film. Wherever light strikes, it changes the silver compounds. The more intense the light, the greater the chemical change. At this stage, the changes are not yet visible. The film must be developed to bring out the image.

3. *Developing* a film involves treating it with chemicals. Only the exposed silver compounds are changed. They turn into dark metallic silver that sticks to the film. The unexposed compounds do not change. They are washed off, leaving a visible image made up of clear patches of film and dark patches of silver. The developed film is called a *negative* because the image is reversed. It is dark in colour where the object was bright, and clear where the object was dark.

4. To get the original light pattern, a positive print must be made. First, light is passed through the negative onto print paper. The paper is coated with light-sensitive chemicals that react wherever light strikes them. These changes are not visible until the print paper is developed.

5. Developing a print involves dipping this coated paper into a series of chemical baths. One bath (the "developer") darkens the exposed parts. Another (the "fixer") prevents further darkening, so the print can be safely taken out of the darkroom. A finished *positive* print is the reverse of the negative. The print is dark where the negative was clear, and white where the negative was dark. The original pattern of light from the object has been restored. A permanent image has been formed.

Photography for Fun and Profit

A surprising number of people earn their living through photography. Not all of them are photographers. For example, real estate agents are often more successful if they display photographs of the homes they want to sell. Camera stores employ staff for sales and repairs. One-hour photo labs employ workers to run huge automatic processing machines. There are many other laboratories in less visible locations. They do work such as enlarging prints, processing colour slides, and developing movie film. Large factories also employ staff to manufacture film, cameras, and accessories.

The big machines used in fast photo labs might make photography seem very complicated.

Figure 21.26 *What makes a picture a winner? Decide which of the photos below and on page 377 should get first prize, and give your reasons. (Your teacher can help you find out what professional judges look for.)*

But developing your own black and white photographs is easy. It is also much cheaper than having it done by professionals. Many of the photographs in this chapter were taken by amateurs and developed in home laboratories.

Photography can be used for taking pictures of friends and nature, or for creating works of art. As a hobby, photography can be enjoyed at many levels. Beginners can have just as much fun as contest winners (Figure 21.26).

1. With your teacher's guidance, carry out a project relating to photography. A few topic ideas are listed here. Your teacher may suggest others.
   (a) A workshop. Arrange for small-group sessions with more experienced amateurs. Get them to explain terms such as f-stop, shutter speed, and film speed. Find out how knowing these meanings can help you take better photos.
   (b) An interview or survey. Do a survey to learn how much money the students and staff of your school spend each year on photography. Do interviews to learn what jobs near your home or school relate to photography. Report your findings to the class as suggested above.
   (c) A guest speaker. Arrange for someone who can provide information not available from books. For example, a professional photographer might tell the class how a beginner gets started, or an experienced amateur could demonstrate the use of camera accessories.

Self-check

1. (a) List three jobs that are directly related to photography.
   (b) List three jobs that are indirectly related to photography.
2. (a) In photography, what is the meaning of "negative"?
   (b) In photography, what is the meaning of "positive"?
3. (a) What do film and print paper have in common?
   (b) How do film and print paper differ?
   (c) How are film and print paper used in photography?

One Picture is Worth a Thousand Words

You take a roll of pictures of your friends and you want your pictures back fast! Your friends can't wait to see the results. What can you do? Thanks to chemistry and technology, you can have your pictures to share with your friends only one hour after you take in your film.

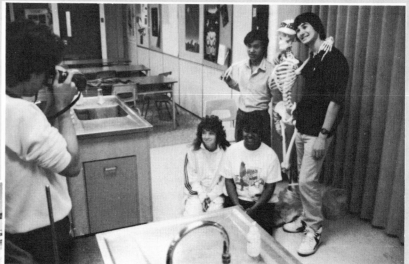

1 *Your film goes into a loading box. The machine pulls the film gradually through separate chemical baths.*

2 *Each chemical bath is responsible for a different chemical reaction. The length of time your film is in each bath, and the temperature of the solution, is precisely controlled. A computer determines when each chemical needs to be replaced. Your negatives have been produced at the end of this step.*

7

6

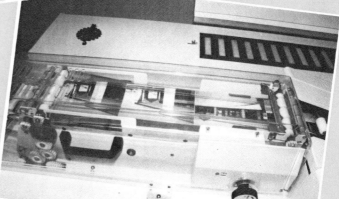

5

3 *A computer helps an operator print pictures from your negatives. Light shines through each negative to expose photographic paper.*

4 *The prints, which will eventually become your pictures, are in one long strip of paper. This strip goes through a series of chemical baths to produce the colour image.*

5 *The strips of photographic paper are automatically cut into your printed pictures.*

6 *Your pictures are separated by a conveyor and packaged.*

7 *Chemical changes and modern technology result in lasting memories.*

3

4

Chapter Objectives

Words to Know

photography
light ray
ray diagram
transmission
transparent
opaque
reflection
image
absorption

light sensitive
photogram
pinhole camera
lens
concave
convex
focal point
focal length

Tying It Together

1. (a) List four scientific facts about light.
   (b) Which is most important to photography?
2. (a) Name three ways light can behave when it meets matter.
   (b) What kind of material does each require?
3. There are two possible results when a material absorbs light. Name them, and give examples.
4. (a) A sheet of print paper is held 10 cm from a lit candle for 5 min. Describe what will happen to the print paper.
   (b) With a diagram, explain why this happens.
5. (a) What part of a pinhole camera forms the image? With a ray diagram, explain how this part does its job.
   (b) What part of a modern camera forms the image? With a ray diagram, explain how this part does its job.
6. If the opening of a pinhole camera is enlarged, the image is destroyed. With the help of a ray diagram, explain why.
7. (a) How could a simple pinhole camera be used to take photographs?
   (b) What problem(s) would you expect?
   (c) Explain what causes the problem(s).

8. In this chapter, you have learned about two image-forming devices. Name them, and list as many uses as you can for each.

9. A pinhole camera and a convex lens are both aimed at the same subject. Which forms the brighter image? Explain.

10. How could you test eyeglass lenses to learn whether the owner's eyes are too short or too long?

11. Which has the longer focal length—a thick convex lens or a thin convex lens? Explain.

12. (a) List all the steps needed to make a photograph, and briefly describe each step.
    (b) Which is the most important step? Explain your answer.

13. (a) What is a photographic negative made from?
    (b) Describe the negative that would result from photographing a hockey puck on ice.
    (c) Describe the positive print that would be made from this negative.

14. Describe three specialized cameras. Be sure to state the purpose of each one, and tell how it differs from a regular camera.

15. Describe the relationship between science and photography. In other words, tell how science depends on photography, and how photography depends on science.

Applying Your Knowledge

1. It is possible to make a photogram on an apple. Describe the method you would use, including the time involved. (Hint: Think about how you would make a photogram on a lawn.)

2. A photographer aims a camera at an outdoor scene. There are flowers in the foreground, mountains in the background, and trees in-between.
    (a) Which will reach the camera first: light from the flowers, light from the trees, or light from the mountains? Explain.
    (b) What must the photographer do to allow for the difference in arrival time? Explain.

3. How could a knowledge of photography help each of the following people?
    police officer
    homeowner
    landscape gardener
    stamp collector
    retired worker
    hockey coach
    student
    car salesperson

4. A photographer wants to shoot a scene that includes an apple tree in a field, but must stand behind a fence 100 m away. Which of the lenses shown in Figure 21.22 would be best for taking a picture of (a) the whole scene with the tree in the middle, (b) the tree with none of the field, and (c) a single apple?

5. What would happen to the resulting photograph
    (a) if the film was placed less than one focal length from the lens? Explain.
    (b) if the film was placed more than one focal length from the lens? Explain.

Projects for Investigation

1. With your teacher's guidance, build a pinhole camera that can be used to take black-and-white pictures. In a darkroom, clip a roll of film into short segments. Label each segment carefully. Load the segments into your camera, one at a time, and experiment with different exposure times. Develop the film. How long must film be exposed in your pinhole camera to photograph an outdoor scene?

2. Without looking at a reference book, list all the non-medical uses you can imagine for an X-ray camera. Then check the library to see how many of your ideas are right.

Unit Seven: The Science of Sports and Leisure

MATCH

In your notebook, write the letters (a) to (f). Beside each letter, write the number of the word in the right column which corresponds to each description in the left column.

(a) repeats ray pattern of original object

(b) records shape of original object, but not its ray pattern

(c) thick in the middle, thin at the edges

(d) place where bent light rays are concentrated

(e) light-sensitive coating on film

(f) distance between camera lens and film

1. concave lens
2. convex lens
3. developer
4. emulsion
5. focal length
6. focal point
7. image
8. photogram
9. shadow

MULTIPLE CHOICE

In your notebook, write the numbers 1 to 10. Beside each number, write the letter of the best choice.

1. Which of the following statements is *not* true?
   (a) Whenever acceleration occurs there is always a force involved.
   (b) Every force causes acceleration.
   (c) Acceleration tells how fast an object is changing speed or direction.
   (d) Negative acceleration occurs when a moving object comes to a stop.

2. Kinesiologists are sports researchers who study
   (a) how exercise affects the body
   (b) how the body learns to run, jump, and throw
   (c) how to design better sports equipment
   (d) the causes and cures of athletic injuries.

3. Gravity affects the path of a projectile by
   (a) making it fall straight down
   (b) making its path curve downward, except for a backspinning ball
   (c) adding a downward force to whatever other forces are acting
   (d) making it impossible for the projectile to move upward.

4. A projectile thrown through air follows a trajectory like that in a vacuum. However, the path through the air is
   (a) shorter and lower
   (b) shorter and higher
   (c) longer and lower
   (d) longer and higher.

5. Which of the following ideas would help a skydiver make a safe landing?
   (a) streamlining surfaces to increase drag
   (b) reducing freefall time to reduce terminal speed
   (c) increasing surface areas to reduce drag
   (d) increasing surface areas to increase drag

6. Lift force is caused by unequal airflow around an object. Which statement is true?
   (a) Lift always pushes toward the faster airstream.
   (b) Lift always pushes away from the faster airstream.
   (c) Lift acts only on symmetrical objects.
   (d) Lift acts only on spinning objects.

7. Which of the following would a pitcher use to make a baseball stay in the air longer?
   (a) no spin
   (b) sidespin
   (c) backspin
   (d) topspin

8. Which of the following is *not* an example of acceleration?
   (a) An athlete runs at a steady pace around an oval track.
   (b) A pitcher throws a non-spinning ball to third base.
   (c) A golfer loses a sidespinning ball in the woods.
   (d) Skydivers fall downward at a steady rate after opening their parachutes.

9. During a football game, the ball is carried 50 m up the field in 20 s. Which statement is false?
   (a) The ball moved at a constant speed of 2.5 m/s.
   (b) The ball moved at an average speed of 2.5 m/s.
   (c) The distance moved by the ball was probably greater than 50 m.
   (d) The ball underwent both positive and negative acceleration.

10. Objects falling in air reach a terminal speed
   (a) that is the same for all objects regardless of shape, mass, or volume
   (b) when the drag force becomes equal to the force of gravity
   (c) because gravity grows smaller as the object gets closer to the earth
   (d) only if they have topspin.

TRUE/FALSE

Write the numbers 1 to 10 in your notebook. Beside each number, write T if the statement is true and F if the statement is false. For each false statement, rewrite it so that it becomes true.

1. Only professional photogaphers can take good pictures.
2. Opaque materials transmit light, while transparent materials absorb light.
3. Light travels in straight lines, unless reflected by a mirror or bent by a lens.
4. The part of a pinhole camera that re-sorts jumbled light rays is the pinhole.
5. The focal length of a convex lens depends on the distance between the lens and the object.
6. A thin convex lens has a shorter focal length than a thick convex lens.
7. For a sharp image on a photographic negative, the lens and the film should be 1f apart.
8. The modern camera is similar to the human eye in all important ways, except that the eye has no adjustable diaphragm.
9. Photographs taken with pinhole cameras require a long exposure time because the image formed is so dim.
10. Modern photography did not develop until after 1850 because so little was known about light-sensitive compounds.

FOR DISCUSSION

Read the paragraphs below and answer the questions that follow.

Knowledge gained through sports research is helping to solve medical problems for one unusual group of non-athletes: workers who earn their living in space. Cosmonauts work in the U.S.S.R.'s space station. Astronauts fly the U.S.A.'s space shuttle. Both share similar working conditions, especially when it comes to gravity.

Earth's gravity pulls down strongly on a shuttle or space station. But this downward motion is balanced by the vehicle's own outward motion. As a result, the vehicle neither falls nor climbs. It simply stays in orbit. For the workers inside, the effect of gravity seems to be neutralized.

Workers in orbit can therefore move about easily in any direction, even upward. They can also lift huge objects with little effort. This makes some parts of the job easier, but there are serious medical drawbacks. Space workers experience a variety of body changes while in orbit. For example, the skeleton loses about 2% of its calcium every month. The spine lengthens by as much as 0.5 cm a month. The muscles get weaker and smaller.

In recent years, the periods spent in space have been increasing in length. Just as this book was being printed, a Soviet citizen returned to earth after setting a record of 327 days in an orbiting space station!

1. Name all the branches of sports research. Which might be helpful in solving the medical problems of space workers? Explain.
2. Suggest what space workers might do to slow down calcium losses. Base your answer on sports research, and explain why you think your method would help.
3. What difficulties would astronauts encounter while carrying out a fitness program in space? How might some of these problems be overcome?
4. The cosmonaut referred to above lost both bone mass and strength, but had "grown" nearly 4 cm taller. Explain why this was not true growth, and explain how it could happen. Hint: The spine contains many cylindrical bones. The spaces between the bones contain fluid-filled sacs.

The Metric System

The metric system of measurement was originally designed by French scientists in 1791. This system is based on multiples and fractions of ten. As measurement techniques improved, many of the original metric definitions of measurement had to be changed. In 1960, the metric system was modernized to become the International System of Units. From the French *Système international d'unités*, it is called the SI. The SI is used by 90% of the world's population in over 100 countries.

There are seven base units in the SI. In this course, you will use only the three base units shown in Table A.1.

All other units are derived from (based on) the SI base units. Units can be multiplied by multiples of ten to give larger units. They can also be divided by multiples of ten to give smaller units. A prefix is used with the name of the unit to identify the derived unit. For example, the prefix *kilo* means one thousand. Thus, kilometre means one thousand metres. The prefix *milli* means one-thousandth. Thus, millisecond means one-thousandth of a second. Table A.2 shows the derived units that you will use in this course.

Table A.1 *Some SI Base Units*

QUANTITY	UNIT	SYMBOL
Length	metre	m
Mass	kilogram	kg
Time	second	s

Table A.2 *Common Derived SI Units Used in This Course*

QUANTITY MEASURED	UNIT PREFIX	milli-	centi-	deci-	UNIT	deca-	hecto-	kilo-
	MULTIPLE	$\frac{1}{1000}$	$\frac{1}{100}$	$\frac{1}{10}$	1	10	100	1000
Length		millimetre mm	centimetre cm	decimetre dm	metre m			kilometre km
Area			square centimetre cm²		square metre m²			
Volume (solid)			cubic centimetre cm³	cubic decimetre dm³	cubic metre m³			
Volume (liquid)		millilitre mL			litre L			kilolitre kL
Mass		milligram mg			gram g			kilogram kg
Energy					joule J			kilojoule kJ

Graphing

A graph is a special kind of diagram that shows information (data) in the form of a picture. Graphs make it easy to see how one factor is related to another. Graphs also show patterns you might not be able to observe if the data were listed in a table or chart. Thus, graphs are used in many different areas of study, such as business, economics, and (of course) science.

Look at the growth record for the teenager named Susan Lee. This record shows measurements of mass as Susan's age changes. Can you easily see a relationship between mass and age in Table B.1? You will graph this information so you can observe more readily the relationship between Susan's age and the change in her mass.

All graphs begin with two number lines. One of these lines is horizontal. The other is vertical. The horizontal line is called the *x*-axis. The vertical line is called the *y*-axis. The plural of axis is axes (pronounced "ak-seez"). Each axis is used to show a different measurement. The *x*-axis is usually used to show the measurement over which the experimenter has control. The measurement graphed on the *y*-axis usually depends on the changes in the other measurements. The experimenter does not have control over the values that are measured and placed on the *y*-axis.

DRAWING A GRAPH

1. The point at which the *x*-axis and the *y*-axis meet is called the origin. On a sheet of graph paper, establish the origin as shown in Figure B.1. The origin should be placed about 3 cm up from the bottom edge of the paper and 3 cm over from the left edge.

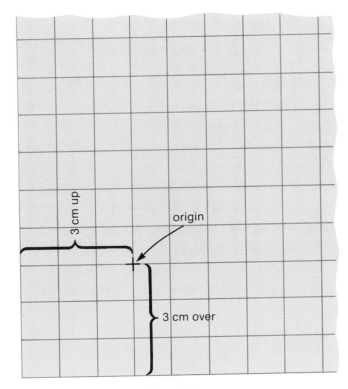

Figure B.1 *Establish the origin.*

Table B.1 *Age and Mass Data for Susan Lee*

Mass In Kilograms	3	10	12	14	16	18	20	22	25	29	33	37	42	46	50	53	56	57	57
Age In Years	0	1	2	3	4	5	6	7	8	9	10	11	12	13	14	15	16	17	18

2. Draw the *x*-axis as shown in Figure B.2. Use a ruler to make certain that your line is directly on the line on the graph paper. Then draw the *y*-axis with a ruler. Neatly print a label on each axis.

3. Title the graph. Near the top of the paper, draw a title block as shown in Figure B.3. Inside it, print a short title for the graph. Include your name and the date.

4. Select an appropriate scale. A scale is the spacing used to separate the numbers. Selecting the right scale for your data is probably the most important skill to learn in graphing. If you choose a scale that is too small, the graph you draw may be too small to read. On the other hand, if you choose a scale that is too large, the graph will not fit on the page. You will have to count the number of squares available and decide what will work best.

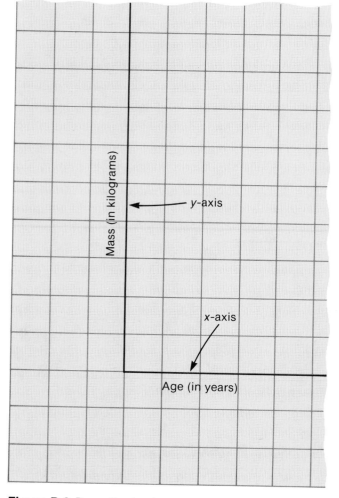

Figure B.2 *Draw the horizontal axis (x-axis) and the vertical axis (y-axis).*

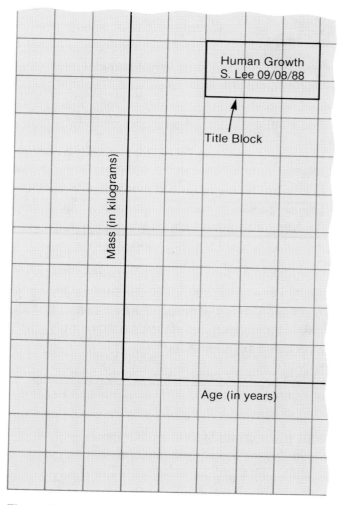

Figure B.3 *Give the graph an appropriate title.*

(a) For the data on Susan's mass and age, we suggest using a scale of 1 cm for every year of age. Thus, in Figure B.4, the number 2 has been printed on the first vertical line away from the origin.

(b) The scale for the *y*-axis in Figure B.4 shows 1 cm for every 5 kg. Thus, the number 15 has been printed on the first horizontal line above the origin.

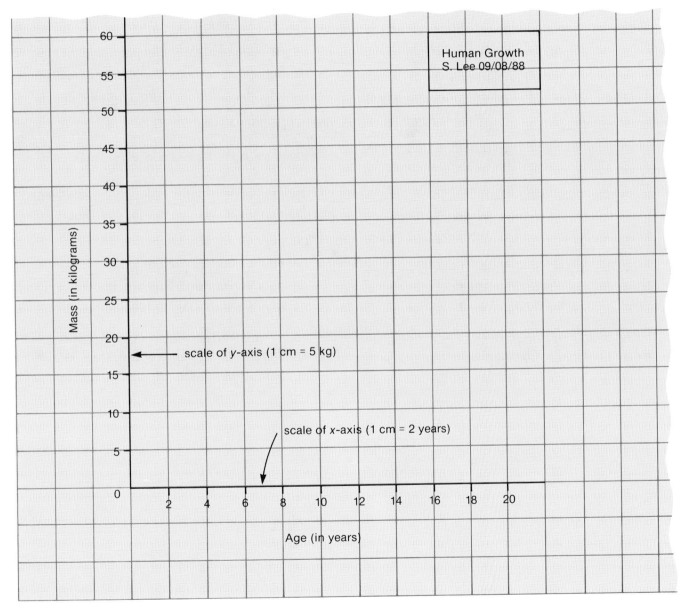

Figure B.4 *Choose an appropriate scale.*

5. Once you have chosen the scale for both axes, put numbers on the axes as shown in Figure B.4. Then plot the data from the table. The numbers in the table come in pairs. There is a mass to go with each age listed. The graph lets you show each number pair, both mass and age, with just one point. The first point will show Susan's mass at birth (3 kg at 0 years).

    To plot the first point, place the tip of your pencil lightly on the age axis at 0 years. Without marking the paper, move the pencil upward until it reaches the 3 kg line. (Note: The first horizontal line above the origin stands for 2 kg and the second line stands for 4 kg. The 3 kg mark lies halfway between these two lines.) Make a small cross at this position with your pencil.

    To plot the second point, place the tip of your pencil lightly on the age axis at 1 year. Without marking the paper, move the pencil upward until it comes to the 10 kg line. Mark the point as before. Continue in this way until you have plotted all of the pairs of numbers.

6. Draw a smooth line from one plotted point to the next, using a ruler and a light pencil line. Note that the origin is not a point on this graph. At birth, age is 0 and mass is 3 kg. This should be the first point on your graph.

Questions

1. (a) At what age did Susan have a mass of 26.5 kg?
   (b) At what age was her mass 35 kg?
2. (a) Describe the pattern of your graph in words.
   (b) Explain why the popular saying, "a picture is worth a thousand words," is appropriate for graphing.
3. (a) What is the period of most rapid increase in mass?
   (b) What is the period of slowest increase in mass?

PRACTICE DRAWING GRAPHS

Use another sheet of graph paper to graph the data in Table B.2. Choose suitable scales and draw your graph.

Questions

1. At what age was Susan 145 cm tall?
2. (a) Do you think Susan will ever grow to a height of 170 cm? Explain.
   (b) What let you make your prediction in answering the previous question?
3. How do you think graphing can be useful in predicting trends?

Table B.2 *Age and Height Data for Susan Lee*

Height In Centimetres	50	74	86	96	102	108	115	120	127	132	138	144	151	157	160	162	163	164	164
Age In Years	0	1	2	3	4	5	6	7	8	9	10	11	12	13	14	15	16	17	18

LINE GRAPHS AND BAR GRAPHS

The graphs you have drawn are called line graphs. Another type of graph you will find useful is called a bar graph. Each bar of a bar graph is labelled to show what it represents. As on a line graph, a suitable scale is selected to show values.

Figure B.5 shows the science examination results of 150 students from one school. A teacher has grouped the results for all grade 9 classes. Each student will be able to see how he or she did on the exam compared with all other grade 9 students in the school. As you can see, the bar graph clearly shows the pattern of results.

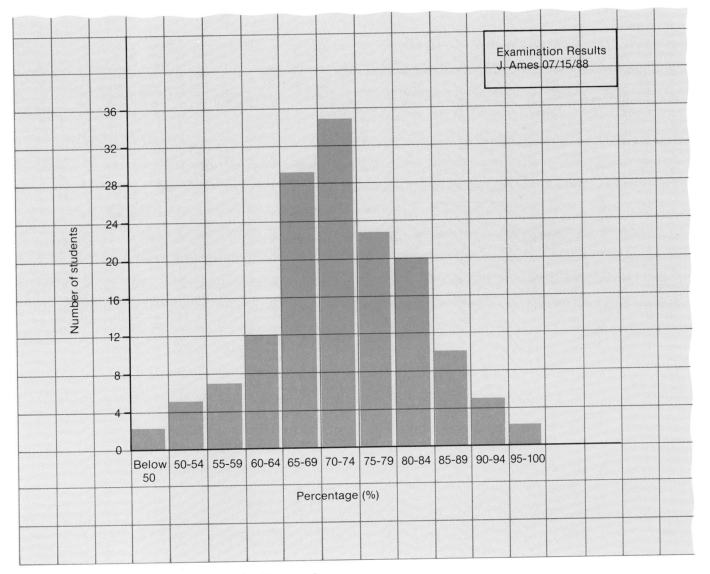

Figure B.5 *A bar graph to show examination results*

A

absorption The process of taking one substance into another

acceleration How fast an object changes speed or direction

acid precipitation Acidic solutions that fall to the ground in the form of rain or snow. These solutions are caused by the reaction of oxides of sulphur and nitrogen with water and oxygen in the atmosphere.

acidic Forming acid; sour; of pH less than 7

adaptation The way an organism's structure is suited to perform its functions efficiently

aesthetics The study or appreciation of beauty in nature or art

air resistance The backward push caused when a moving object collides with air molecules

allergic reaction An inflammation or irritation caused by a particular substance

anorexia nervosa A disease in which the person is convinced he/she is too fat and refuses to eat

aperture Opening that lets light enter a camera

asexual reproduction Reproduction involving only one parent

aspartame An artificial sweetener

atherosclerosis A disorder of the circulatory system that occurs when cholesterol builds up on the inside of blood vessels and makes them narrow

atom The smallest particle of an element

average speed Total distance travelled divided by total time taken

B

bacteria Microscopic organisms that can be helpful but can also cause diseases in plants and animals

basic Forming a base; of pH greater than 7

binocular microscope A microscope that has two eyepieces

biodegradable Able to be broken down by decay organisms (bacteria and fungi) into simpler substances

blade The broad, thin part of a leaf

blight A type of plant disease caused by a fungus

body tube The portion of a microscope that holds the lenses in position

boiling Rapid change of state from a liquid to a gas

boiling point The temperature at which a substance boils

bubble tower A commercial device for separating petroleum into its component fractions by fractional distillation

bulb An underground stem surrounded by leafy layers containing food

by-products Other substances, in addition to the main product, that result from a chemical reaction

C

calcium A mineral that helps form bones and teeth, aids in blood clotting, and helps muscles to work

cambium The actively dividing cells of a plant root or stem

cancer Growth formed when body cells divide in an uncontrolled manner

carbohydrate A nutrient substance, such as sugars and starches, that is the primary source of energy for humans

catalyst A substance that changes the rate of a chemical reaction without being used up in the reaction

catalytic converter A device in an automobile that changes dangerous compounds in the exhaust into safer materials that will not pollute the air so much

cell The basic unit of which all living things are made

cell membrane The outer living surface of a cell

cell plate Formed between dividing plant cells and made of cellulose

cell theory The idea that living organisms are made up of cells and that all new cells are produced from cells

cell wall The non-living layer that surrounds most plant cells

centrifuge A separation device that spins a mixture at high speed to cause settling

centrifugation The process of separating the parts of a mixture of different densities by spinning at high speed

centriole A cell organelle that is involved in cell division and forms spindle fibres

centromere The structure which attaches the spindle fibres to chromosomes during cell division

change of state A change from one state (solid, liquid, or gas) to another

chemical change A process in which new substances with new properties are formed

chemical formula A combination of symbols of atoms representing the composition of a compound

chemical property A characteristic behaviour of a substance

chemical reaction A chemical change in which some substances (reactants) act on one another to produce new substances (products)

chlorophyll The green pigment in plants, algae, and cyanobacteria able to convert the energy of light into chemical energy (photosynthesis)

chloroplast An organelle containing chlorophyll located in the cytoplasm of green cells and some protists

cholesterol A type of fat that can build up on the inside of blood vessels

chromatid One of the double structures appearing during mitosis that later forms a chromosome

chomatin The substance of which chromosomes are composed, DNA and protein

chromosome Thread of DNA within the nucleus of a cell that contains hereditary information

cilium (plural: cilia) Hairlike projection on the surface of cells and protists

clay A type of soil consisting of very small particles that do not allow much drainage

climate The kind of weather that occurs in an area over the years

coarse adjustment The knob on a microscope that produces large focussing movements

colloid A mixture containing tiny visible particles that are too small to settle

combustion The process of burning

compost Decayed plant material

compound A pure substance made up of two or more elements

compound light microscope An instrument that magnifies the view of an object and contains at least two lens systems

concave Hollow; the surface of a mirror or lens that is curved like the inside of a sphere

concave lens A lens that is thin through the middle and thick at the edges

concentration Strength of a solution

condensation The change of state from a gas to a liquid

condensation point The temperature at which a substance condenses from a gas to a liquid

conducting system (of a plant) The continuous series of cells that carry water and dissolved materials from the roots through the stem to the leaves, flowers, and fruit. It consists of xylem vessels and phloem tubes (fibrovascular bundles).

conductivity The ability of a substance to transmit electricity or heat

conductor Substance that can transmit electricity or heat

contractile vacuole A vacuole that expels its contents into the environment

contraction A decrease in volume

control A set of standard values or measurements in an experiment with which changes are compared

converge To come closer together (for example, light rays after they pass through a convex lens)

convex Bulged out; the surface of a mirror or lens that is curved like the outside of a sphere

convex lens A lens that is thick through the middle and thin at the edges

cotyledon Part of the seed of a flowering plant, containing stored food. In most plants the two cotyledons emerge from the ground when the seed sprouts and serve as temporary "seed leaves."

coverslip Small piece of glass used in preparing a wet mount slide

crop rotation The process of growing different crops in a field after a given length of time

cuticle A waxy coating that is secreted by the upper epidermis of a leaf to prevent loss of water by reducing evaporation

cutting A portion of a plant (stem, leaf, or root) that can be used to produce a new plant

cytoplasm Fluid within which all cell organelles are found. It is the basic substance in cells.

D

daughter cell A product of cell division. When a cell divides, it produces cells that are identical to the original parent cell.

DDT A chemical pesticide that was used to kill insects but is now known to be harmful to other organisms

decompose To break down chemically

deficiency disease A disease resulting from a lack of a certain vitamin or mineral

deforestation The clearing away of trees

density The amount of mass (matter) in a given volume

deoxyribonucleic acid (DNA) The molecule of chromosomes that contains the code for the inherited chemical activities of the cell

destructive distillation The process of breaking down a complex material such as wood into its products by heating strongly and then cooling the vapours

developing (1) Treating a film with chemicals to make a negative (exposed silver compounds on the film change to dark patches of silver)
(2) Treating coated paper with chemicals to make a permanent positive print (exposed light-sensitive compounds on print paper change colour)

diaphragm The portion of a microscope below the stage that regulates the size of the opening that admits light

differential permeability The ability of a cell membrane to allow only certain molecules to pass into and out of the cell

differentially permeable membrane A cell surface that will allow some substances to pass through while keeping other materials in (or out)

diffusion Movement of molecules from an area of high concentration to an area of low concentration

dissolve To mix completely a solute with a solvent to form a solution

distillate The liquid produced by distillation

distillation The process of separating the parts of a mixture by heating a liquid until it vaporizes, and then condensing the gas

diverge To spread farther apart (for example, light rays spreading out from a source)

dormancy A resting period in which an organism is alive but carries on its life processes at a very slow rate

downy mildew Fungal disease of grapes and corn

drag A name for the braking force caused by air resistance

E

ecosystem The living and non-living environment in a region

elasticity Springiness caused when an object is temporarily pushed out of shape and then springs back to its original shape

electron microscope A microscope in which the beam of light is replaced by a beam of electrons

element The simplest pure substance that contains only particles of that substance

embryo The immature plant found inside a seed

emulsifying agent A substance that keeps one liquid mixed with another, preventing settling

emulsion (1) A mixture in which one liquid is dispersed in another, such as oil in water
(2) A thin paste made from light-sensitive compounds and used to coat film

energy The ability to do work

enzyme A molecule that increases the rate of chemical reactions

epidermis The outermost layer(s) of thin flat cells that cover the parts of a plant

ethylene A chemical that increases the rate at which fruit ripens

eyepiece (ocular) lens The lens of a microscope that is nearest the eye

expansion An increase in volume

F

fats A nutrient group that includes butter, oils, and cholesterol

fertilizer A material that contains substances to improve the nutrient content of the soil

fibrovascular bundle A "vein" in a leaf or stem made up of xylem vessels, phloem tubes, and supporting fibres. It is part of the conducting system of a plant.

field of view Area observed through the eyepiece of the microscope

film Clear plastic coated with light-sensitive emulsion and used in cameras to record photographic images (must be developed to give a visible image)

filtrate The liquid that passes through a filter

filtration The process of separating the parts of a mixture by screening out the larger particles

fine adjustment focus The knob on a microscope that produces focussing movements to make the image clear and sharp

flagellum (plural: flagella) A whiplike projection from a cell

floc formation The process of attracting tiny suspended particles to a substance such as alum so that they form larger particles that will settle

flotation The process of separating a mixture where a difference in density allows one of the parts to float while the other sinks

fluid Any material that flows, such as a gas or a liquid

focal length Distance between the centre of a lens and its focal point

focal point The place where light rays meet after passing through a convex lens

focus To bring a number of light rays together at one point

food additive Something added to a food product to improve its appearance, act as a preservative, or add flavour

force Any push or pull that can cause acceleration

fraction A portion of something, such as the parts of a mixture of petroleum, where one of the fractions is gasoline

fractional distillation The separation of different parts (fractions) of a mixture such as crude oil (petroleum) by heating and then condensing the gases at specific temperatures

freezing point The temperature at which a substance solidifies

fruit The ripened ovary of a plant

fungus (plural: fungi) An organism that cannot make its own food but must get its nutrients from other organisms. Mushrooms, moulds, and wheat rust are examples of fungi

G

galvanometer Equipment used to measure conductivity

glucose A simple sugar produced in photosynthesis and converted into starch for storage

goiter A swelling in the neck area caused by a lack of iodine in the diet

grafting The process of taking a stem from one plant and joining it to the stem of another plant

gravity A natural force that pulls all objects toward earth's centre

greenhouse effect The warming of the earth's atmosphere by the trapping of heat by the carbon dioxide in the air

green manure Vegetation that is ploughed under and allowed to rot to increase the quality of the soil

guard cells Paired cells in the epidermis of plant leaves and green stems that surround a stoma (opening), which they open and close

H

herb A plant used in cooking to enhance the flavour of the food

herbicide A chemical used to kill plants, especially weeds

heterogeneous Not of uniform composition

homogeneous Of uniform composition throughout

hormone A chemical substance that controls activities in a living organism

humus Decayed plant or animal material

hybrid A product of selective breeding crossing different varieties

hydrocarbon Compound, such as gasoline or paraffin, that is composed of only carbon atoms and hydrogen atoms

I

image A picture that looks like the original object because it reflects light in the same way. An image may be upright (right-side up) or inverted (upside down).

inference A conclusion based upon observations

insecticide A substance that kills insects

insoluble Not able to dissolve

insulator Substance that does not conduct electricity or heat

iodine A mineral element needed to help the thyroid gland function

iris Part of the eye which controls the amount of light entering the eye

iron A mineral element that is necessary for the functioning of red blood cells

irrigation Any method used to increase artificially the amount of water reaching a crop

J

joule (J) The unit used to measure energy. It is the work done when one newton of force moves an object one metre, or one watt of power is used for one second.

K

kilojoule (kJ) 1000 joules

L

layering A method of vegetative propagation in which a branch of an existing plant is bent over and the tip covered with soil to allow root growth at that point

legume A plant belonging to the family that includes peas, beans, clover, and alfalfa. Legumes are important food plants and a good source of protein.

lens A transparent device that bends the path of light passing through it

lift Upward force on a forward-moving object, caused when airflow around top of object is faster than airflow beneath

light A form of energy that can be detected by the human eye

light ray A straight path of light represented in drawings by a thin line with an arrowhead

light sensitive Any material that undergoes a chemical change when exposed to light (for example, human skin or camera film)

light system Part of a microscope that provides the light to view objects clearly

lipid Fat

M

magnification The act of magnifying; a measure of the power by which something has been magnified

magnifying power The number engraved on the side of the lenses of a microscope. It shows the degree of enlargement of the image produced by a lens.

Magnus force Extra force pushing across the path of a spinning object. It may push up, down, left, or right depending on spin direction

Marquis wheat A variety of hybrid wheat produced in Canada in 1904 that was valuable in resisting disease and maturing early

mass The amount of matter in a substance or object

matter Anything that has mass and occupies space

mechanical mixture A mixture made up of two or more easily identifiable parts

mechanical system Part of a microscope that allows the lenses to move and consists of the frame and adjusting knobs

megajoule (mJ) 1000 kilojoules

melting The change of state from solid to liquid

melting point The temperature at which a substance melts

mesophyll A plant tissue consisting of the cells in the middle of a leaf, often divided into two layers, palisade (upper) and spongy

midrib The main "vein" in the centre of a leaf blade

mineral A type of nutrient containing chemical elements needed in small quantities in the diet

mitochondrion (plural: mitochondria) An organelle responsible for releasing energy from food

mitosis The process of division of the cell nucleus in which the parent cell passes on all its characteristics to the daughter cells

mixture Material made up of two or more substances

model Something used to explain known behaviour or to predict future behaviour

molecule The smallest particle of a compound, composed of two or more atoms

mulching A method used to keep plant roots cool and prevent the growth of weeds by covering the soil with straw, paper shreds, etc.

N

negative A reversed image that is dark where the original object was bright, and clear where the original was dark

neutral Neither acidic nor basic; a substance, such as pure water, with a pH equal to 7

nitrogen An element that is necessary for making proteins and encourages the growth of healthy plant leaves and stems

non-renewable resource See **resource**

nosepiece (revolving) The rotating portion of a microscope that has the objective lenses attached to it

no-spray interval The time that must elapse between the spraying of a pesticide on a food crop and the eating of that crop

nuclear fluid The solutions within the nucleus of a cell

nuclear membrane The boundary around the outside of the nucleus of a cell

nuclear pores Openings through a nuclear membrane, through which materials may enter or leave the nucleus

nucleolus (plural: nucleoli) An organelle within the nucleus

nucleus The control organelle of a living cell

nutrient A substance that an organism can use to provide energy or materials for any of its daily activities

O

objective lens The lens of a microscope that is nearest to the object

observation Careful study using the five senses

opaque Able to prevent light from passing through (for example, black plastic or metal)

optical system The eyepiece, the body tube, revolving nosepiece, and the objective lenses of a microscope

ore Rock containing a metal

organ Two or more tissues organized to do a particular job

organelle A structure within a cell that has a specific function

organ system Two or more organs that work together to perform a special function

osmosis The diffusion of water from an area of high concentration to an area of low concentration through a differentially permeable membrane

P

paper chromatography The process of separating different solutes from a solution by having it flow up a strip of paper

particle model An important scientific model used to explain and predict the behaviour of matter in terms of particles

perspiration A loss of water from the surface of the skin

pesticide A general name given to any chemical that kills pests, including weeds, insects, etc.

petiole The stalk of a leaf, joining the leaf blade to the stem

petroleum A mixture of many liquid compounds of carbon and hydrogen, such as gasoline, lubricating oils, and paraffin

phloem A plant conducting tissue consisting of thin-walled cells joined end-to-end, forming a "conveyor belt" to carry the products of photosynthesis from the green cells to all parts of the plant

phosphoric acid A chemical found in fertilizers that contains phosphorus

phosphorus A mineral needed by plants to help roots grow and fruit ripen

photogram Outline made by exposing film or other light-sensitive materials to light shining past an object

photograph A picture made by exposing film to an image of an object

photography The making of pictures using a camera and film

photosynthesis The process of converting the energy of light into chemical energy of sugars and starch. It occurs in the chlorophyll of green plants, algae, and cyanobacteria.

pH scale A series of numbers indicating how acidic or basic a substance is

physical change A change of properties that does not change the type of substance

pinhole camera A sealed container with a small hole at one end and a screen at the other. The hole forms images that fall on the screen.

positive print A permanent photograph whose image repeats the original pattern of light and dark

potash A substance found in fertilizers that contains potassium

potassium A mineral needed by organisms for cell division and growth

potato scab A bacterial disease of potatoes

potential chemical energy The energy contained in food that can be released by chemical reactions

producer In an ecosystem, an organism capable of photosynthesis

product The substance(s) formed in a chemical reaction

projectile Any object that can be thrown, fired, or shot (either into a fluid or into space)

property Characteristic of a substance

protein A nutrient necessary for growth and maintenance of cell parts and as an energy source

protist A single-celled organism containing a nucleus and other organelles

pupil Opening that lets light enter the eye

pyrethrum A plant related to daisies and chrysanthemums that is poisonous to insects; a natural insecticide

Q

qualitative Observed with the senses

qualitative property A characteristic of a substance that may be described but not measured

quantitative Measured using equipment

quantitative property A characteristic of a substance that can be measured

R

random motion Motion without definite direction or order

ray diagram A drawing that uses rays to show or predict the path of light

reactant One of the substances that takes part in a chemical reaction

reactive Able to combine readily with other substances

reflection Occurs when light bounces off of a surface

renewable resource See **resource**

residue The part remaining after a separation process such as filtration or distillation

resource Any substance that fills a need. Such substances are often classified as renewable if they can be used and then restored for use again (such as forests), or non-renewable if they cannot be restored (such as coal and petroleum).

retina Light-sensitive surface inside the human eye

ribosome A cell organelle containing RNA, involved in the manufacture of proteins

RNA (ribonucleic acid) Nucleic acid contained in ribosomes within the cell

root hair A slender extension of a cell in the epidermis of a root, just back of the growing root tip. Root hairs greatly increase the surface area of the root for absorption.

root pressure The force created by osmosis in roots that helps to push water upwards through the stem to the upper parts of a plant

rotenone A substance that is poisonous to insects. Rotenone is made from a plant, *Derris*. It is a natural insecticide.

runner A stem that grows along the ground and can send down roots to start new plants

rust A plant disease caused by fungus

S

safiole A natural flavouring derived from sassafras that was once used to flavour root beer. It is now known to be unsafe for humans.

sand Large soil particles with large spaces between them that allow for rapid water drainage

saturated Not able to dissolve any more solute at a particular temperature

scanning electron microscope (SEM) An instrument for seeing an enlarged view of tiny objects in three dimensions by reflecting a beam of electrons off them

scurvy A deficiency disease caused by a lack of vitamin C

sedimentation A separation process involving the settling of the denser particles of a mixture

seed A structure consisting of an embryo plant, a food source, and a protective coat used to start many plants

seed coat The protective layer surrounding the embryo plant and food source

shade tolerant A type of plant that is capable of growing in low light conditions

shutter Camera part that controls amount of light striking the film

silt Soil particles of medium size that allow for adequate drainage

simple microscope A single lens or magnifying glass

smut A plant disease caused by fungus

solidification The change of state from a liquid to a solid

solubility The property of being able to dissolve. More specifically, it refers to the mass of a solute that can dissolve in a given amount of solvent to form a saturated solution at a particular temperature.

solubility curve A graph representing the solubility of a substance at a range of temperatures

solute The substance that dissolves in a solvent to form a solution

solution The product formed when a solute dissolves in a solvent

solvent The substance that dissolves a substance to form a solution

soot Black carbon in smoke resulting from combustion of fuels

speed How fast an object moves at one particular point on its path. Speed is expressed in units such as m/s.

spindle The arrangement of fibres that separates and moves the chromosomes during mitosis

spores Reproductive cells produced by algae, bacteria, fungi, mosses, and ferns

stage The part of a microscope where the slide is placed

standard test A technique for identifying a certain chemical compound or element by a characteristic behaviour that differs from all other chemicals

starch A chemical compound (carbohydrate) produced by plants to store the energy obtained from photosynthesis. Starch is made of many similar units of glucose, a simple sugar.

state of matter Solid, liquid, or gas

stoma (plural: stomata) An opening through the epidermis of a leaf or a green stem through which gases can diffuse. A pair of guard cells surrounds each stoma, controlling its size.

streamlining Polishing or reshaping an object to reduce drag

sublimation The change of state from a solid to a gas or from a gas to a solid without becoming liquid

sugar A type of carbohydrate that supplies energy

supersaturated solution A solution that contains more dissolved solute than can normally be dissolved in a given mass of solvent at a particular temperature

suspension A mixture containing floating particles that will settle after some time

synthetic Something that has been made artificially and is not natural

T

terminal speed Steady speed reached by a falling object when upward drag force becomes equal to downward force of gravity

thermostat Temperature-sensing device

tissue A group of similar cells that perform the same function

toxic Poisonous

trajectory The path followed by a projectile as it moves through a fluid or through space

transmission electron microscope (TEM) An instrument for greatly enlarging small objects by passing a beam of electrons through them, and focussing the beam using magnetic "lenses"

transparent Able to allow light to pass through it

transpiration The loss of water from the exposed parts of a plant by evaporation from cell walls

transpiration pull A force created in the leaves that helps pull water upward through stems

tuber An underground stem from which new plants may develop

Tyndall effect The scattering of a beam of light by reflection from tiny suspended particles

U

upper epidermis The upper "skin" of a leaf, a thin layer of cells

V

vacuole Fluid-filled organelle within a cell

vaporization The change of state from a liquid to a gas

vegetative propagation A method of producing plants without the use of seeds

vessel A tube in the xylem of a plant through which water rises to the cells of leaves, stems, flowers, and fruit

viability rate Percentage of seeds that are able to grow when planted

viable Able to grow

viewfinder A device that allows a photographer to see the image that has been formed by a camera lens

virus A microscopic particle capable of causing disease

viscosity A measure of how much a liquid resists flow

vitamin A nutrient that does not provide energy for the organism but is essential for daily functions

volume The amount of space an object or substance occupies

W

wet mount A slide prepared, using water, for viewing under a microscope

word equation A statement representing the reactants and products of a chemical reaction, often showing the energy relationships

working distance The distance between the lens and the stage on a microscope

X

xylem Thick-walled, hollow conducting vessels that carry water and dissolved materials through a plant

Photographs not specifically credited to another source are courtesy of the authors. Those photographs in the Science and Technology in Society (STS) features, which are not specifically credited to another source, are courtesy of Jane L. Forbes.

We are particularly grateful to the following for providing the photographs that appear in the colour section of the book: Canada Packers, Imperial Oil Limited, Jack Peakes, D. Cowl.

Unit Openers—One: Ford Motor Company of Canada, Limited; Two: Mary Kay Winter; Three: Derek Bullard; Four: E. Kassner; Five: Donald I. Galbraith; Seven: Derek Bullard. 1.7: Marion Gadsby. 1.11(a): Gordon R. Gore. 1.19(b): Imperial Oil Limited. Chap. 1 Science and Technology in Society: Steve Nash, with sincere thanks and gratitude. Chap. 2 opener: Canada Blower and Pumps Limited. 2.5(a): Marion Gadsby. 2.8: Halton Hills Herald. 2.11: Ontario Ministry of Natural Resources. 2.13: Ontario Ministry of Transportation and Communications. Chap. 4 STS: with the kind permission of the Metropolitan Toronto Zoo; Maurice Nelischer, University of Guelph. Chap. 6 opener: Derek Bullard. 6.4: Derek Bullard. 6.7(a): G. Kaustinen and P. Senkowski. Chap. 7 opener: Dofasco. 7.2: Paul Crucini, PetroCanada Station. 7.6: Metropolitan Toronto Works Department. 7.11: Imperial Oil Limited, Chapter 8 opener: Canapress Photo Services; Gordon R. Gore. 8.3: Allen Wootton. 8.6: Allen Wootton. 8.9: Bill Hause. 8.10: Allen Wootton. 8.13: Allen Wootton. 8.14: J.D. Rising. 8.15: Derek Bullard. Chap. 8 Science at Work: University of California at Berkeley. Chap. 8 STS: Domtar Chemicals Groups; Ministry of Northern Development and Mines; Ontario Ministry of Transportation and

Communication; Schnieders. Chap. 9 opener: Goodyear Canada Inc. 9.2: S.T. Fox. 9.3: Interprovincial Pipe Line Limited. 9.4: Bill Hause. 9.8: Canapress Photo Services. 9.10: Mississauga Diving Services Limited. Chap. 10 opener: Miller Services. 10.2: Derek Bullard. 10.4: Derek Bullard. 10.5: Institute for Hydrogen Systems. 10.9: COMINCO. 11.3: H.M. Lang. 11.5 H.M. Lang. 11.6: H.M. Lang. 11.12: Health and Welfare Canada. 11.15: Dr. J. Lai-Fook. 11.16: Dr. J. Lai-Fook. Chap. 11 Science at Work: Ontario Science Centre. 12.2: Dr. J. Lai-Fook. 12.9: Dr. J. Lai-Fook. 12.10: Dr. J. Lai-Fook. 12.11: Dr. J. Lai-Fook. 12.12: Dr. J. Lai-Fook. 12.16: Ontario Science Centre. Chap. 13 opener: the family of Terry Fox. 13.3: Richard A. Boolootian, Science Software Systems, Inc. 13.11: H.M. Lang. 13.13: Canadian Cancer Society. 14.4: Donald I. Galbraith. 14.8: British Columbia Ministry of Forests. 14.9: Donald I. Galbraith. 14.11: Donald I. Galbraith. 14.14: Donald I. Galbraith. 15.11: Ontario Ministry of Natural Resources. 16.13: Canadian Forest Products Limited, 16.20: Pollution Probe. Chap. 16 STS: with the kind permission of Rick Doyle, Mississauga Recreation and Parks Department. Chap. 17 opener: Paul Thompson. 17.1: Paul Thompson. 17.5: Paul Thompson. 17.9: Paul Thompson. 17.13: Stokes Seed, St. Catharines, Ontario. 17.14: Stokes Seed, St. Catharines, Ontario. Chap. 18 STS: Ontario Ministry of Agriculture and Food (all photos). 19.1: Robin Caulderwood. 19.10: Robin Caulderwood. Chap. 20 opener: Derek Bullard. 20.1: Vancouver Sun. 20.8: Derek Bullard. 20.19: Derek Bullard. Chap. 21 opener: Derek Bullard; Gordon R. Gore 21.4: Derek Bullard. 21.6: Derek Bullard. 21.13: University of Texas at Austin. 21.20: Derek Bullard. 21.21: Derek Bullard. 21.22: Derek Bullard. 21.26: Derek Bullard. Chap. 21 STS: Kent Kawasaki, Japan Camera Centre.

We are particularly grateful to the following for providing the photographs that appear in the colour section of the book and on the cover: Canada Packers Inc., Canadian Standards Association, Coca-Cola Limited, Scott K. Edgington, Imperial Oil Limited, Ontario Hydro, Jack Peake, The Canadian Red Cross Society and Thomas R. Shultz.

Jon. D.
&
Paul. A